Ecological Modelling and Ecophysics

Agricultural and environmental applications

Ecological Modelling and Ecophysics

Agricultural and environmental applications

Hugo Fort
Department of Physics, Republic University, Montevideo, Uruguay

IOP Publishing, Bristol, UK

ISBN 978-0-7503-2432-8 (ebook)
ISBN 978-0-7503-2430-4 (print)
ISBN 978-0-7503-2433-5 (myPrint)
ISBN 978-0-7503-2431-1 (mobi)

DOI 10.1088/978-0-7503-2432-8

Version: 20200601

IOP ebooks

British Library Cataloguing-in-Publication Data: A catalogue record for this book is available from the British Library.

Published by IOP Publishing, wholly owned by The Institute of Physics, London

IOP Publishing, Temple Circus, Temple Way, Bristol, BS1 6HG, UK

US Office: IOP Publishing, Inc., 190 North Independence Mall West, Suite 601, Philadelphia, PA 19106, USA

To Silvia, José and Rodrigo

Contents

Preface

The title of this book highlights that its focus is on quantitative methods applied to practical problems. Indeed, this is a book oriented toward use-inspired basic science. It falls into the so-called Pasteur's quadrant, *sensu* Stokes (1997), where the motivation is both expanding understanding and increasing our abilities to solve **practical issues** with rigorous **quantitative methods**. This is why this book is organized in such a way that each chapter devoted to methods in community/population ecology is followed by a companion **Application chapter**, containing a practical application of the presented methods. A main goal of the latter chapters is to engage the reader interested in developing tools and strategies to solve their own problems. Two of the applications are about production optimization in agriculture: livestock production and polyculture crops. The focus of the other two applications is more towards environmental issues: the dynamics of tree species in tropical forests and the development of early warning signals of catastrophic shifts in lakes. The common theme underlying all these problems is that they are approached through population dynamics models. There are few textbooks connecting theoretical methods and mathematical developments in ecology with the real world and its practical problems either in production or conservation. This book attempts to fill such a gap.

In the title also appears the neologism 'Ecophysics' to stress that, the spirit as well as the methods of physics permeate this book. However, a clarification is in order. I do not want to promote the 'ecology as physics' viewpoint. Rather, I advocate for adopting certain general principles which have repeatedly demonstrated their effectiveness in physics.

One of these principles is not to be afraid of relying on partial aspects and partial relations of things which we only have an incomplete understanding. Indeed, abstracting and isolating certain relevant features of phenomena and forgetting about the rest of the Universe is a recipe that has yielded great dividends since Galileo! More recently, Enrico Fermi was celebrated for his ability to make fast, excellent approximate calculations with little or no concrete data. One of the best well-known examples is when the first atomic bomb was detonated during the Manhattan Project. To estimate the power of the blast Fermi, who was standing at base camp 10 miles away from the explosion, dropped a few scraps of paper as the shock wave from the detonation passed. Using the distance covered by the pieces of paper, after some coarse calculations, Fermi estimated a power of 10 kilotons of TNT, which is remarkably close to the now-established value of 20 kilotons. ***Fermi problems***, or *order estimation problems*, have become quite standard in physics and engineering courses to develop the students' skills in solving complex problems using simple shortcuts to make approximate, but meaningful, calculations. Since Fermi problems illustrate the 'no fear principle', I briefly discuss them in the Introduction, and readers who wish to dig deeper can find more material in appendix II.

Occam's razor, aka parsimonious modelling approach, is another fruitful principle: always start with the simplest possible model capable of capturing the phenomenon or problem you want to describe, explain or predict. This *keep it simple* principle has demonstrated to be crucial to make progress in several disciplines, it prevents us from introducing parameters in our models which are often very difficult, or even impossible, to measure. A large number of unsupported or free parameters are more likely to induce overfitting and decrease our trust in the model. Complex models are often perceived as more reliable than simpler ones, although they can be as intractable as the real systems they aim to model; too many parameters in a model can easily lead to confusion rather than insight. Furthermore, it is quite common that experimental errors preclude distinguishing predictions from these more sophisticated models to those from simpler models (hopefully the Application chapters will help to illustrate this point). So our philosophy here is that the complexity of proposed models should try to match the complexity of the problem that they seek to address. Therefore, we prefer to aim at simplicity at the outset, and, if necessary, make our model more complex or extend it by adding more model parameters.

The first part of this book is devoted to methods in population/community ecology that have become classical. These methods, originally devised by Alfred Lotka and Vito Volterra in the 1920s, were later developed and extended by theoretical ecologists such as Robert MacArthur, Robert May and many others.

The second part aims to introduce the reader to certain tools and techniques from different branches of physics—like thermodynamics, statistical mechanics and complex systems—and their applications to address questions in ecology and environmental sciences. Connecting ecological problems with well-studied phenomena in physics allows exploiting analogies to gain deeper insight into these problems, to identify novel questions and problems, and to get access to alternative quantitative methods and tools from physics.

What this book does not cover

This book does not cover many ecological issues, such as delay models, age-structured models, disease ecology, migration, or the analysis of the complexity–stability problem. These important topics are well explained in classical textbooks on mathematical ecology—the books by Pastor (2008) or Kot (2003) are good examples—, or in volumes devoted to mathematical biology—for example Keshet-Edelstein's (1988) or Murray's (1989). Indeed a possible criticism is that the topics covered in this book might seem rather eclectic and strongly tailored towards my own personal tastes and experience. However, hopefully, it will include material the reader will not readily find elsewhere.

References

Keshet-Edelstein L 1988 *Mathematical Models in Biology* (NewYork: Random House)

Kot M 2003 *Elements of Mathematical Ecology* (Cambridge: Cambridge University Press)

Murray J D 1989 *Mathematical Biology(Biomathematics series* vol 19) (Berlin: Springer)

Pastor J 2008 *Mathematical Ecology of Populations and Ecosystems* (Singapore: Wiley-Blackwell)

Stokes D E 1997 *Pasteur's Quadrant: Basic Science and Technological Innovation* (Washington, DC: Brookings Institution Press)

Acknowledgements

I want to thank several people. First of all Raul Donangelo, Clive Emary, Carla Kruk, Jorge Pullin and Angel 'Marley' Segura, who thoroughly went over drafts of the chapters of this manuscript and corrected many typographical and other errors, also pointing out where more explanation was required, and details I had missed.

I am grateful to colleagues with whom we have collaborated in carrying out the work reported in the Applications chapters: Francisco Dieguez (Application chapter 1), Valentín Picasso (Application chapter 2), Tomás Grigera (Application chapter 3).

I want to make a special mention of several of my paper co-authors: Raul Donangelo, Ariel Fernández, Pablo Inchausti, Néstor Mazzeo, Boon Leong Lan, Muhitin Mungan, Marten Scheffer, Angel Segura, Egbert van Nes and Diego Vázquez. I have collaborated with them over the years on different topics of mathematical modelling and ecology. Much of the material of this book corresponds to research developed in collaboration with them.

I also want to thank Stefano Allesina, György Szabó, Attila Szolnoki, Donald Waller and Fredy R Zypman for interesting discussions.

It has been a pleasure working with the Institute of Physics (IOP) team; particularly with Ashley Gasque (Senior Commissioning Editor), Robert Trevelyan (ebooks Editorial Assistant), Daniel Heatley (ebooks Editorial Assistant) and Caroline Mitchell (Commissioning Editor). All of them provided valuable help and assistance.

Most of these chapters were written during 2018–9, while I was on sabbatical leave across Europe and the US thanks to the generous support from ANII through project ERANET-LAC R&I2016–1005422 and the University of the Republic (UdelaR), Uruguay. I am also indebted to Sebastian Ibanez, Ferenc Jordan and Valentín Picasso for their hospitality at, respectively, Chambery (France), Budapest (Hungary) and Madison (US).

Finally, above all, I must thank Silvia Etchamendi, my wife, as well as my two sons, José and Rodrigo, for their patience and support during the writing of this book.

Montevideo, Uruguay
Hugo Fort

Author biography

Hugo Fort

Dr Hugo Fort is Professor at the Physics Department of the Faculty of Sciences of the Republic University (Montevideo, Uruguay) and Head of the Complex System Group. After earning his PhD in Physics from the Autonomous University of Barcelona in 1994 he conducted research on quantum field theory. Since 2001 his scientific interests evolved from theoretical physics to complex systems and mathematical modelling applied to problems in biology, with focus in ecology and evolution.

A main goal of his research is to develop quantitative methods and tools for a wide variety of practical problems in fields ranging from agro-economy (crop mixtures for overyielding, scientific or precision livestock production) to environmental sciences (early warnings of catastrophic shifts, forecasting biodiversity dynamics) and real-time evolution (*quasispecies* theory for RNA viral dynamics, bacteria evolutionary experiments). Professor Fort is currently involved in several international research collaborations pursuing used-inspired basic science. A central aim is to connect ecological and evolutionary problems with well-studied phenomena in physics to gain deeper insight into these problems, to identify novel questions and problems, and to get access to alternative powerful computational tools.

Author of over a hundred articles in scientific journals and book chapters in diverse fields: agriculture sciences; applied mathematics; biology; ecology; physics and social sciences modelling, Professor Fort has taught several courses in mathematical modelling, complex systems, non-linear dynamics and statistical physics. He has also been collaborating in different projects with national agencies as a senior scientific consultant.

IOP Publishing

Ecological Modelling and Ecophysics
Agricultural and environmental applications
Hugo Fort

Chapter 0

Introduction

'In time, those unconscionable maps no longer satisfied, and the cartographers guild drew a map of the empire whose size was that of the empire, coinciding point for point with it. The following generations, who were not so fond of the study of cartography saw the vast map to be useless and permitted it to decay and fray under the Sun and winters.'

Jorge Luis Borges *On Exactitude in Science* (1946).

'*But everything takes a different shape when we pass from abstractions to reality. In the former, everything must be subject to optimism, and we must imagine the one side as well as the other striving after perfection and even attaining it. Will this ever take place in reality?*'

Carl von Clausewitz *On War* (1832).

0.1 The goal of ecology: understanding the distribution and abundance of organisms from their interactions

Let us start from the beginning. Ecology can be defined as the scientific study of the distribution and abundance of organisms (Andrewartha 1961) or, more precisely, as the scientific study of the interactions that determine the distribution and abundance of organisms (Krebs 1978). Krebs' definition is both clear and comprehensive. And, in contrast to other definitions, avoids ambiguities or vagueness. Species distribution is the manner in which species are spatially arranged. Species distribution patterns depend on biotic (living) and abiotic (non-living) factors and change depending on the scale at which they are viewed. Abundance is an ecological quantity of paramount importance when making management and conservation decisions (Andrewartha and Birch 1954, Krebs 1978, Gaston 1994). Estimates of future

doi:10.1088/978-0-7503-2432-8ch0

abundance are essential either for species of value to humans for regulating current and future harvests, or for species of conservation concern since their future numbers will be an important determinant of their extinction risk, or for harmful/toxic species to anticipate their blooms.

To analyze the relationships between living organisms and their environments, the interaction of organisms with one another, and the resulting patterns of abundance and distribution of organisms, ecology has progressed through the incorporation of sophisticated computational methods. The first major use of mathematics in ecology can be traced to the seminal models of Alfred Lotka (1925) and Vito Volterra (1926). They produced basic ecological models for competition and predation—now referred to as the Lotka–Volterra models—that have become the cornerstone for much of ecology today and particularly for two of its sub-disciplines: *community ecology* and *population ecology*. Community ecology is the study of the organization and functioning of communities, which are assemblages of interacting populations of different species coexisting at a particular location. Population ecology focuses on the dynamics of populations of single-species groups (or even a particular population within the same species). Since then, the above stylized mathematical models and their descendants have been used mostly to provide *qualitative* explanations for patterns in nature. An example is the use of competition models to explain species diversity (May 1974, Diamond and Case 1986). Simple competition models served to show that species that utilize the same resource can coexist and avoid competitive exclusion.

On the other hand, modelling efforts to provide *quantitative* results in applied fields have generally rejected simple mathematical models in favor of giant hyper-realistic simulation models (Onstad 1988). Individual-based models (DeAngelis and Grimm 2014) constitute an example of this modelling approach whose use in ecology has been growing rapidly in the last two decades. They simulate populations and communities by following individuals and their properties. These *individuals* might represent plants and animals in ecosystems. Individual-based models have been playing an increasingly important role in questions posed by complex ecological systems. For example, to understand the causes of vegetation change, an important long-term goal of ecology, we can link individual-based simulations of populations to models of detritus composition and nutrient release. These models provide important information on plant community processes, constraints over selection and biogeochemistry (Levin 1992). Individual-based models have also been used for addressing theoretical questions like understanding how traits of individual organisms are connected with the assembly of communities and food webs (DeAngelis and Grimm 2014). The problem with these more realistic simulation models is that they sacrifice understandability for ecological realism and can have hundreds of parameters and state variables.

However, in recent decades, an increased interest in applied questions among ecologists, and the resulting research has begun to suggest a different use for simple mathematical models, both for Lotka–Volterra models (Fort 2018a, 2018b, Fort and Segura 2018) as well as a general approach (Ginzburg and Jensen 2004, Hilborn and Mangel 1996, Ishida 2007, Lonegran 2014). In the last section of this chapter we will comment on this well-known trade-off in mathematical modelling between realism and usability, which is captured by the saying 'All models are wrong, but some are useful' (Box 1976). But before this, in the section, we will review some of the basics of mathematical modelling, including why models are of central importance in scientific research, its principles and the classification of mathematical models.

0.2 Mathematical models

0.2.1 What is modelling?

Models describe our beliefs about how the world functions. They are idealizations that simplify reality and never provide a completely accurate representation of the real-world phenomena. Among the uses of models are: to describe the behavior or results observed; to explain why that behavior and results occurred as they did; to predict future behaviors or results that are as yet unseen or unmeasured.

modelling constitutes a central piece of the *scientific method* that roughly can be viewed as having four stages: observation, modelling, prediction and verification. The observation stage consists in measuring what is happening in the real world. Here we gather empirical facts and real-world data related to the phenomenon or phenomena we are interested in understanding/explaining. The modelling stage includes, first analyzing and filtering the above observations to only retain a reduced set of relevant data associated with this phenomenon and then to construct a hypothesis—an educated guess about how things work—that explains the phenomenon. For example, in physics this often means proposing a mathematical relation or a causal mechanism. In the prediction stage we use models to tell us what will happen in a yet-to-be-conducted experiment or in an anticipated set of events in the real world. These predictions are then followed by observations in the verification stage that serve either to validate the model or to suggest reasons why the model fails and then it either has to be changed or completely rejected. Actually, the above process is iterated so that scientists build up a better and better representation of the phenomena they want to explain or use for prediction as time goes on.

There is a large element of compromise in modelling. The majority of systems in the real world are far too complicated to model in their entirety. So it is crucial to identify the most important parts of the system. These will be included in the model, the rest will be excluded. This compromise between realism/accuracy and usefulness/practicality is well illustrated by the *map–territory relation*: the most accurate map possible would be the territory itself, and thus would be perfectly accurate and perfectly useless. The quotation from Jorge Luis Borges at the beginning of this chapter describes the tragic uselessness of the perfectly accurate, one-to-one map.

0.2.2 Why mathematical modelling?

Models can be of different types; conceptual models (abstractions of things that exist in the real world) to better understand, graphical models for visualizing purposes, mathematical models, etc. Formulating models into the language of mathematics has many advantages. It allows:

- **Precision and non-ambiguity.** Mathematics is a very precise language with well-defined rules for manipulations.
- **Algorithmic compression.** Mathematical models can condense vast arrays of data into compact formulas. According to Motz (1987), modern science began with Galileo and Newton and the quest for algorithmic compressibility: Newton's fundamental discovery that all the information recorded about the motion of bodies in the heavens or on Earth could be encapsulated in three simple rules he called 'the laws of motion' plus his law of gravitation. Algorithmic compression is crucial to understand why things are the way they are as well as to predict how they are going to be in the future.
- **To discover non-apparent relationships and develop scientific understanding.** Besides providing brevity and formality of description, quantitative expressions of current knowledge of a system also allow manipulation of the model and provide the opportunity of discovering emergent properties not apparent without mathematical reasoning and that are required to understand complex systems.
- **To use computers to perform numerical calculations.** Although mathematics has the potential to *prove* general results, i.e. of analytical proof, most systems of interest in ecology are described by equations whose analytical solutions (in terms of closed mathematical formulas) are unknown. However, these equations can be solved numerically by using computers, i.e. manipulating numbers directly to produce a numerical result rather than a formula.
- **To simulate virtual experiments.** Related with using computers, simulation allows one to perform virtual experiments that would be difficult or impossible to perform in the real world, for either practical or ethical reasons. This is particularly the case for many ecological systems for which experimentation is impossible. In fact, the complexity, cost or risk of many experiments make the use of simulations unavoidable. Furthermore, simulation can be useful when the real system does not yet exist (e.g. artificial ecological communities, like mixture of crops) or when the real system works too fast (e.g. an electrical network) or too slow (e.g. geological processes or evolution or speciation) precluding direct analysis. The systematic use of computer simulations—from climate science to high energy physics (Massimi and Bhimji 2015)—is undeniable.
- **Aid decision making**. Quantitative information, e.g. in the form of standardized indices, is helpful for making both tactical decisions by managers as well as strategic decisions by planners.

The importance of the application of mathematical methods to ecology has been discussed in several studies (Hallam and Levin 1986, Levin *et al* 1989, May 1974, 2004, Vandermer 2010). Mathematical models have been fundamental in describing many practical issues. A non exhaustive list covers from the fate and transport of pollutants in the environment, the spread of agricultural pests, the dynamics and control of epidemics, to the management of renewable and nonrenewable resources, the response of ecological systems to global climate change, and fisheries management.

Indeed, there are few areas of science which cannot benefit from mathematical modelling. Our aim is that the methods covered in this book allow the reader to start developing her/his own models and provide insight into useful methods with a wide range of applications, including developing early warnings of catastrophic shifts in ecosystems, sustainable harvesting and conservation management.

0.2.3 What kind of mathematical modelling?

Mathematical models can be classified in several ways, according to their use (description, explanation, prediction, optimization), accuracy (qualitative or quantitative), degree of randomness (deterministic and stochastic), degree of specificity (specific or general), etc. Another possible distinction is between deductive and inductive models. A deductive model is a logical structure based on a theory. An inductive model arises from empirical findings and generalization from them. Mathematical models in ecology can also take many forms, including but not limited to dynamical systems, statistical models, partial differential equations, etc.

This book deals with models based on *differential equations* or their discrete counterpart, *finite difference equations* (and, in part III, with *cellular automata* resulting from spatial partial differential equations commonly used to describe systems such as vegetation patterns). These equations determine how a system changes from one state to the next and/or how one variable depends on the value or state of other variables (*state equations*). For instance, they describe the rates at which populations change due to birth, death, migration, etc. The majority of the models we will consider will be deterministic models, which ignore random variation, and so always predict the same outcome from a given starting initial condition. However, when needed, we will include stochastic variables to model uncertainty or randomness of some environmental condition. Take, for instance, a population of lynxes and hares. The lynxes need to eat hares in order to produce offspring. So the number of young each lynx produces is a function of the number of hares. But as the lynxes eat the hares, the number of hares decreases, thus the hare death rate is related to the number of lynxes. A model known as the *Lotka–Volterra predator–prey model* describes this kind of interacting system with a pair of coupled nonlinear differential equations. As we will see in the next chapter, this is an example of a *dynamical system*. More specifically, a deterministic dynamical system. That means that if you start with the same inputs, you always get the same prediction for the number of lynxes and hares over time.

There are several excellent textbooks on mathematical modelling. A completely non exhaustive list includes as general introductions to mathematical modelling the text by Giordano *et al* (2009) which offers a solid introduction to the entire modelling process across several fields with balance between theory and practice, and Bender (2000) using a practical 'learn by doing' approach. Mathematical biology is extensively covered by these two classics: Edelstein-Keshet (1988) and Murray (1989). Highly valuable introductions to mathematical models, methods, and issues in population ecology can be found in Kot (2003), May (1974) and Pastor (2008).

0.2.4 Principles and some rules of mathematical modelling

Mathematical modelling has both principles behind it and rules that are either necessary to warrant logical consistency or helpful to get insight with your model. The principles can be phrased as questions about the intentions and purposes of mathematical modelling (Dym 2004). We outline such principles in box 0.1, and we illustrate them with a practical problem. Next, we briefly review some of the rules and discuss them in the context of this illustrative problem.

Suppose our **goal** is to estimate when the population of yeast *Saccharomyces cerevisiae* inoculated into a culture medium will reach 80% of its maximum equilibrium value under the same experimental conditions used by Gause (1932) in his famous experiments. So **we want to know** the volume occupied by yeast as a function of the time after inoculation.

What do we know? Gause's experimental setting and details are as follows. As a nutritive medium the so-called 'yeast fluid' is employed; it was prepared in the following manner: 20 g of dry pressed beer yeast were mixed with 1 liter of distilled water, boiled for half an hour in Koch's boiler, and then filtered through infusorial earth. To this mixture 5 per cent of sugar was added, and then the medium was sterilized in an autoclave. The sterile medium was aseptically poured into the test-tubes which had been stopped by cotton-wool and first sterilized by dry heat. The

Box 0.1. A list of useful questions for guiding the process of mathematical modelling.

Question	*Task*
• **Why?** What is the goal?	Identify the need for the model.
• **Find?** (output) What do we want to know?	List the data we are seeking.
• **Given?** (Input) What do we know?	Identify the available relevant data.
• **Assumptions?** What can we assume?	Identify the circumstances that apply.
• **How?** How should we look at this model?	Identify the governing rules or principles and the equations that will be used.
• **Predictions?** (Output) What will our model predict?	Identify the calculations that will be made, and the answers that will result.
• **Valid?** Are the predictions valid?	Check the accuracy of the model's representation of the real system.
• **Improve?** How can we improve the model?	Identify parameter values that are not adequately known, variables that should have been included, and/or assumptions/restrictions that could be lifted.

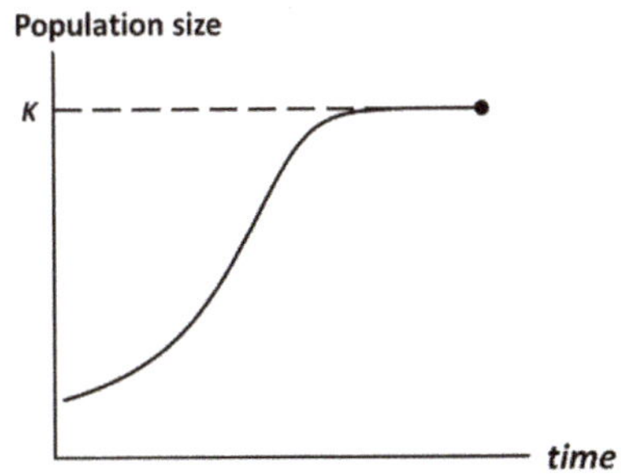

Figure 0.1. Logistic curve for the population size as a function of time, $N(t)$.

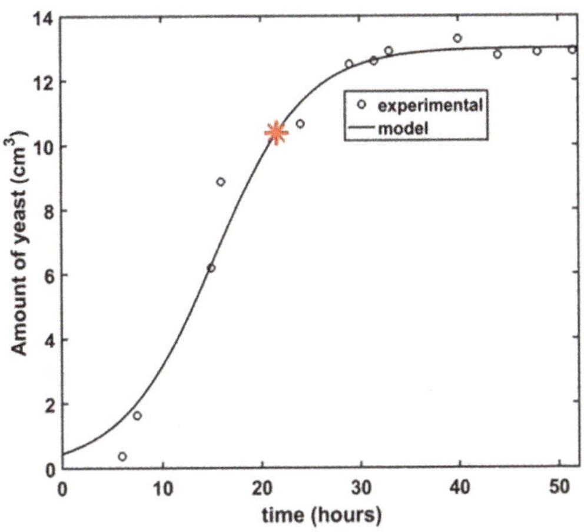

Figure 0.2. The growth of the volume of *Saccharomyces* (o) and the logistic curve (filled). The * symbol denotes when the yeast reaches 80% of its carrying capacity.

experiments were made in a thermostat at a temperature of 28 °C. The volume of yeast was measured and the number of cells were counted in a counting chamber under the microscope for 6, 7.5, 15, 16, 24, 29, 31.5, 33, 40, 44, 48 and 51.5 h after inoculation (circles in figure 0.2).

We should list any relevant **assumptions**, so any assumptions made about population growth should be spelled out. The reader can find a list of relevant assumptions in table 1 of the next chapter. For example, that there is a continuous overlap of generations so that a continuous time variable can be used, the population is isolated (i.e. there are no in or out fluxes), the environment is considered homogeneous (i.e. there are no sharp gradients of resource distribution), etc. It turns out that all these assumptions are consistent with the experimental conditions.

Which rules or principles apply to this model? As we will see in the next chapter, the growth of a homogeneous population within a limited environment exhibits a sigmoid behavior, i.e. at first the population N growths exponentially with time until it reaches an inflection point where it changes concavity and finally it saturates to a maximal equilibrium population or carrying capacity $\mathcal{K}$ (figure 0.1). Mathematically the population growth is described by the logistic equation (see next chapter)[1]

$$dN/dt = \mathcal{r}N(1 - N/\mathcal{K}), \tag{0.1}$$

where $\mathcal{r}$ denotes the population growth rate. Solving this equation we get the logistic curve

$$N(t) = \frac{\mathcal{K}}{1 + (\mathcal{K}/N_0 - 1)e^{-\mathcal{r}t}}, \tag{0.2}$$

where N_0 is the initial population. In our problem, the variable will be the volume $V(t)$ rather than the number of cells $N(t)$.

Predictions. Our model will predict the time t_{80} when the fraction $x = N/\mathcal{K}$ reaches 80%. From equation (0.2), this function as a function of the elapsed time t can be expressed as

$$x(t) = \frac{1}{1 + (1/x_0 - 1)e^{-\mathcal{r}t}}. \tag{0.3}$$

The parameters $\mathcal{K}$ and $\mathcal{r}$ as well as N_0 can be obtained from the experimental values by a simple least square regression (see exercise 1.2) producing $\mathcal{K} = 13.0$, $\mathcal{r} = 0.218$ and $N_0 = 0.45$ ($x_0 = 0.45/13 = 0.035$).

Therefore t_{80} can be obtained by solving equation (0.3) for $x = 0.8$, which gives:

$$t = \frac{1}{\mathcal{r}} \ln\left(\frac{1/x_0 - 1}{1/0.8 - 1}\right) = \frac{1}{0.218} \ln\left(\frac{1/0.035 - 1}{1/0.8 - 1}\right)$$
$$= 21.6 \text{ h (indicated as an asterisk in figure 0.2)}.$$

Validation. From figure 0.2 we can see that the theoretical logistic curve, with estimated parameters $K = 13.0$, $r = 3.328$, agrees pretty well with the experimentally measured amounts of yeast (circles).

If we find that our model is inadequate or that it fails in some way, we then enter an iterative loop in which we cycle back to an earlier stage of the model building and re-examine our assumptions, our known parameter values, the principles chosen, the equations used, the means of calculation, and so on. This iterative process is essential because it is the only way that models can be improved, corrected, and validated.

Now we will review some rules that need to be followed as well as some recipes that are generally useful.

A. ***Abstraction and the value of idealized systems.*** An important decision in modelling is choosing an appropriate level of detail for the problem at hand. This abstraction process requires a thoughtful approach to identifying those phenomena on which we want to focus. Finding the right level of detail is therefore intimately connected with finding the right level of abstraction. With this aim, using ideal systems, i.e. simplified or stylized representations

[1] A note on notation, as we will see at the end of this chapter, variables (like 'N' and 't') are denoted in regular italic while parameters (such as '$\mathcal{r}$' and '$\mathcal{K}$') are denoted by manuscript font.

of real systems, as a starting point to model a phenomenon and then complicating things if this simplest model fails in capturing important aspects of this phenomenon is a common practice in many fields of science that has demonstrated its usefulness. An example of such idealized systems in physics is the *ideal gas*, a theoretical gas composed of many randomly moving point particles whose only interactions are perfectly elastic collisions, like the ones between billiard balls. For a mol of an ideal gas the well-known equation of state $PV = \mathcal{R}T$ (where P, V and T denote, respectively, pressure, volume and temperature, and $\mathcal{R}$ is the gas constant 8.314 J K^{-1} mol^{-1}), the *ideal gas law*, can be derived using Maxwell's kinetic theory of gases. There are no 'ideal gases' in reality, however, the ideal gas law, is a good approximation. Under usual conditions, most real gases behave qualitatively like an ideal gas. Furthermore, many gases such as, noble gases, nitrogen, oxygen, and hydrogen can be treated like ideal gases within reasonable tolerances. For example one mol of nitrogen, the most common pure element in the Earth and making up 78.1% of the entire volume of the atmosphere, occupies at standard temperature and pressure (a temperature of 20 °C = 293.15 K and an absolute pressure of 1 atm = 101 325 Pa) a volume of 24.00 l. And, under the same standard conditions, the ideal gas law predicts a volume of $8.314 \times 293.15/101\,325 = 0.024\,05\ m^3 = 24.05$ l, i.e. a relative difference of 0.21%. However, it turns out the ideal gas model tends to fail at lower temperatures or higher pressures, when intermolecular forces and molecular size becomes important. It also fails for most heavy gases, such as many refrigerants, and for gases with strong intermolecular forces, notably water vapor. At high pressures, the ideal gas law generally underestimates the volume of real gases and at low temperatures overestimates the pressure of real gases. Most importantly, at some point of low temperature and high pressure, real gases undergo a phase transition, such as to a liquid or a solid. The model of an ideal gas, however, does not describe or allow phase transitions. These must be modelled by more complex equations of state.

History has shown, not only in physics, but also in ecology, that one can make considerable progress with simple models. For example, in population ecology a well-known ideal system is a population growing at a constant rate, i.e. governed by the Malthus equation $dN/dt = \mathfrak{r}N$ (see next chapter), which can be obtained as a limiting case of the more realistic logistic equation (0.1) when N is much smaller than $\mathcal{K}$. And, although it is very unrealistic as a general description of population growth, bacteriologists have used it for differentiating species by restricting the model to the initial stage of exponential growth. Indeed, as we will discuss in chapter 4, there is an interesting formal analogy between the equation of state of a real gas and its ideal gas limit for one side and the logistic equation and its Malthus limit for the other. In the next section we will get back to the value of using idealized systems and situations to get insight into the phenomena we want to understand.

B. ***Aggregation and compartmental modelling.*** Compartmental modelling can help the above abstraction process. Compartmental models are composed of sets of interconnected 'compartments' or 'chambers'. Each compartment of the system is considered to be homogeneous (perfectly mixed, with uniform concentration or density). Compartments can be used as 'black' boxes without explicit internal structure, connected by 'flows' that simulate how individuals in different 'compartments' interact. In population ecology, the plants or animals in each compartment are assumed to be the same as all the other plants or animals in that compartment. Flows between compartments of the model can represent interactions. For instance, predators eating prey is described as a 'biomass flow' from the prey compartment into the predator compartment. This flow represents the fact that the number of predators that can survive is related to the number of prey available for them to eat and, vice versa, the survival of prey depends on the number of predators around to eat them. Another example is live individuals that flow to a 'dead' compartment with a certain rate. The rates of flow between compartments, and interaction rates between compartments, are parameters of the model. For instance, a compartmental model could be used to simulate the number of animals over time in a predatory/prey system like lynxes and hares. One compartment would be the predators (the lynxes), and they would be assumed to all be the same, with the same death and birth rates, ability to catch prey, no distinction between ages, etc. The other compartment would be the hares, and they would also be assumed to all be the same and have the same probability of being caught by lynxes. Actually, this model of two boxes, one for prey and the other for predator, is the basis behind the Lotka–Volterra predator–prey model we already mentioned.

Aggregation of species which are similar regarding traits or function into sets or groups can also be helpful. For example, functional groups are non-phylogenetic, aggregated units of species sharing an important ecological characteristic and playing an equivalent role in the community (Cummins 1974). In the second part of this book we will consider a compartmental model obtained by aggregation of species of plants and pollinators in plant–pollinator networks.

C. ***Dimensional homogeneity and consistency.*** Biology, physics and chemistry, rest on a number of measurable entities or *quantities* such as length, mass or time. To each such quantity a unit of measurement is assigned. For example, length measured in meters, mass in kilograms and time in seconds define the MKS (meter–kilogram–second) *system of units*. We associate to each of these three quantities the *physical dimensions* of length, mass and time, respectively. The concept of physical dimension was introduced by the physicist/mathematician Joseph Fourier in 1822. Physical quantities that are of the same kind (called *commensurable*) have the same dimension (length, time, mass) and can be directly compared to each other, even if they are originally expressed in differing units of measure (such as yards and meters). If physical quantities have different dimensions (such as length versus mass),

Table 0.1. Dimensions for standard units in the SI system.

Quantity	Unit	Abbreviation	Dimension
Length	Meter	m	$[L]$
Mass	Kilogram	kg	$[M]$
Time	Second	s	$[T]$
Temperature	Kelvin	K	$[\theta]$
Amount of substance	Mole	mol	$[N]$
Charge	Coulomb	C	$[Q]$
Luminous intensity	Candela	cd	$[I]$

they cannot be expressed in terms of similar units and cannot be compared in quantity (also called *incommensurable*). The dimensions of length, time and mass—denoted, respectively, by L, M, and T—are *fundamental* in the sense that the dimension of any mechanical quantity can be written as a product of powers of them. Thus, the dimension of an area is $[L^2]$, of a speed $[LT^{-1}]$, of an acceleration $[LT^{-2}]$, and of a force $[MLT^{-2}]$. For phenomena outside mechanics, more dimensions are required. For instance, in thermodynamics and bioenergetics the additional dimension is temperature, denoted by $[\theta]$. The standard unit is 1 degree on the Kelvin scale, denoted as K. In the same way, if we want to address electromagnetic phenomena we need to add the charge dimension $[Q]$ for electromagnetic quantities, etc. Table 0.1 shows the dimensions for each of the seven standard units of the International System of Units (SI, abbreviated from the French *Système International*).

Likewise, an extra dimension, the population $[N]$ is needed for biological entities. The corresponding base unit is a biological entity # rather than a mole. Examples of biological entities are individuals, species, or cells. Therefore, growth rates have dimension of population divided by time $[NT^{-1}]$, densities of individuals have dimension of population divided by area $[NL^{-2}]$, etc.

A requirement that is central to mathematical modelling, is that every equation we use must be dimensionally homogeneous or dimensionally consistent. It is quite logical that every term in the second law of Newton, force = mass × acceleration, has total dimensions of force. In the same way, a population dynamic equation with the temporal variation of a population density on the left-hand side, must have total dimensions $[NT^{-1}]$. Therefore, equations will involve four quantities:

i. Dimensional variables, e.g. yield (biomass density), of dimensions $[ML^{-2}]$ and measured in g m^{-2}.
ii. Dimensional parameters, e.g. yield growth rates, of dimensions $[ML^{-2}T^{-1}]$ and measured g m^{-2} t^{-1}.

iii. Dimensionless variables, e.g. relative yields, defined as yields in polyculture divided by yields in monoculture.
iv. Dimensionless parameters.

D. ***Constructing linear approximations.*** Linearity is a concept of central importance in mathematical modelling. A model or system is said to be linear when its basic equations are such that the magnitude of its behavior or response produced is directly proportional to the excitation or input that drives it. Even though most systems in nature are more fully described by nonlinear models, their behavior can often be approximated by linearized models whose equations are, in general, easier to solve analytically than nonlinear equations. Additionally, a linear system obeys the *principle of superposition*: the response of that system to the sum of many individual inputs is obtained by adding or superposing the separate responses of the system to each individual input. Therefore, as engineers do all the time, we can use this principle to predict the response of a system to a complicated input by decomposing or breaking down that input into a set of simpler inputs that produce known system responses or behaviors. For example, we know that the relation between force F and relative extension x of a simple coiled spring, e.g. an automobile spring, is nonlinear, i.e. $F(x) = -k_1x + k_2x^2 + \cdots$. However, civil engineers frequently use, as a first approximation, a simple linear elastic spring ($F(x) = -k_1x$) to model the static and dynamic behavior of a tall building, from wind loading to how the building would respond to an earthquake (Dym 2004). Indeed, this linear spring is an example of a *harmonic oscillator*, which, like the ideal gas, is another fruitful idealized system. Notice that if we write the logistic equation as $dN/dt = rN - (r/\mathcal{K})N^2$, its right-hand side term looks similar to the one of the nonlinear spring. So a Malthusian population would be like the 'linear spring' of population ecology (while the logistic equation would correspond to a more realistic nonlinear spring). As we will see in the next chapter, linearization is also crucial to study the stability of equilibria of dynamical population equations.

E. ***Heuristics for problem-solving.*** *Heuristic* as an adjective means 'serving to discover'. A heuristic method, often called simply a heuristic, is any approach to problem-solving or self-discovery that employs a practical method, not guaranteed to be optimal, perfect, logical, or rational, but instead sufficient for reaching a quick estimation. Heuristics are 'rules of thumb', educated guesses, intuitive judgments or simply common sense. Here are a few other commonly used heuristics, from George Pólya's book, *How to Solve It* (Pólya 2014):
   - If you are having difficulty understanding a problem, try drawing a picture.
   - If you can't find a solution, try assuming that you have a solution and see what you can derive from that ('working backward').

- If you cannot solve the proposed problem try to solve first some related and more accessible problem or look for a similar problem you already solved or know the solution for.
- If the problem is abstract, try examining a concrete example.

Another example of useful heuristic methods in science and engineering are *back-of-the-envelope calculations*, i.e. rough calculations, typically scribbled on any available piece of paper (such as an envelope). It is more than a guess but less than an accurate calculation or mathematical proof. This estimation technique is also called *Fermi order estimate*, after physicist Enrico Fermi as he was known for his ability to make good approximate calculations with little or no actual data. Fermi estimations typically involve making justified guesses about quantities and their variance or lower and upper bounds. Depending on the difficulty of the problem, and the number of estimation sub-problems required, one can usually hope to be correct to within a factor of 2 or 3, and other times to within the correct order of magnitude (i.e. the closest power of 10).

Why bother with Fermi estimates, if they are likely to be off by a factor of 2 or even 10? Well, often, getting an estimate within a factor of 10 or 20 is enough to make a decision. So Fermi estimates can save you a lot of time. Actually, scientists often look for Fermi estimates of the answer to a problem before turning to more sophisticated methods to calculate a more accurate answer. While the estimate is almost certainly incorrect, it provides a useful and simple first check on the results. By contrast, precise calculations can be extremely complex and time consuming. The far larger number of factors and operations involved can obscure very significant errors, either in mathematical process or in the assumptions the modelling equations are based on. Without a reasonable benchmark to work from it is unclear if a result is acceptably precise. The Fermi estimation gives a quick, simple way to obtain this benchmark for what might reasonably be expected to be the answer, giving context to the results.

In addition, practicing Fermi estimations enables you to:

i. Develop important modelling skills, like choosing relevant information, making reasonable estimates of size and quantity and combine bits of information, your estimates and rates appropriately.
ii. Get you comfortable estimating, visualizing and working with unknown quantities or incomplete information.
iii. Rapidly and approximately calculate an unfamiliar quantity.
iv. Learn dimensional analysis and master approximation techniques.

One famous instance came during the first atomic bomb test in New Mexico on 16 July 1945. As the blast wave reached him, Fermi dropped bits of paper from a known height (his outstretched arm). Measuring the distance they were blown, gave him an approximation of wind speed which, together with knowing the distance from the point of detonation, provided an estimation of the energy of the blast. Fermi concluded that the blast must be greater than 10 kilotons of TNT. His guess was remarkably accurate for having so little data: the true answer turned out to be 18.6 kilotons (Kelly 2004). Another illustrative example of a Fermi estimation is in box 0.2.

When getting started with Fermi practice, it is recommended estimating quantities that you can easily look up later, so that you can see how accurate your Fermi estimates tend to be. You might allow yourself to look up particular pieces of the problem—e.g. the world electric energy consumption, or the gross domestic product of a country—but not the final quantity you're trying to estimate. In appendix I we develop some other examples of Fermi estimations.

Box 0.2. Fermi estimation: 'How many piano tuners are there in Chicago?'

A problem Fermi posed to his to students is 'How many piano tuners are there in Chicago?' A typical solution to this problem involves multiplying a series of estimates that yield the correct answer if the estimates are correct. For example, we might make the following assumptions:

1. The population of Chicago is approximately 3 000 000.
2. On average, there are two persons in each household in Chicago.
3. Roughly one household in twenty has a piano that is tuned regularly.
4. Pianos that are tuned regularly are tuned on average about once per year.
5. Each piano tuner works eight hours in a day, five days in a week, and 50 weeks in a year.
6. Each day a piano tuner tunes 3 pianos (it takes about two hours to tune a piano, plus travel time).

From these assumptions, we can compute that the number of piano tunings in a single year in Chicago is

(3 000 000 persons in Chicago) ÷ (2 persons/household) × (1 piano/20 households) × (1 piano tuning per piano per year) = 75 000 piano tunings per year in Chicago.

We can similarly calculate that the average piano tuner performs

(50 weeks/year) × (5 days/week) × (3 pianos per day) = 750 piano tunings per year.

Dividing gives (75 000 piano tunings per year in Chicago) ÷ (750 piano tunings per year per piano tuner) = 100 piano tuners in Chicago. Depending on the specific values you chose, you would probably get answers in the range 30–300. In 2009, the actual number of piano tuners in Chicago was about 290 (Wolfram alpha 2019).

A couple of lessons from the piano tuners example. First, suppose the precise answer tells you there are many thousands of piano tuners in Chicago. Then you know you need to find out why there is this such a divergence from the expected result. Maybe you missed something important. For example, Does Chicago have a number of music schools or other places with a disproportionately high ratio of pianos to people? Or perhaps, most piano tuners have other jobs and they only work part-time? etc. Second, although this Fermi calculation is not very accurate, it may be good enough for practical purposes. Imagine we want to start a store in Chicago that sells piano tuning equipment, and we calculate that we need 10 000 potential customers to stay in business. So we quickly conclude that we should consider a different business.

0.3 Community and population ecology modelling

0.3.1 Parallelism with physics and the debate of the 'biology-as-physics approach'

Much of the rationale of mathematical population theory is explicitly derived from physical science. In fact, its major proponents in the 1920s, Alfred Lotka and Vito Volterra, have been physicists and mathematicians. For instance, Lotka analyzed populations as chemical systems with exchanges of energy and matter governed by the second law of thermodynamics and treated organisms as engaged in competition for energy. According to Lotka, evolution would increase the flow of energy and matter through the system. Furthermore, it seems that the audience which Lotka hoped to reach consisted of physicists, and it was the ecologist C C Adams who pointed out to him the connection of his work with ecology (Kingsland 1981). Volterra, although he was interested in explaining the dynamics of fish populations in the Adriatic, used similar physical analogies to Lotka. The physicist V A Bailey in the 30s modelled organisms as gas molecules obeying the gas laws. Beginning in the 1940s, concepts like 'energy flow', 'trophic levels' and 'ecosystem' became popular in the leading ecology journals, and they indicated a view of nature shaped more by physics than by botany (Worster 1993).

This similarity with physics may be regarded as a virtue for those seeking to turn ecology into a 'hard' science but also as a vice for those who believe that ecology has specificities that make it intrinsically different from physics. From the very beginning mathematical population theory received criticism. Reviewing Lotka's book Charles Elton wrote a common complaint of many later ecologists: 'Like most mathematicians he takes the hopeful biologist to the edge of a pond, points out that a good swim will help his work, and then pushes him in and leaves him to drown.' A common criticism was that so many assumptions have been made in order to simplify the mathematical treatment that the entities considered can nowhere be found in the roster of living organisms (Stanley 1932), i.e. organisms do not behave like gas molecules, a common assumption of population theory (McIntosh 1985). Another concern, common to many ecologists working with natural populations was that the diversity and complexity of natural populations could not be expressed in simple mathematical formulations, i.e. the authors of mathematical population ecology generalized too much, simplified too much (Nice 1937).

This debate between those leaning towards the biology-as-physics approach and those leaning towards the natural history tradition has continued until the present. Those supporting the first viewpoint think that biology would be best served by emulating the physical sciences in selecting simple models that generate testable hypotheses with research questions structured as branching logical trees (Platt 1964). They consider that the fundamental reason of why there is little progress in ecology is that biologists do not follow the four rules of reasoning formulated by Newton (Murray 1992). The first rule is Occam's Razor or the principle of parsimony for the determination of the causes of natural phenomena. The second is the causality principle, i.e. the same cause always produces the same effect. The third corresponds to the generalization by induction in order to allow the extension of the range of applicability of a theory to all natural things. 'The process of induction is the process of assuming the simplest law that can be made to harmonize with our experience' (Wittgenstein 1922). The fourth rule states that a hypothesis is only as good as the extent to which its predictions conform to fact. Murray (1992) goes on by developing a metaphor in terms of the motion of a cart pushed along the street. To understand this motion it is wise to begin with the simplest possible situation, and proceed gradually to the more complicated ones (Einstein and Infeld 1938). For example, if we give a cart a push along a level road it will move some distance before coming to rest. How can we increase this distance? There are various ways, like oiling the wheels or making the road smoother to reduce friction both in the wheels and between the wheels and the road. The more the friction is reduced, the longer the cart will travel before stopping. One significant step farther is to imagine this simplest idealized situation: the road is perfectly smooth, and wheels with no friction at all. Then there would be nothing to stop the cart, so that it would move at a constant speed in a straight line forever. But, we cannot actually perform this experiment since it is practically impossible to eliminate friction completely. It is a *gedankenexperiment* (from the German, a thought experiment), we can only think it. Therefore, 'the first law of motion or principle of inertia cannot be derived directly from experiment, but only by speculative thinking consistent with observation. The idealized experiment can never be actually performed, although leads to a profound understanding of real experiments' (Einstein and Infeld 1938). Moreover, it allowed Newton to predict from how far an apple would fall near the Earth's surface to the elliptical shape of planetary orbits. Murray (1992) regrets biologists do not approach problems in this way. Rather, 'biologists called in to study the physics of motion would be designing experiments to determine the role of those factors impeding the forward progress of the cart, such as the nature of the ball bearings, the viscosity of the lubricating oil, the shape of the cart, wind speed and direction, temperature and humidity, the road materials, and so on'. And he concludes: biologists shouldn't be 'distracted by the biological equivalents of friction, if we are interested in discovering the biological equivalents of inertia'.

Many ecologists reply that biologists should not think like physicists since biology has developed in a way that is distinctly different from the physical sciences (Aarsen 1997). For example, biology is largely context dependent; often deals of necessity with small sample sizes; and for a broad range of organisms, the most interesting phenomena occur over longer timescales than the feasible long-term studies (Anderson 2017). Naturalists have been fascinated by biological diversity and do not view organisms merely as models, or vehicles for theory, but rather as the thing itself that excites their admiration and their desire for knowledge, and understanding (Schmidley 2005). A common claim of ecologists that deplore the biology-as-physics approach is that because of the nature of biological science, progress in ecology has been hampered by an excessive focus on simple models, which fail to adequately capture important processes driving ecosystem dynamics. Rather, complex, system-specific models are often needed to provide the ability to predict the current and future behavior of systems. These models need to incorporate all relevant processes and then be tested, perhaps by simulation, to assess which processes most influence the predictions of the model (Evans 2013). It was also argued that biology needs historical explanations which do not systematically follow the methods of Newtonian physics (Quenette and Gerard 1993). On both sides of the debate, there is a widespread belief that ecology is different from physics because (A) it lacks general laws, and (B) it is not a predictive and, therefore, not a 'hard' science (Turchin 2003).

Freeman Dyson summarized the reason of this 'clash of cultures' (1988). According to him, there are two kind of scientists, the unifiers and the diversifiers. The driving passion of unifiers is to find general principles which will explain everything. They are happy if they can leave the Universe looking a little simpler than they found it. Diversifiers are people whose passion is to explore details and are in love with the heterogeneity of nature. They are happy if they leave the Universe a little more complicated than they found it. Whereas physicists are unifiers, many ecologists are diversifiers.

0.3.2 Trade-offs and modelling strategies

The main trade-offs we face when modelling population dynamics as well as the different choices we have were analyzed by Richard Levins in an essay of 1966 that became a classic (Levins 1966). He stated that an optimally general, precise, and realistic model would require using a very large number of simultaneous partial differential equations involving a huge number of parameters to estimate from measurements. These equations would be analytically insoluble, and even if soluble (in the form of quotients of sums of products of parameters) would be uninterpretable, i.e. they would have no meaning for us. It is true that, from 1966 to today, we have augmented our computational abilities and power by simulations, including individual-based models, for addressing both applied and theoretical questions. However, as we already mentioned, these more realistic simulation models (a) can have hundreds of parameters which are impossible to measure in practice by limitations of time and resources, and (b) by including so many variables they sacrifice understandability for ecological realism. Therefore, there is an unavoidable trade-off for us between the generality, precision, and realism of the mathematical models if they are to be of any use to ecologists. Levins thus concludes the above trade-offs suggest three alternative strategies for building models:

Type I models sacrifice generality to precision and realism.
Type II models sacrifice realism to generality and precision.
Type III models sacrifice precision to generality and realism.

Type I models are large models preferred in applied fields, like Fisheries, or by *systems ecology* practitioners who advocate for a holistic approach of ecology that takes all the various components into account (Odum 1994, Watt 1968). The stylized type II models would correspond to the biology-as-physics approach we already reviewed. Type III models are favored when ecologists are concerned with qualitative rather than quantitative results.

This book is structured around practical applications and ours will be an instrumentalist viewpoint, regarding models primarily as operational tools to make quantitative predictions. However we will attempt to maintain balance between the simplicity of the unifiers and the complexity of the diversifiers. In that sense, a general recipe that has worked very well in science is the so-called *principle of parsimony* aka *Occam's razor*: 'plurality should not be posited without necessity', meaning that all things being equal, the simpler theory is more likely to be correct. Simpler models are less prone to overfitting occurring when, in the effort to refine the model to the maximum, we end up adjusting the model to the imperfections of the data we have, worsening our predictive capacity and confusing signal with noise. Therefore, we will adopt a parsimonious approach for model-building. This implies that, since most of the applications we will consider require quantitative output, we will mainly deal with type II models (much simpler than type I models and more adequate than type III models for quantitative analysis).

Besides being parsimonious, our approach will also be pragmatic since we want to overcome the existing trade-off between efficiency and robustness. Something is efficient if it performs optimally under ideal circumstances, while something is robust if it performs pretty well under non-ideal circumstances. It turns out that robust (simple) models are often more efficient than efficient (complex) models are robust. When facing practical problems the conditions are in general far from ideal. There are model parameters that are unknown, let alone errors in the parameters. Moreover, the data are usually incomplete and/or contaminated and/or the statistics is not clearly sufficient, etc. Under these conditions complex efficient models can fail spectacularly because the assumptions they're based on don't hold. We will show how simple models can contribute to solve practical problems in fields like agriculture and environmental sciences. However, since no single model can meet all the requirements of generality, realism, precision, and manageability, for those specific applications for which type II models are unfit we will consider models that could be considered of type I or of type III.

Framing ecology as physics is probably neither possible nor desirable. Nevertheless, there are practices that have paid off. Firstly, borrowing ideas from other sciences, especially physics, and incorporating them into ecology. This is a tradition that has existed at least since the days of Lotka (1925). Secondly, embracing the spirit of physics, do not be afraid of relying on partial aspects and partial relations of things which we only understand partially. Amazingly, in spite of all our ignorance, we can make rather precise statements and predictions about certain aspects of phenomena. Thirdly, whether you are a diversifier or a unifier, it is undeniable that using idealized systems has been crucial in providing us with insight and understanding of several natural phenomena. Therefore, we will adopt these three practices throughout this book.

Definitions and tips for non mathematicians

Many people, including agronomists and ecologists, often dislike reading books full of equations. One reason is they are not used to mathematics and find it hard, but so was any language before we learned it. Comprehension and eventual proficiency in handling equations results from repeated encounter and practice. Another reason is that they don't know how to relate equations to anything real. This lack of clear meaning is a source of frustration to the reader. If that happens, you're doing it wrong. Math is nothing more than a concise language for describing relationships between objects. So if you are not a mathematician (and maybe even if you are) but someone who sees calculus as a tool, rather than an end in itself, the definitions and tips below hopefully will be useful to finding meaning in equations, to manipulating equations in order to facilitate understanding, and to making you feel comfortable reading/writing equations.

Basic definitions and notation

The first step is to specify the different entities coexisting in an equation. Most equations contain *variables*, denoted by letters in italic font signifying things that are known or unknown. That is, variables can be classified as dependent and independent. The values of **dependent** variables depend on the values of **independent** variables. The dependent variables represent the output or outcome whose variation is being studied. The independent variables represent inputs or causes. In other words, the dependent variable is a ***function*** of the independent variable. For example, the population density of a species N is a function of time t. And, using the notation first introduced by Leonhard Euler in 1734, this *functional dependence* is denoted explicitly as $N(t)$, i.e. by a symbol consisting generally of a single letter in italic font, followed by the independent variable between parentheses. The notation reads: 'N of t'.

Some widely used functions are represented by a symbol consisting of several letters (usually two or three, generally an abbreviation of their name). By convention, in this case, a roman type is used, such as 'sin' for the sine function, in contrast to italic font for single-letter symbols.

Besides variables there are quantities which are assumed to be constant or fixed and are called *parameters* (from the Greek word *para-*, meaning 'beside,' and *metron*, meaning 'measure'). To distinguish parameters from variables we will denote parameters using manuscript font. For example the per-capita growth rate and the carrying capacity of a population are denoted, respectively, by '$\mathscr{r}$' and '$\mathscr{K}$'.

When parameters are present, they actually define a whole family of functions, one for every valid set of values of the parameters.

Historically, the concept of function was elaborated with the infinitesimal calculus at the end of the 17th century, i.e. the functions that were considered were differentiable (that is, they had a high degree of regularity). So besides algebraic equations, relations only involving variables (and parameters), we have differential equations. A **differential equation** is a mathematical equation that relates some function with its derivatives. The first equation that appears in this book, equation (0.1), is an example of a differential equation, that contains all the ingredients mentioned up to now:

- an independent variable (the time t),
- a dependent variable or a function of t (the population density $N(t)$),
- the derivative of this variable (dN/dt),
- and two parameters ($\mathscr{r}$ and $\mathscr{K}$).

To emphasize that the population density N is a dependent function of t, equation (0.1) can be re-written as

$$dN/dt = rN(t)[1 - N(t)/K]$$

Finally, sets of variables or sets of parameters can be arranged into *arrays*, for example one-dimensional arrays or *vectors*, and two-dimensional arrays or *matrices*. We will denote vector and matrices as bold symbols. Unless we specify otherwise, we will use lowercase for vectors and uppercase for matrices. For example, if we have S species coexisting in a community with densities N_1, N_2, ..., N_S, we can group them into a column vector denoted as the column of densities between square brackets:

$$\mathbf{n} = \begin{bmatrix} N_1 \\ N_2 \\ \vdots \\ N_S \end{bmatrix}.$$

Likewise, we will arrange the pairwise interaction coefficients a_{ij} between these species (which are parameters) into an $S \times S$ square matrix:

$$\mathbf{A} = \begin{pmatrix} a_{11} & \cdots & a_{1S} \\ \vdots & \ddots & \vdots \\ a_{S1} & \cdots & a_{SS} \end{pmatrix}.$$

Reading and writing equations

When reading an equation it becomes crucial to understand what the equation is telling you; you should try to 'inhabit' the situation this equation describes. Translating equations into words is a crucial step for understanding them. Then you should answer the following questions (Pólya 2014):

- What is the unknown?
- What are the data?
- How are the various items connected?
- How is the unknown linked to the data?

Let us suppose we have a differential equation for the growth of a certain population N living in a given area. Something of the form:

$$\frac{dN}{dt} = i(N(t);\ \{c_\alpha\}) - o(N(t);\ \{c_\beta\}),$$

where $i(N(t);\{c_\alpha\})$ and $o(N(t);\{c_\beta\})$ correspond, respectively, to 'input' and 'output' terms (corresponding for instance to birth and immigration rates and death and emigration rates) which are functions of the variable population N (t) and of sets of parameters denoted by curly brackets as $\{c_\alpha\}$ and $\{c_\beta\}$, $\alpha = 1,2,\ldots$, $\beta = 1,2,\ldots$, (we use Greek indices for parameters).

A helpful recipe to qualitatively analyze this kind of differential equation is to break it down into chunks and visualize what the relationship between variables looks like. You can follow these steps:

1. Check that all terms in both sides of the equation have the same dimension. (In our case $[NT^{-1}]$).

2. Identify which of the quantities are constants and which are variables. (Variables: t, N; constants: $\{c_\alpha\}$ and $\{c_\beta\}$).
3. As a first approximation, ignore the constants; you can just set them all to 1.
4. Break each term down and visualize what it would look like when plotted on an x–y graph. (Plot $i(N(t);\{1\})$ and $o(N(t);\{1\})$ versus $N(t)$).
5. Add or subtract the terms as the equation says, and visualize the resulting curve. $(i(N(t);\{1\}) - o(N(t);\{1\}))$.
6. Go back and look at the constants to see how this curve scales.
7. If you are interested in the behavior of $N(t)$, remember that to solve the above differential equation it is necessary to integrate it. And, that the geometrical interpretation of the integral is the 'area under the curve' between the initial and final times t_i and t_f. (The area under $i(N(t);\{c_\alpha\}) - o(N(t);\{c_\beta\})$ varying t_i and t_f.)

References

Aarsen L W 1997 On the progress of ecology *Oikos* **80** 177–8

Anderson J G T 2017 Why ecology needs natural history *Am. Sci.* **10** 290

Andrewartha H G 1961 *Introduction to the Study of Animal Populations* (Chicago, IL: University of Chicago Press)

Andrewartha H G and Birch L C 1954 *The Distribution and Abundance of Animals* (Chicago, IL: University of Chicago Press)

Bender E A 2000 *An Introduction to Mathematical modelling (Dover Books on Computer Science)* 1st edn (New York: Dover)

Box G E P 1976 Science and statistics *J. Am. Stat. Assoc.* **71** 791–9

Cummins K W 1974 Structure and function of stream ecosystems *BioSciences* **24** 631–41

DeAngelis D L and Grimm V 2014 Individual-based models in ecology after four decades *F1000Prime Rep.* **6** 39

Diamond J M and Case T J (ed) 1986 *Community Ecology* (New York: Harper and Row)

Dym C L 2004 *Principles of Mathematical modelling* 2nd edn (Amsterdam: Elsevier)

Dyson F 1988 *Infinite in All Directions* (New York: Harper and Row)

Edelstein-Keshet L 1988 *Mathematical Models in Biology* (New York: Random House)

Einstein A and Infeld L 1938 *The Evolution of Physics: From Early Concepts to Relativity and Quanta* (New York: Simon and Schuster)

Evans M R *et al* 2013 Do simple models lead to generality in ecology? *Trends Ecol. Evol.* **28** 578–83

Fort H 2018a Quantitative predictions from competition theory with an incomplete knowledge of model parameters tested against experiments across diverse taxa *Ecol. Model.* **368** 104–10

Fort H 2018b On predicting species yields in multispecies communities: Quantifying the accuracy of the linear Lotka–Volterra generalized model *Ecol. Model.* **387** 154–62

Fort H and Segura A 2018 Competition across diverse taxa: quantitative integration of theory and empirical research using global indices of competition *Oikos* **127** 392–402

Gaston K J 1994 *Rarity* (London: Chapman and Hall)

Gause G F 1932 Experimental studies on the struggle for existence *J. Exp. Biol.* **9** 389–402

Ginzburg L R and Jensen C X J 2004 Rules of thumb for judging ecological theories *Trends Ecol. Evol.* **19** 121–6

Giordano F R *et al* 2009 *A First Course in Mathematical modelling* (Belmont, CA: Cengage Learning)

Hallam T G and Levin S A (ed) 1986 *Mathematical Ecology: An Introduction to BiomathematicsBiomathematics series* vol 17 (Berlin, Heidelberg: Springer)
Hilborn R and Mangel M 1996 *The Ecological Detective: Confronting Models with Data* (Princeton, NJ: Princeton University Press)
Ishida Y 2007 Patterns, models, and predictions: Robert MacArthur's approach to ecology *Philos. Sci.* **74** 642–53
Kelly C K (ed) 2004 *Remembering The Manhattan Project: Perspectives on the Making of the Atomic Bomb and its Legacy* (Singapore: World Scientific)
Kingsland S E 1981 Modelling nature: theoretical and experimental approaches to population ecology 1920–1950 *PhD Thesis* University of Toronto
Kot M 2003 *Elements of Mathematical Ecology* (Cambridge: Cambridge University Press)
Krebs C J 1978 *Ecology: the Experimental Analysis of Distribution and Abundance* 2nd edn (New York: Harper and Row)
Levin S A, Hallam T G and Gross L J (ed) 1989 *Applied Mathematical Ecology(Biomathematics series* vol 18) (Berlin: Springer)
Levin S A (ed) 1992 *Mathematics & Biology: The Interface Challenges and Opportunities* (Berkeley, CA: Lawrence Berkeley Laboratory)
Levins R 1966 The strategy of model building in population biology *Am. Sci.* **54** 421–31
Lonegran M 2014 Data availability constrains model complexity, generality, and utility: a response to Evans *et al Trends Ecol. Evol.* **29** P301–2
Lotka A J 1925 *Elements of Physical Biology* (Baltimore, NY: Williams and Wilkins)
Massimi M and Bhimji W 2015 Computer simulations and experiments: the case of the Higgs Boson *Stud. Hist. Philos. Sci. Part B* **1** 71–81
May R M 1974 *Stability and Complexity in Model Ecosystems* (Princeton, NJ: Princeton University Press)
May R M 2004 Uses and abuses of mathematics in biology *Science* **303** 790–3
McIntosh R 1985 *The Background of Ecology: Concept and Theory, Cambridge Studies in Ecology* (Cambridge: Cambridge University Press)
Motz L 1987 Introduction *The World of Physics* ed J H Weaver (New York: Simon and Schuster) pp 17–27
Murray J D 1989 *Mathematical BiologyBiomathematics series* vol 19 (Berlin: Springer)
Murray B G 1992 Research methods in physics and biology *Oikos* **64** 594–6
Nice M M 1937 *Studies in the Life History of the Song Sparrow* (NewYork: Dover)
Odum H T 1994 *Ecological and General Systems: An Introduction to Systems Ecology* (Niwot, CO: University Press of Colorado)
Onstad D W 1988 Population-dynamics theory: The roles of analytical, simulation, and supercomputer models *Ecol. Model.* **43** 111–24
Pastor J 2008 *Mathematical Ecology of Populations and Ecosystems* (Singapore: Wiley-Blackwell)
Platt J R 1964 Strong inference *Science* **146** 347–53
Pólya G 2014 *How to Solve It* (Princeton, NJ: Princeton University Press)
Quenette P Y and Gerard J F 1993 Why biologists do not think like Newtonian physicists *Oikos* **68** 361–3
Schmidley D J 2005 What it means to be a naturalist and the future of natural history at American universities *J. Mammol.* **86** 449–56
Stanley J 1932 A mathematical theory of the growth of populations of the flour beetle *Tribolium confusum*, Duv *Can. J. Res.* **6** 632–71

Turchin P 2003 *Complex Population Dynamics: A Theoretical/Empirical Synthesis* (Princeton, NJ: Princeton University Press)

Vandermer J 2010 *The Ecology of Agroecosystems* (Burlington, MA: Jones and Bartlett)

Volterra V 1926 Fluctuations in the abundance of a species considered mathematically *Nature* **118** 558–60

Watt K E F 1968 *Ecology and Resource Management: A Quantitative Approach* (New York: McGraw-Hill)

Wittgenstein L 1922 *Tractatus Logico-Philosophicus* (London: Routledge and Kegan Paul) fifth impression 1951

Wolfram alpha 2019 https://wolframalpha.com/input/?t=crmtb01&f=ob&i=how+many+piano+tuners+are+in+chicago

Worster D 1993 *The Wealth of Nature: Environmental History and the Ecological Imagination* (New York: Oxford University Press)

Part I

Classical population and community ecology

IOP Publishing

Ecological Modelling and Ecophysics

Agricultural and environmental applications

Hugo Fort

Chapter 1

From growth equations for a single species to Lotka–Volterra equations for two interacting species

'The importance of the method is this: if we know certain variables, mostly desired by ecologists and in some cases already determined by them, we can predict certain results which would not normally be predictable or even expected by ecologists. The stage of verification of these mathematical predictions has hardly begun; but their importance cannot be under-estimated, and we look forward to seeing the further volumes of Lotka's studies.'

Charles Elton 1935 review of A J Lotka's *Théorie analytique des associations biologiques*.

Summary

We start this chapter in section 1.1 with the simplest growth equation for a single isolated population, the Malthus equation. The problem with this equation is that it assumes a constant per capita growth rate, leading to exponential growth. Then we will show that if we assume resource limitation, which means replacing the constant per capita growth rate by a ***density dependent*** per-capita growth rate, the resulting ***logistic equation***, saturates the population to an equilibrium value called the ***carrying capacity***.

Before we move to two interacting species, in section 1.2, we consider general models for single species populations and show how to analyze the ***local equilibrium stability***, both algebraically and geometrically.

In section 1.3 we present the Lotka–Volterra equations for a predator and its prey. We first consider the original Lotka–Volterra predator–prey model.

We introduce the useful concept of ***phase plane*** for representing the dynamical behavior of a two-species system as a trajectory followed by a point representing the system in this plane. We show that this model gives rise to periodic oscillations in the abundances of predator and prey populations, resembling the empirically observed behavior. However, the model is unrealistic and it has conceptual problems, mainly its ***structural instability***. Namely, oscillations of predator and prey abundances correspond to closed trajectories in the phase plane and any small perturbation could make the system jump onto another orbit which can be quite different from the original one. We next show that the Lotka–Volterra model can be modified in simple ways to make it more biologically realistic. These modifications, consisting in making the prey populations density-dependent (e.g. logistically) and making the predator death rate depend inversely on prey density, cure the problem of the structural instability: the modified predator–prey equations give rise to robust periodic orbits, termed stable ***limit cycle*** oscillations.

In section 1.4, we introduce the Lotka–Volterra competition equations for a pair of species and analyze the possible equilibria: competitive exclusion of one species or species coexistence. We also show how this descriptive or phenomenological model (i.e. without specifying either the limiting resources the two species are competing for or the mechanism of competition) can be transformed into a mechanistic model by expressing the competition coefficients and carrying capacities in terms of rates of utilization and renewal of resources.

Finally, in section 1.5, we discuss a system formed by a pair of species engaged in a mutualistic relationship. Two types of mutualism between two species, obligate and facultative, are introduced as well as the corresponding Lotka–Volterra equations. We conclude by analyzing the possible equilibria.

1.1 From the Malthus to the logistic equation of growth for a single species

1.1.1 Exponential growth

Describing how a population of a given species changes with time has a long history. For example, the Italian mathematician Leonardo of Pisa (1175–1250), better known as Fibonacci, in his book *Liber Abaci* (1202) posed this problem involving the growth of a population of rabbits:

A certain man had one pair of rabbits together in a certain enclosed place. How many pairs of rabbits can be created from the pair in one year if we suppose that each month each pair begets a new pair, which on the second month on becomes productive?

Fibonacci used a simple mathematical model based on idealized assumptions to solve this problem (see exercise 1.1).

Here we will devote some space to explain step by step the population dynamics approach as well as analyzing its implicit underlying assumptions. So imagine a population of a given species, rabbits if you like, in a given area. This population changes in a given period due to births, deaths, immigration and emigration that occur in this period. We can express this as a population balance (inputs minus outputs) relationship:

$$\text{change in the population} = \text{births} - \text{deaths} + \text{immigration} - \text{emigration}, \quad (1.1)$$

Let us assume the simplest situation without migration in which the above relation simplifies to:

$$\text{change in the population} = \text{births} - \text{deaths}, \quad \text{(isolated population)} \qquad (1.2)$$

and see how we can traduce the above conceptual relationship into a mathematical equation.

Let N_t be the population density (number of individuals in a specific area) of the species at time t. Hence, the left-hand side of equation (1.2) can be written as a **difference equation**

$$N_{t+1} - N_t = B_t - D_t. \qquad (1.3)$$

where B_t and D_t denote, respectively, the number of births and deaths that occurred during the interval from t and t +1. This interval between consecutive integer time values may denote, depending on the problem and the species involved, from minutes (in the case of bacteria or protozoa) to years (in the case of say humans).

Suppose now that the population N is a large number and we are interested in changes of this population for an arbitrary small interval of time Δt. In particular, if we take Δt infinitesimally small, equation (1.3) can be rewritten in terms of instantaneous rates of births and deaths, b_t and d_t (measured as individuals per unit of time), as

$$\frac{N_{t+\Delta t} - N_t}{\Delta t} = \frac{B_t}{\Delta t} - \frac{D_t}{\Delta t} = b_t - d_t. \qquad (1.3')$$

Actually, N_t and $N_{t+\Delta t}$ are discrete variables so they cannot differ in less than one individual, and if Δt is taken small enough, they will be equal and equation (1.3′) breaks down. The ‘trick’ is to promote the discrete (integer number) population variable N_t to a continuous (real number) $N(t)$ (to emphasize we are using continuous variables, t appears between parentheses rather than as a sub-index). The left-hand side of equation (1.3′) defines the derivative of N with respect to time:

$$\lim_{\Delta t \to 0} \frac{N_{t+\Delta t} - N_t}{\Delta t} = \frac{dN}{dt}$$

—which sometimes is also denoted as $\dot{N}$—, so that equation (1.3′) becomes a **differential equation**:

$$\frac{dN}{dt} = b(t) - d(t). \qquad (1.3'')$$

So, what should we use, a continuous or discrete time description? From a biological point of view, if births occur continuously with overlapping generations in relatively seasonal environments, continuous time is a sound choice. However, for many species births occur in regular time-intervals so they have no overlap whatsoever between successive generations and therefore population growth is better-suited for a discrete-time description. For primitive organisms such discrete steps can be quite short in which case a continuous time model may be a reasonable approximation. In fact, the step lengths can vary widely from species to species.

Biological considerations aside, the advantage of using differential equations is that qualitative insight is usually gained from simple model problems that may be solved using analytical methods. That is, as we shall see in a moment, calculus often helps in uncovering functional relationships between the relevant problem variables. However, most problems of interest lead to differential equations that cannot be solved easily using analytic techniques. In such cases numerical methods allow us to use the power of a computer to obtain quantitative insight. Since a computer is limited to finite combinations of the four arithmetic operations, $+$, $-$, $\times\_$, $\div\_$, and logical operations, numerical methods require discrete time for running such computations. Therefore, hereafter in Part I, we will resort to the continuous time description and to equation (1.3′) as the starting point to introduce the formalism of population dynamics. On the other hand, for most practical applications, it seems more natural to build the model as a *discrete* difference equation from the start, without going through the doubly approximative process of first, during the modelling stage, finding a differential equation to approximate a basically discrete situation, and then, for numerical computing purposes, approximating that differential equation by a difference scheme (for an interesting discussion see van der Vaart 1973). Thus, in the companion practical application chapters, involving more advanced models that need to be approached by numerical solution techniques, we will use the discrete time representation[1].

The birth and death rates, $b(t)$ and $d(t)$, in equation (1.3′) are still unspecified functions of the time t. Obviously, both rates depend on population size; the larger $N(t)$ the larger the number of births and deaths per unit of time. The simplest assumption we can make is that both rates are proportional to $N(t)$:

$$b(t) = \mathscr{b}N(t), \tag{1.4a}$$

$$d(t) = \mathscr{d}N(t), \tag{1.4b}$$

where $\mathscr{b}$ and $\mathscr{d}$[2] are ***constant*** per-capita rates, measured, respectively, as births and deaths per individuals per unit of time. Therefore the equation (1.3′) becomes

$$\frac{dN}{dt} = (\mathscr{b} - \mathscr{d})N(t). \tag{1.5}$$

It is customary to group $(\mathscr{b} - \mathscr{d})$ into a net constant per-capita growth rate $\mathscr{r}$ and write equation (1.5) as (Malthus 1798):

[1] A note of caution is in order here. As was shown by Robert May (1976), some of the simplest nonlinear difference equations can exhibit a wide spectrum of dynamical behavior. From stable equilibrium points, to stable cyclic oscillations between two population points, four points, eight points, etc, through to a chaotic regime in which (depending on the initial population value) cycles of any period, or even totally aperiodic but bounded population fluctuations, can occur.

[2] Warning: $\mathscr{d}N$ in equation (1.4b) is NOT the same as the numerator of the temporal derivative dN/dt!

$$\boxed{\frac{dN}{dt} = r\,N.} \tag{1.6}$$

(where we have omitted the temporal dependence of N for compactness of notation.)

We know from calculus that the solution of this equation is

$$N(t) = N_0 e^{rt}, \tag{1.7}$$

where $N_0 = N(0)$ is the initial population. Thus if $r > 0$ (i.e. $b > d$) the population grows exponentially while if $r > 0$ (i.e. $b < d$) it dies out. The graph of this exponential function (figure 1.1) shows the behavior of the population over time.

The differential equation (1.6), due to Malthus in 1798, is fairly unrealistic. It says that the rate of change in population size over time (dN/dt) increases by a proportional rate of growth (r) multiplied by the current population size (N). This is an example of a density-independent biological force, because it depends only on the population N, not on external forces such as crowding or food supply. We already mentioned some of the simplifications leading to equation (1.6). In addition, there are several underlying implicit assumptions and it is worth making them explicit because it is important for modellers to be aware of the approximations behind the models they use so that they understand how these models can fail. Box 1.1 summarizes the assumptions behind equation (1.6) as well as the corresponding mathematical implications.

In principle, all the above assumptions would be only justified for an isolated, well mixed in a homogeneous habitat, sexless *parthenogenetic* (i.e. exhibiting asexual reproduction) population in which individuals are immediately reproductive when they are born, with no resource limitation and completely controlled conditions (no stochasticity). A growing population of bacteria or protozoa most closely approximates this situation for a while (until it faces resource limitations).

Unrealistic as it is, equation (1.6) is the foundation for all of the population models that we will discuss in this book. In particular, the mean-field approximation IV together with the approximations V and VI of a dynamical autonomous system is common to all population ecology, the subject of part I. As we will see in the

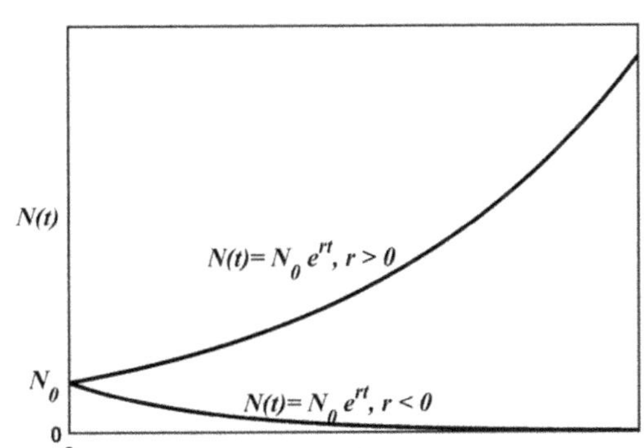

Figure 1.1. Exponential growth ($r > 0$) and decay ($r < 0$).

Box 1.1. Assumptions behind Malthus equation (1.6).

	Assumption	Mathematical implication
I	**Continuous overlap of generations.**	Use of differential equations.
II	**Geographical isolation.**	No migration rates in the differential equation.
III	**Interactions with individuals of other species are negligible.**	The rates $b(t)$ and $d(t)$ do not depend on other species neither through parameters (exercise 1.4) nor through population variables (sections 1.3 and 1.4).
IV	***Mean-field approximation*** (MFA): the spatial nature of interactions is ignored; only changes in mean quantities, such as global densities, are tracked. The MFA in fact implies two approximations (Morozov and Poggiale 2012): **IVa** The approximation of ***well mixed population***, i.e. the individuals in the habitat are well mixed and the probability of interaction of a randomly taken individual with any other individual does not depend on the individual chosen. **IVb** The environment is considered to be **homogeneous** (e.g. no sharp gradients of resource distribution).	$N(t)$, $b(t)$ and $d(t)$ do not depend on space coordinates. No spatial derivatives in the differential equation.
V	***Dynamical system.*** There is a branch of mathematics developed specifically to deal with dynamics: *dynamical systems theory.*	No higher time derivatives than the first appear in the differential equation[3].
VI	***Autonomous system.***	The rates $b(t)$ and $d(t)$ do not *explicitly* depend on t, their dependence is only *implicit* through $N(t)$ i.e. they are functions of $N(t)$.
VII	**Non interacting individuals.** Equivalent to assume that resources for growth and repro-duction are unlimited (see next subsection).	$b(t) = \mathscr{b}N(t)$, $d(t) = \mathscr{d}N(t)$, $\mathscr{b}$, $\mathscr{d}$ constants
VIII	**No age or size structure.**	There are no differences in the per-capita rates $\mathscr{b}$ and $\mathscr{d}$ among individuals due to age or body size.
IX	**Deterministic.** The population is determined solely by N_0 and $\mathscr{r}$: If we started over with the same set of conditions, the population would grow to precisely the same size.	No noise or stochastic term in the differential equation.

[3] Notice however that many higher-order differential equations can be written as first order systems of differential equations by the introduction of derivatives as new dependent variables (see exercise 1.12).

Application chapters, very simple models based on most of the above assumptions still produce reliable results for more complex species under much less restrictive conditions.

1.1.2 Resource limitation, density dependent per-capita growth rate and logistic growth

In deriving equation (1.6) we unrealistically assumed that per-capita rates b and d were constant, independent of the population density (assumption VII in box 1.1), and thus the same happens for $r = b - d$. However, when resources (nutrients, water, space, etc) are limited, if crowding is increased, we expect the per capita birth rate to decrease and the per capita death rate to increase because fewer resources are available for organisms to use for reproduction and survival, respectively. Hence, these per-capita rates should really be a function of the population density of the species, so that, as population density increases to sufficiently high levels, the death rate rises and the birth rate falls, bringing about a fall in r toward zero and below, into negative values. In other words, more realistic per-capita rates b and d exhibit *density dependence*. The simplest formula for a decreasing net growth rate $r(N)$ is a linear one—straight line with a proportionality constant c—as shown in figure 1.2:

$$r(N) \equiv r - \mathrm{c}N = r - \mathrm{c}N = r\left(1 - \frac{N}{r/\mathrm{c}}\right). \tag{1.8}$$

If we define $\mathcal{K} \equiv r/\mathrm{c}$ and replace r in equation (1.6) by (1.8), we re-obtain the logistic equation of Pearl and Reed (1920):

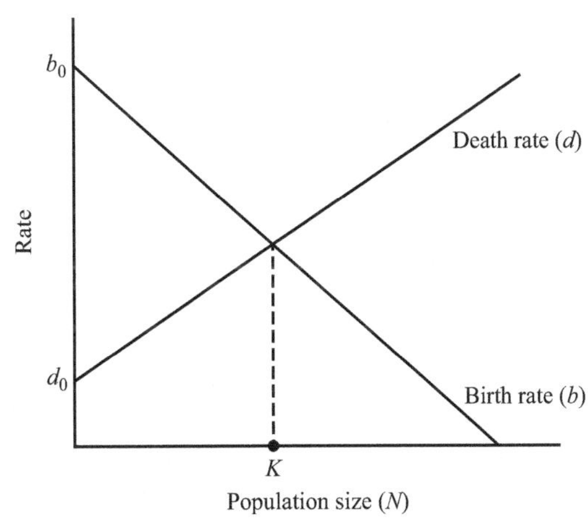

Figure 1.2. Linear density-dependent birth and death rates for the logistic model illustrating how the per capita rates of birth and death change as a function of crowding. The population reaches a stable equilibrium ($N = \mathcal{K}$) at the intersection of the curves, where birth and death rates are equal.

$$\boxed{\frac{dN}{dt} = r N\left(1 - \frac{N}{\mathcal{K}}\right).} \tag{1.9}$$

The logistic equation which was originally described as a model for human population growth by Pierre-Francois Verhulst (1838), is a simple descriptive model of how competition limits the growth of populations. The model assumes that a maximum sustainable population density, the carrying capacity $\mathcal{K}$, exists where $dN/dt = 0$. The per capita population growth rate $(1/N)(dN/dt)$ is at its maximum value of r when N is close to zero, then declines linearly to zero when N reaches $\mathcal{K}$. If N exceeds $\mathcal{K}$, the per capita growth rate becomes negative.

The logistic model is the simplest nonlinear differential equation (box 1.2). Most nonlinear differential equations cannot be explicitly solved, that is, there are no methods to integrate them to obtain an explicit equation for $N(t)$ in terms of initial conditions, $N(0)$, and the parameters—or else the explicit, time-dependent solutions are too unwieldy to be of much help. Fortunately, the logistic equation has an explicit simple solution. Let us integrate this equation to review, at least once, how differential equations are solved analytically by the *separation of variables* method. So the first step is to separate the terms in N on the lhs and those in t on the rhs:

$$\frac{dN}{N(1 - N/\mathcal{K})} = r dt$$

Then integrate the lhs from $N(0)$ to $N(T)$ and the rhs from $t = 0$ to $t = T$:

$$\int_{N(0)}^{N(T)} \frac{dN}{N(1 - N/\mathcal{K})} = \int_0^T r dt$$

Box 1.2. Logistic model assumptions.

Because the logistic model is derived from the Malthus model it shares all its assumptions except # VII of box 1.1, i.e. we have seen that resources are limited in the logistic model, which implies a density dependent per capita growth rate. So assumption VII is replaced by two additional assumptions:

VIIa *Constant carrying capacity*. In order to achieve the S-shaped logistic growth curve, we must assume that $\mathcal{K}$ is a constant: resource availability does not vary through time. In the companion Application chapter, we will relax this assumption, and consider a periodically forced carrying capacity.

VIIb *Linear density dependence*. Each individual added to the population causes an incremental decrease in the per capita rate of population growth. This is illustrated in figure 1.2 which shows the per capita population growth rate $(1/N)(dN/dt)$ as a function of population density.

The integral of the rhs is the simplest one and we immediately get $\mathfrak{r}t$. To integrate the lhs we will use the identity:

$$\frac{1}{N(1 - N/\mathcal{K})} = \frac{1}{N} + \frac{1}{1 - N/\mathcal{K}},$$

and then we integrate each part separately:

$$\int_{N(0)}^{N(T)} \frac{dN}{N(1 - N/\mathcal{K})} = \int_{N(0)}^{N(T)} \frac{dN}{N} + \int_{N(0)}^{N(T)} \frac{dN}{1 - N/\mathcal{K}}.$$

Recall that the integral of $1/N$ is $\ln N$ and the integral of the second term on the lhs can be obtained, by making a change of variable $N' = 1 - N/\mathcal{K}$, as $-\ln(1 - N/\mathcal{K})$.[4] To find definite integrals we have to evaluate at the upper and lower bounds and subtract the expression for the lower bound from that of the upper. Therefore, we get:

$$\ln(N(t)) - \ln(N(0)) - \ln(1 - N(t)/\mathcal{K}) + \ln(1 - N(0)/K) = \mathfrak{r}t.$$

Finally, after some algebraic manipulation we get:

$$\boxed{N(t) = \frac{\mathcal{K}}{1 + \left(\dfrac{\mathcal{K}}{N(0)} - 1\right)e^{-\mathfrak{r}t}}.} \tag{1.10}$$

We can easily check that equation (1.10) produces for $T = 0$, $N(0)$ and for very large times the asymptotic value $N(\infty) = \mathcal{K}$.

Of particular importance for models like the logistic equation is the question of whether the model has an equilibrium, that is, whether there is a value N^* of N such that $dN/dt = 0$, and if the equilibrium is stable or not. The right-hand side becomes 0 either for the trivial case where $N = 0$ or, as we just have seen, for $N = \mathcal{K}$. So we have two equilibria: $N^* = 0$, i.e. the population is extinct, and $N^* = \mathcal{K}$, i.e. the population density reaches its maximum sustainable value or carrying capacity. We will see in the next section different methods to find out whether an equilibrium is stable or unstable.

1.2 General models for single species populations and analysis of local equilibrium stability

1.2.1 General model and Taylor expansion

Both the Malthus equation (1.6) and the logistic equation (1.9) are examples in which the growth rate depends on population density. We can generalize them and write a general growth equation for a single population as

$$\frac{dN}{dt} = f(N), \tag{1.11}$$

[4] Or, you can also obtain it from a table of integrals.

where $f(N)$ is a general function of the population density N. Notice that for equations (1.6) and (1.9) the function/is the polynomial

$$f(N) = a_o + a_1 N + a_2 N^2,$$

where $a_0 = 0$; for equation (1.6) $a_1 = r$ and $a_2 = 0$; for equation (1.9) $a_1 = r$ and $a_2 = -r/\mathcal{K}$, i.e.

$$\begin{cases} \text{Malthus model: } f(N) = rN & (1.12a) \\ \text{Logistic model: } f(N) = rN(1 - N/\mathcal{K}) = rN - (r/\mathcal{K})N^2 & (1.12b) \end{cases}$$

More generally, if the function f is sufficiently smooth or 'well behaved', it is possible to write f as a Taylor infinite power series around a fixed value of N, $N = N^*$:

$$\begin{aligned} f(N) = \sum_{n=0}^{\infty} f^{(n)}(N^*)\frac{(N - N^*)^n}{n!} &= f(N^*) + f'(N^*)\frac{(N - N^*)}{1!} \\ + f''(N^*)\frac{(N - N^*)^2}{2!} &+ f'''(N^*)\frac{(N - N^*)^3}{3!} + \cdots, \end{aligned} \tag{1.13}$$

where $n!$ denotes the factorial of n and $f^{(n)}(N^*)$ denotes the nth derivative of f evaluated at the point N^*, i.e. $f^{(0)} = f$, $f^{(1)} = f' = df/dN$, $f^{(2)} = f'' = d^2f/dN^2$ and so on. Substituting equation (1.13) into (1.11) we get the general (infinite) series expression for the growth of a single population:

$$\begin{aligned} \frac{dN}{dt} = f(N^*) &+ f'(N^*)\frac{(N - N^*)}{1!} + f''(N^*)\frac{(N - N^*)^2}{2!} \\ &+ f'''(N^*)\frac{(N - N^*)^3}{3!} + \cdots, \end{aligned} \tag{1.14}$$

Remark. Thus any growth function may be written as a (possibly infinite) polynomial

$$f(N) = \sum_{n=0}^{\infty} a_n N^n = a_0 + a_1 N + a_2 N^2 + a_3 N^3 + \cdots. \tag{1.15}$$

1.2.2 Algebraic and geometric analysis of local equilibrium stability

We have seen that the equilibrium points N^* satisfy

$$f(N^*) = 0, \tag{1.16}$$

so that $dN/dt = 0$ when evaluated at these N^*. Thus, to find the equilibrium points we have to solve for all N^* that satisfy equation (1.16).

To perform the algebraic stability analysis of an equilibrium point N^* we consider a perturbation or departure x from the equilibrium value N^*,

$$x(t) = N(t) - N^*, \tag{1.17}$$

which we initially assume is much smaller than 1 ($x(0) << 1$)[5]. Substituting equations (1.15) and (1.17) into equation (1.14) we get:

$$\frac{dx}{dt} = f'(N^*)x + f''(N^*)\frac{x^2}{2} + f'''(N^*)\frac{x^3}{6} + \cdots,$$

and neglecting terms of order x^2 and higher, we obtain an approximate *linear* differential equation for the perturbation $x(t)$:

$$\frac{dx}{dt} \approx f'(N^*)x. \tag{1.18}$$

Since $f'(N^*)$ is equal to constant a1, we know that the solution of equation (1.18) is given by an exponential:

$$x(t) = x(0)e^{at}. \tag{1.19}$$

Therefore, if $a < 0$ this perturbation dies away exponentially, whereas for $a > 0$ the disturbance grows unbendingly (the special case $a = 0$ corresponds to neutral stability). In summary, this neighborhood stability analysis gives the equilibrium point at N^* as

$$\begin{cases} \textbf{stable if } a = f'(N^*) < 0 \\ \textbf{unstable if } a = f'(N^*) > 0 \end{cases}.$$

More formally, we say that if and only if $f'(N^*) < 0$ the equilibrium is **locally stable**. This means, starting at the equilibrium value of N, N^*, if the population changes slightly in size, will it tend to return to its equilibrium value. **Global stability** is a more general property that implies that a system will return to the equilibrium point *from any initial population value*. We will illustrate the difference between local and global stability by considering the Malthus and logistic models.

Let us start by considering the simplest Malthus model. According to equation (1.12a) there is only one equilibrium solution to this model, $N^* = 0$. To determine the stability of this equilibrium, we take the derivative of $f(N) = rN$ with respect to N:

$$f'(N) = r \text{ for all } N. \quad \text{(Malthus equation)} \tag{1.20}$$

That is, it is a constant—the parameter r. Therefore, 0 is an unstable equilibrium point when $r > 0$ and a stable equilibrium when $r < 0$. Furthermore, *any perturbation of any size* x away from $N^* = 0$ will decrease to 0 if $b < d$ ($r < 0$) or increase unbounded and diverge to infinity if $b > d$ ($r > 0$) (figure 3.1). This is because the differential equation is a linear equation and N has disappeared after we took the derivative with respect to it. Linear models are therefore *globally stable* or *globally unstable*. That is, if the model is globally stable, $N(t)$ will converge to the single equilibrium point N^* from any starting value of N. If the model is globally unstable, $N(t)$ will be repelled away from the single unstable equilibrium point N^* for any value of N.

[5] Remember that we are assuming that N is a continuous variable so that x can be as small as we want.

In the case of the logistic equation, by equation (1.12b),

$$f'(N) = r(1 - 2N/K). \quad \text{(logistic equation)} \tag{1.21}$$

Hence, inserting the two equilibria: $N^* = 0$ and $N^* = \mathcal{K}$, we have

$$\begin{cases} f'(0) = r(1 - 2 \times 0) = r > 0, & (1.22a) \\ f'(\mathcal{K}) = r(1 - 2 \times 1) = -r < 0. & (1.22b) \end{cases}$$

Therefore,

$$\begin{cases} N^* = 0 \text{ is an } \textbf{unstable equilibrium} \\ N^* = \mathcal{K} \text{ is a } \textbf{stable equilibrium} \end{cases}$$

In the logistic model, unlike the Malthus linear model, the equilibria are locally stable. The simplest way to see this is by using the geometric analysis (see below).

For the geometric stability analysis, we graph $f(N)$ against N, thus showing how the rate of change of the population varies with population density, N. In the case of the logistic equation[6], since its rhs is a quadratic expression, the graph is simply a parabola that crosses the N-axis at the two equilibrium points, 0 and $\mathcal{K}$, where $f(N) = dN/dt = 0$ as shown in figure 1.3. We now place arrows in this figure to the right or left to each side of each equilibrium point in the direction of population change specified by the sign of the derivative. Notice that arrows point away from $N^* = 0$ on either side but they both point towards $N^* = \mathcal{K}$ from either side. This means that by introducing or removing at least one individual (e.g. by emigration or immigration) in the vicinity of $\mathcal{K}$ will always result in the population returning to $\mathcal{K}$. Doing the same around 0 will always result in the population moving away from 0, so long as $r > 0$ (a point to which we shall return shortly). Therefore, $N^* = \mathcal{K}$ is a

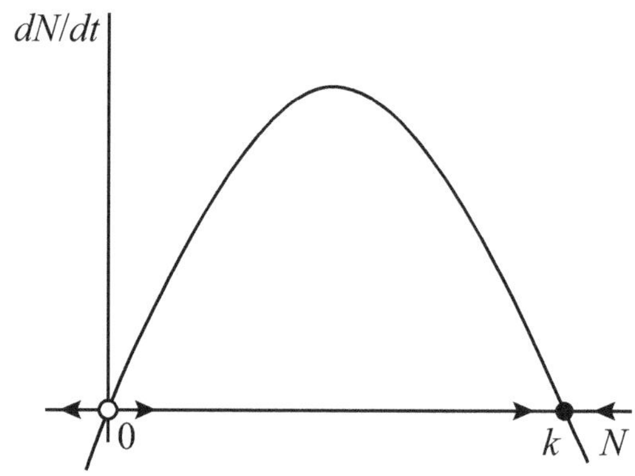

Figure 1.3. The inverted parabola of the logistic equation and its two equilibria: $N^* = 0$ (unstable) and $N^* = \mathcal{K}$ (stable).

[6] We will discuss only the case of the logistic equation because it includes the simplest case of Malthus equation.

stable equilibrium point while $N^* = 0$ is an unstable equilibrium point, in complete agreement with the algebraic stability analysis. For the Malthus equation, instead of an inverted parabola, we have a straight line by the origin and therefore only one intersection point, $N^* = 0$, which is an unstable equilibrium.

Notice that, when $\mathfrak{r} > 0$, $N^*=\mathcal{K}$ is stable only for $N > 0$: if we start from a $N < 0$,[7] figure 1.3 shows that the trajectory will be repelled towards the left and will never reach $N^*=\mathcal{K}$. Likewise, if we start from $N > \mathcal{K}$ the trajectory will move towards the left (towards $\mathcal{K}$) rather than be repelled towards the right by $N^* = 0$. Therefore, $N^*=\mathcal{K}$ is locally stable rather than globally stable and $N^* = 0$ is locally unstable rather than globally unstable.

Remarks.

(i) The algebraic stability analysis, through equation (1.19), tell us that $\mathfrak{r}$ controls *quantitatively* the fate of perturbations near equilibria for the Malthus and logistic equations: they grow unboundedly as $\exp(\mathfrak{r}t)$ near $N^* = 0$ for both equations and they die out as $\exp(-\mathfrak{r}t)$ near $N^*=\mathcal{K}$ for the logistic model. On the other hand, the geometric analysis only provides *qualitative information*: it tell us whether the equilibria are stable or unstable.

(ii) In the logistic model, unlike the Malthus linear model, the equilibria are locally stable. The reason is because nonlinear models, which typically have powers of N, can have several equilibrium solutions. Because these models are in powers of N, taking the derivative with respect to N and evaluating it at an equilibrium point results in an expression for the eigenvalue which is itself a function of N^*. For nonlinear models, the sign of f' and the stability of the equilibrium point will depend on which equilibrium point they are evaluated at. In general, unlike globally stable or unstable linear models, nonlinear models are typically stable or unstable only within certain local values of N in the neighborhood of a particular N^*.

1.3 The Lotka–Volterra predator–prey equations

1.3.1 A general dynamical system for predator–prey

A general dynamical system for modelling predator–prey involves two species. One of them (the predators) feeds on the other species (the prey), which in turn feeds on some third food resource available around, i.e. the model assumes *implicitly* a third trophic level below the prey. The population densities of prey and predator at a reference time t are denoted by $N(t)$ and $P(t)$, respectively, and obey two autonomous differential equations:

$$\dot{N} = Nf(N, P) \tag{1.23a}$$

[7] Of course, from a mathematical point of view, a negative population density does not make sense. But the global stability is a mathematical concept.

$$\dot{P} = Pg(N, P), \tag{1.23b}$$

i.e. the time t does not appear explicitly in the functions $f(N,P)$ and $g(N,P)$ which denote the respective per capita growth rates of the two species. It is assumed that these functions are *continuously differentiable* and that $df(N,P)/dP < 0$ and $dg(N,P)/dN > 0$, that is, the prey abundance is negatively affected by the predator abundance and vice versa, the predator abundance is positively affected by the prey abundance. A standard example is a population of Canadian lynxes and snowshoe hares. However, depending on their specific settings of applications, predator–prey models can take the forms of resource–consumer, plant–herbivore, parasite–host, etc.

1.3.2 A first model for predator–prey: the original Lotka–Volterra predator–prey model

In 1926, the Italian mathematician Vito Volterra proposed a differential equation model to explain the observed simultaneous increase in predator fish and decrease in prey fish in the Adriatic Sea during World War I (when fishing was largely suspended). At the same time in the United States, the equations studied by Volterra were derived independently by Alfred Lotka (1925) to describe a hypothetical chemical reaction in which the chemical concentrations oscillate. The Lotka–Volterra model is the simplest model of predator–prey interactions. It is based on linear per capita growth rates, written as

$$f(N, P) = r - eP \tag{1.24a}$$

$$g(N, P) = aN - d, \tag{1.24b}$$

where the parameter r denotes, as usual, the growth rate of the prey species in the absence of interaction with the predator species (a positive quantity), the parameter e is a consumption rate of prey by predators that measures per-capita loss of prey due to predators, the parameter a denotes the net rate per capita of growth of the predator population in response to the size of the prey population, and the parameter d is the per-capita loss (by death or emigration) rate of predators in the absence of interaction with prey. Substituting the equations (1.24) into (1.23) we get:

$$\dot{N} = rN - eNP \tag{1.25a}$$

$$\dot{P} = aNP - dP. \tag{1.25b}$$

Let us interpret what these equations are telling us. The idea of the first equation is that in the absence of predators ($P = 0$), the prey would grow in a Malthusian way at a constant rate r, but decreases linearly as a function of the density P of the predators. Similarly, in the absence of prey ($N = 0$), the density of predators would decrease but the rate increases proportional to the density of the prey. A necessary condition for predation to occur is that an individual predator and an individual prey physically meet. The product of the two densities, NP, represents the expectation that an individual predator will meet an individual prey (assuming random movement of the two through a homogeneous landscape). It is equivalent to the chemistry ***law of mass action***, which states that the rate of molecular collisions of two chemical species in a dilute gas is proportional to the product of the two concentrations of the reactants. Hence the parameter $\mathscr{e}$ can be interpreted as the probability that, upon meeting a prey, the predator will successfully kill it. The total prey kill is therefore $\mathscr{e}NP$. Since not all the biomass of a killed individual is transformed into predator biomass, so the growth of the predator is itself proportional to the total prey harvest, or $\beta\mathscr{e}NP$, where β represents the conversion efficiency of prey biomass into predator biomass. Thus we can write $\mathscr{a} = \beta\mathscr{e}$ and the NP term can be thought of as representing the conversion of energy from one species to another: $\mathscr{e}NP$ is taken from the prey and $\mathscr{a}NP$ accrues to the predators. This first predator–prey model inherits all the nine assumptions of the Malthus equation (box 1.1). Three additional assumptions are listed in box 1.3.

We shall see that this model has serious drawbacks. Nevertheless, it has been of considerable value in posing relevant questions and as a starting point for more realistic models that we will briefly review in the next subsection. But before doing this we are going to analyze the solution of the Lotka–Volterra predator–prey model and compare it against empirical data.

As a first step in analyzing the Lotka–Volterra model it is convenient, for simplicity, to nondimensionalize the system by introducing a change of variables:

$$\tau = \mathscr{r}t,\ x(\tau) = \frac{\mathscr{a}N(t)}{\mathscr{d}},\ y(\tau) = \frac{\mathscr{e}P(t)}{\mathscr{r}},\ \alpha = \mathscr{d}/\mathscr{r}, \tag{1.26}$$

so that the Lotka–Volterra equations (1.25) can be rewritten as:

$$\frac{dx}{d\tau} = x(1 - y), \tag{1.27a}$$

$$\frac{dy}{d\tau} = \alpha y(x - 1). \tag{1.27b}$$

The system of differential equations (1.27) has two immediate advantages over that of (1.25). First, it involves just a single parameter α. It is much easier to explore how the solutions of these equations change when we move in this single-dimensional 'space of parameters' than in the original four-dimensional one (exercises 1.7 and 1.8

Box 1.3. Additional assumptions of the original Lotka–Volterra predator–prey model.

X The effect of the predation is to reduce the prey's per capita growth rate by a term proportional to the prey and predator populations; this is the $-\mathscr{e}N\ P$ term.

XI In the absence of any prey for sustenance the predator population would experiment an exponential decay, that is, the $-\mathscr{d}P$ term in (1.25b).

XII The ***functional response***, or per capita consumption rate of prey per predator, increases linearly with prey abundance as $\mathscr{e}N$.

illustrate this point clearly). Second, the two equilibrium states now become very simple, either

$$\begin{cases} x^* = 0 = y^* & \text{or} \quad (0, 0), \qquad (1.28a) \\ x^* = 1 = y^* & \text{or} \quad (1, 1). \qquad (1.28b) \end{cases}$$

In spite of its simplicity, the system of differential equations cannot be solved analytically. However, we will first show that we can get useful qualitative insights using calculus. Next, to obtain the model quantitative predictions we will solve these equations numerically. Therefore, dividing equation (1.27b) by equation (1.27a) we get:

$$\frac{dy}{dx} = \alpha \frac{y(x-1)}{x(1-y)}, \qquad (1.29)$$

We can separate variables and express (1.29) as

$$\frac{dy}{y} - dy = \alpha\left(dx - \frac{dx}{x}\right), \qquad (1.29')$$

so that we can integrate exactly the lhs and the rhs of (1.29′) to get:

$$\ln y - y = H = \alpha(x - \ln x),$$

where H is an undetermined constant. Or equivalently

$$ax + y - \ln x^a y = H. \qquad (1.30)$$

To visualize the meaning of equation (1.30), a useful representation is provided by the *phase plane*; a coordinate plane with axes x and y (see box 1.4).

The direction field or phase portrait corresponding to equation (1.27) is plotted in figure 1.4. There are two singular points at which both the numerator and denominator of equation (1.29) become 0: at $x = y = 0$ and $x = y = 1$ (i.e. the two equilibrium states). We have also drawn the predator and prey ***zero-growth isoclines*** or ***nullclines***, along these zero growth lines population densities do not change. We can see from equation (1.27a) that the prey zero growth isoclines are given by $x = 0$ and $y = 1$. likewise, the predator zero growth isoclines are given by $x = 1$ and $y = 0$. Obviously, equilibria occur at the intersections of the predator and prey zero-growth isoclines. If predator and prey numbers are both low, i.e. the particle representing the system is in quadrant $[x < 1, y < 1]$, predator numbers decrease while prey numbers increase (the arrows point towards the right and downward directions). If prey numbers are high but predator numbers are low, i.e. the particle is in quadrant $(x > 1, y < 1)$, both predators and prey increase (the arrows point towards the right and upward). As predator numbers increase so that we move to quadrant $[x > 1, y > 1]$, prey now begin to decrease (the arrows point towards the left and upward directions). Finally, in quadrant $[x < 1, y > 1]$, when predator numbers are high but prey numbers are low, both predators and prey decrease (the arrows point towards the left and downward directions).

Box 1.4. Phase plane.

For analyzing the dynamics of pairs of interacting populations, governed by coupled pairs of differential equations, a useful visual representation is provided by the *phase plane*; a coordinate plane with axes being the values of the two state variables, say (x, y), or (q, p) etc (any pair of variables). In this plane, the evolution of a dynamical system governed by a couple of differential equations of the form $dx/dt = f(x,y)$ and $dy/dt = g(x,y)$ can be visualized as a the 'flow' of a 'particle' with coordinates $x(t)$ and $y(t)$.

Let us provide some definitions.

- *Phase curves or phase trajectories* of this system in the x–y phase plane are a family of curves solutions of

$$\frac{dy}{dx} = \frac{f(x, y)}{g(x, y)}.$$

- *Phase portrait*: two-dimensional vector field whose vectors have components $(dx/dt, dy/dt)$. Each point in phase space has one phase trajectory going through it and therefore can be associated with only one vector. This is a consequence of the fact that continuously differentiable equations such as equations (1.23) have unique solutions at each point, a unique solution guaranteeing one and only one vector at each point.
- Through any point (x_0, y_0) there is a unique curve except at *singular* or *fixed points* (x_s, y_s) where

$$f(x_s, y_s) = g(x_s, y_s) = 0.$$

Therefore, a particular path taken along a flow line (i.e. a path always tangent to the vectors of the phase portrait) is a phase trajectory. The flows in the vector field indicate the time-evolution that the pair of differential equations describes. With enough of these arrows in place the system evolution over a region of the plane phase can be visualized as the flow of a particle representing the state of the system.

In this way, phase planes are useful in visualizing the behavior of dynamical systems; phase trajectories can 'spiral in' towards zero, 'spiral out' towards infinity, or reach neutrally stable situations called centers where the path traced out can be either circular, elliptical, or any other closed curve (see appendix I). This is useful in determining if the dynamics are stable or not.

We can get additional qualitative information by applying a useful modelling tip we mentioned in the previous chapter, namely constructing linear approximations. Hence we will linearize equations (1.27a) and (1.27b) about each one of the two equilibrium points: (0, 0) and (1, 1). Near (0, 0), we may neglect the nonlinear terms and consider

$$\frac{dx}{d\tau} \approx x, \tag{1.31a}$$

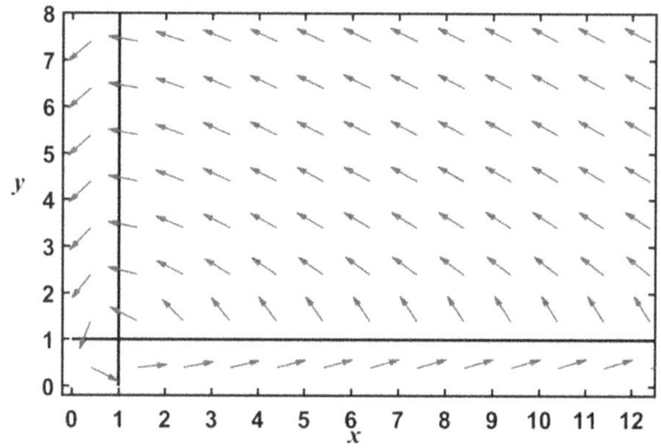

Figure 1.4. Phase portrait for the Lotka–Volterra predator–prey equation (1.27) with nullclines for prey and predator.

$$\frac{dy}{d\tau} \approx -\alpha y. \tag{1.31b}$$

Thus, the prey increase exponentially fast close to the origin, while the predators decrease, corresponding to the horizontal arrows at the bottom of figure 1.4. The equilibrium at the origin has one stable direction (y) and one unstable direction (x) and, as such, is referred to as a *saddle point* (see appendix I).

Near the nontrivial equilibrium (1,1), we introduce new variables u and v that measure the distance from (1, 1),

$$u = x - 1, v = y. \tag{1.32}$$

Therefore, equations (1.27a) and (1.27b) become

$$\frac{du}{d\tau} = v(1 + u), \tag{1.33a}$$

$$\frac{dv}{d\tau} = \alpha u(1 + v). \tag{1.33b}$$

Since u and v are small close to equilibrium (1,1), we again can linearize equations (1.33) and obtain:

$$\frac{du}{d\tau} \approx -v, \tag{1.34a}$$

$$\frac{dv}{d\tau} \approx \alpha u. \tag{1.34b}$$

Deriving both equations with respect to t we can get a second order differential equation for u:

$$\frac{d^2u}{d\tau^2} + \alpha u = 0 \tag{1.35}$$

For this linear, constant-coefficient equation, we may try an exponential solution of the form

$$u(t) = ce^{\lambda\tau}. \tag{1.36}$$

We thus obtain a characteristic equation,

$$\lambda^2 + \alpha = 0, \tag{1.37}$$

that has two purely imaginary roots,

$$\lambda = \pm i\sqrt{\alpha}. \tag{1.38}$$

And, substituting into equation (1.36) we get for the small departure $u(t)$ around $x = 1$:

$$u(t) = ce^{\sqrt{\alpha\tau}} = c\cos\sqrt{\alpha\tau} + ic\sin\sqrt{\alpha\tau}. \tag{1.39}$$

That is, the linearized system has sinusoidal solutions in τ, which have period $2\pi/\sqrt{\alpha}$. Let us see what this means in terms of ecological parameters like intrinsic growth and death rates. From equation (1.26), this period is $T = 2\pi\sqrt{r/d}$; that is, it increases with the ratio of the linear growth rate, r, of the prey to the death rate, d, of the predators. Hence, either an increase in the growth rate of the prey or a decrease of the predator death rate will increase the period. We can see that this makes sense. in the limit in which r/d tends to infinity, then $\alpha = d/r$ tends to 0, so equation (1.27b) can be approximated as $dy/dt \approx 0$, this implies that y can be replaced by a constant value, and therefore equation (1.27a) would become a Malthusian equation, with infinite period (i.e. non periodic).

Unfortunately, we cannot conclude that the fully nonlinear system has these same simple periodic solutions. Purely imaginary roots, like equation (1.38), imply that the linearized system is on the knife-edge between instability and asymptotic stability and on the edge between oscillatory solutions that increase in amplitude and those that decrease in amplitude (see appendix I). For the fully nonlinear system, the nonlinear terms that we have neglected are now critical, since they may tip the nonlinear system one way or another with regard to stability. Three qualitatively different types of phase trajectories that would be, in principle, consistent both with the information from the linearized system and the direction field are shown in figure 1.5. All these three types of phase

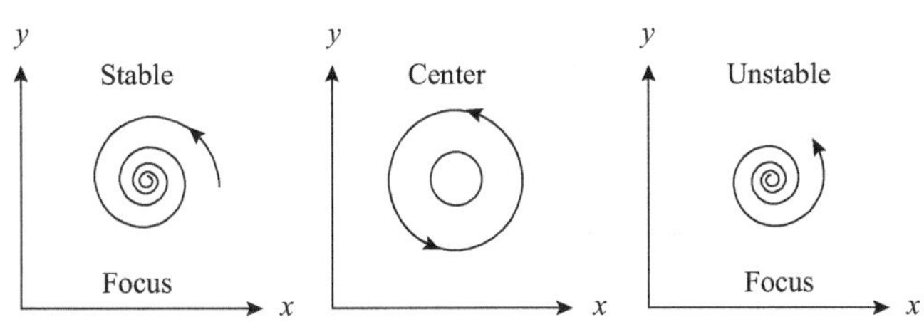

Figure 1.5. Possible scenarios for the equilibrium point (1,1).

trajectories are tangent to the arrows representing the direction field which rotate as we move through the plane. To elucidate which of these scenarios is the correct one we will use equation (1.30) obeyed by the family of phase trajectories, with one phase trajectory for each value of H. Firstly, it can be shown that the minimum possible value for H is $H_{\min} = 1 + \alpha$ and it occurs at the singular point $x = y = 1$ (exercise 1.5). Secondly, equation (1.30) can be solved graphically (exercise 1.6) and we conclude that the nontrivial equilibrium is a ***center*** surrounded by a family of periodic orbits, and not a stable or unstable focus (aka spiral). This is also termed ***neutral stability***.

Hence, we have seen that although the Lotka–Volterra predator–prey equations cannot be solved analytically, a lot of enlightening qualitative information can be obtained using analytical calculations. In general, before embarking on solving a problem by numerical methods it is advisable trying to approach it by analytical calculations which are more efficient to understand the fundamentals, easier to visualize and also serve for verification purposes (numerical calculations are more prone to errors).

Now we will integrate numerically the system of differential equations (1.27) (exercise 1.7) and draw the phase trajectories. For a given $H > 1 + \alpha$, the trajectories (1.30) in the phase plane are closed as illustrated in figure 1.6. The larger the H the larger is the area enclosed by the phase trajectory. The initial conditions, $x(0)$ and $y(0)$, determine the constant H in equation (1.30) and hence the phase trajectory in figure 1.6.

A closed trajectory in the x-y plane implies periodic solutions in τ for x and y. Typical periodic solutions $x(\tau)$ and $y(\tau)$ are shown in figure 1.7. From equations (1.27) we can see immediately that x has a turning point when $y = 1$ and y has one when $x = 1$.

How do the oscillatory solutions compare with predator–prey abundances observed in nature? It turns out that there is a classical set of data on a pair of interacting

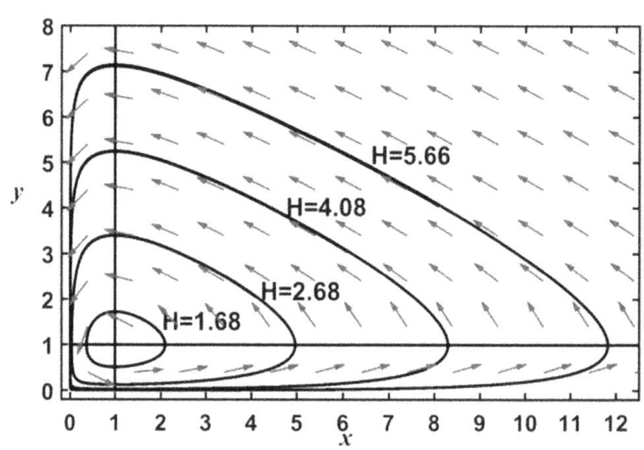

Figure 1.6. Phase portrait for the Lotka–Volterra predator–prey equations (1.27), nullclines for prey ($x = 0$, $y = 1$) and predator ($x = 1$, $y = 0$) and four orbits surrounding (1,1) for different values of H.

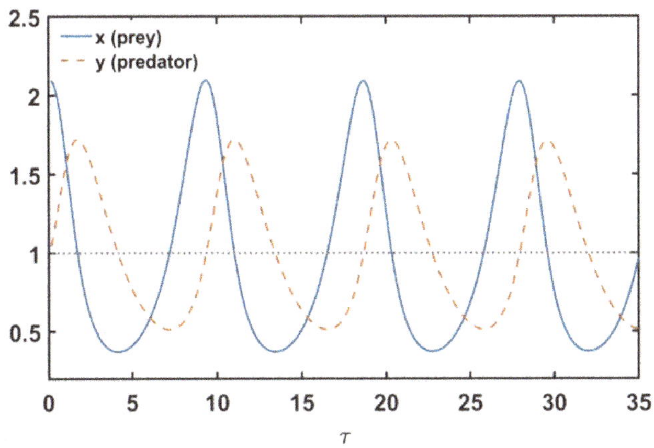

Figure 1.7. Predator–prey cycles corresponding to the closed orbit with $H = 1.68$ of figure 1.6.

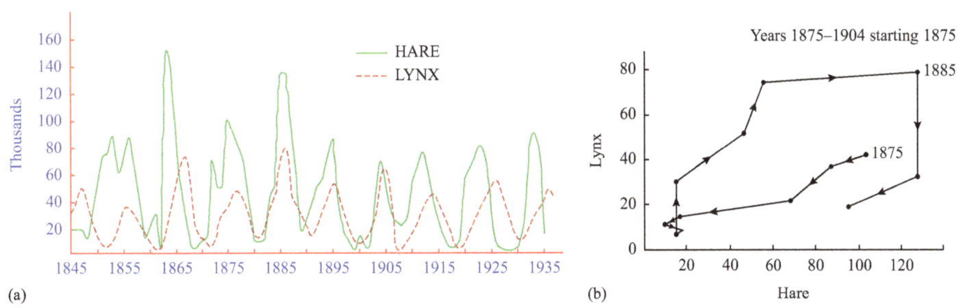

Figure 1.8. (a) Fluctuations in the number of pelts sold by the Hudson Bay Company. (Redrawn from Odum 1953.) (b) Phase plane plot of the data represented in (a) from 1875 to 1904. (After Gilpin 1973.)

populations that come close: the Canadian lynx and snowshoe hare pelt-trading records of the Hudson Bay Company over almost a century. Figure 1.8 (adapted from Odum 1953) shows a plot of that data. We can see they look similar to figures 1.6 and 1.7. However, we shall see in a moment that there is an important difference.

Let us conclude this subsection with a criticism on this first predator–prey model.

A major inadequacy of the Lotka–Volterra model is clear from figure 1.6 which shows that all phase trajectories pass close to the x and y axes. Suppose, for example, $x(0)$ and $y(0)$ are such that x and y for $\tau > 0$ are on the phase trajectory defined by a given value of H (e.g. $H = 4.08$). Then, when the system is 'traveling' along the horizontal or vertical segment of an orbit, any small perturbation could make it jump onto another orbit which does not lie *everywhere* close to the original one. Thus no single orbit is stable in the sense that the system converges to it after perturbations in either direction: perturb the system away from any cycle (for example by harvesting or stocking) and the system will automatically move to another cycle with different amplitude and period. Such a set of equations are called

structurally unstable[8]. Structural instability is a consequence of the fact that the equilibrium is a center, i.e. of neutral stability.

Another inconsistency of this simple model is that figure 1.8(a) shows reasonable periodic fluctuations and figure 1.8(c) a more or less closed curve in the phase plane as we now expect from a time-periodic behavior in the variables. Nevertheless, if we examine the results given in figure 1.8 a little more carefully we note that: first, the *direction* of the time arrows in figure 1.8(b) is clockwise in contrast to that in figure 1.6. This is reflected in the time curves in figure 1.8(a) where the lynx oscillation precedes the hare's. The opposite is the case in the predator–prey situation illustrated in figure 1.7. We have a severe interpretation problem; figure 3.3 would imply that the hares are eating the lynx! Different ways to solve this nonsense were proposed by Gilpin (1973): for example the hares could kill the lynx if they carried a disease which they passed on to the lynx. Or, a more probable candidate is *Homo sapiens*, i.e. the Canadian fur trappers. That is, trappers might sit out poor years and return to the trap lines only when the hare again became abundant. Since lynx were more profitable to trap than hare they could turn a disproportionately large share of their efforts toward catching the more profitable lynx.

1.3.3 Realistic predator–prey models: logistic growth of prey and Holling predator functional responses

The original Lotka–Volterra predator–prey model, unrealistic though it is, suggests that simple predator–prey interactions can give rise to oscillations of the abundances, i.e. to periodic orbits in the phase space. This oscillatory behavior is expected since if the prey population increases, it promotes growth of its predator. Since more predators consume more prey, the prey population starts to decline. With less available food the predator population, in turn, declines and when it is low enough, this allows the prey population to recover and the whole cycle starts over again. We have seen that the problem with these periodic orbits is that they are unstable.

The Lotka–Volterra model can be modified in simple ways to make it more biologically realistic. Reasonable modifications include making the prey populations density-dependent (e.g. logistically), and making the predator death rate depend inversely on prey density. Remarkably, these changes, inspired by biological arguments, solve the problem of the structural instability: the modified predator–prey equations give rise to robust periodic orbits, termed stable ***limit cycle*** oscillations. Here we will briefly present this modified model version. A very pedagogical treatment discussing by stages the buildup of such a more realistic and well behaved model is provided in chapter 8 of the book by Pastor (2008).

We have seen that one of the unrealistic assumptions in the Lotka–Volterra predator–prey equations (1.25a) and (1.25b), is that the prey growth is unbounded in

[8] Note that structural instability is different as a change in stability due to a ***bifurcation*** when a parameter passes through a critical value. In a bifurcation, the stability of an equilibrium of the model changes because of a change in the value of a parameter, but the form of the equations remains unchanged. In a structurally unstable equation, either slight modifications of the initial conditions or of the form of the model equations alter the stability.

the absence of predation. We already know that a more realistic growth rate is provided by the logistic function. So, for example, a more realistic prey population equation might take the form

$$\frac{dN}{dt} = r N\left(1 - \frac{N}{\mathcal{K}}\right) - NPR(N). \tag{1.40a}$$

where $R(N)$ is the predator's functional response. C S Holling (1959) proposed three different functional response types:

$$\begin{cases} R(N) = c & (1.41a) \\ R(N) = \dfrac{c}{N + k} & (1.41b) \\ R(N) = \dfrac{cN}{N^2 + k^2} & (1.41c) \end{cases}$$

which are called, respectively, Holling's type I, II, and III (see figure 1.9). The constant c is the maximum per capita *consumption rate* and the parameter k is the *half-saturation constant* or the value of N which gives half the per capita consumption rate $c/2$.

Type I functional response is used in the Lotka–Volterra predator–prey model. It corresponds to a linear increase in intake rate with food density and assumes that the time needed by the consumer to process a food item is negligible, or that consuming food does not interfere with searching for food.

On the other hand, type II and III functional responses, are characterized by a decelerating intake rate, which follows from the assumption that the consumer is limited by its capacity to process food. Type II functional response describes the observed behavior of insects and parasitoids, as well as modelling microbial growth rates in a limiting nutrient environment, which is known as the Monod equation (Monod 1949). This functional response is also termed the Michaelis–Menten (1913) equation in the context of enzymatic reactions. Type III functional response is similar to type II in that at high levels of prey density, saturation occurs. But it has an inflection point when concavity changes. So, first at low prey density the rate of prey capture accelerates, and then it decreases with increasing prey density. This

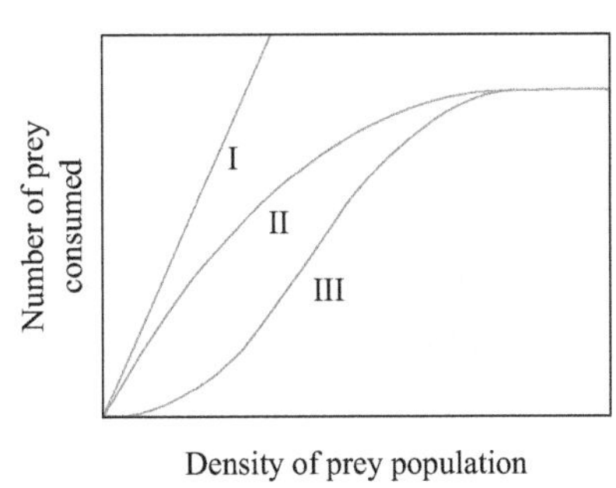

Figure 1.9. Holling type I, II and III response functions.

'sigmoidal' behavior has been attributed (Real 1977) to the existence of 'learning behavior' in the predator population. That is, predators learn more specialized techniques for hunting or focusing their search in particular places within the environment. Type III works well in general for vertebrates. For example, Holling (1959) identified this mechanism in deer mice feeding on sawflies. At low numbers of sawfly cocoons per acre, deer mice especially experienced exponential growth in terms of the number of cocoons consumed per individual as the density of cocoons increased, until a saturation point was reached.

The equation for predator dynamics can take different forms. For example,

$$\frac{dP}{dt} = \beta P R(N) - dP, \tag{1.40b}$$

where β, as before, represents the conversion efficiency of prey biomass into predator biomass.

However, there are several other possible choices for the predator equation. For example, in the so-called Holling–Tanner model (Tanner 1975) the predator equations are:

$$\frac{dN}{dt} = rN\left(1 - \frac{N}{\mathcal{K}}\right) - \frac{cN}{N + k}P \tag{1.40a'}$$

$$\frac{dP}{dt} = sP\left(1 - \eta\frac{P}{N}\right). \tag{1.40b'}$$

It turns out that the above modified Lotka–Volterra predator–prey equations with functional responses of Holling type II or III exhibit stable ***limit cycle*** oscillations (see exercises 1.10 and 1.11). A limit cycle solution is a closed trajectory in the predator–prey space which is not a member of a continuous family of closed trajectories such as the solutions of the original Lotka–Volterra predator–prey depicted in figure 1.6. Rather, it is an isolated periodic orbit (the thick closed curve shown in figure 1.10), such that any small perturbation from this trajectory decays to zero. This limit cycle persists for a range of parameter values. And, its name is because it also attracts orbits that occur at nearby initial conditions, as shown in figure 1.10 for the Holling–Tanner model. For two different initial conditions, one inside and the other outside the limit cycle, the phase trajectories spiral into it (see exercise 1.11). In the application chapter companion to this chapter we will consider a modified Lotka–Volterra predator–prey model with a variable carrying capacity and a Holling type III consumption to model extensive livestock farming.

1.4 The Lotka–Volterra competition equations for a pair of species

1.4.1 A descriptive or phenomenological model

Lotka (1925) and Volterra (1926) independently proposed the simplest dynamical system for modelling two-species competition using extensions of the logistic equation. The Lotka–Volterra competition model is an *interference* competition

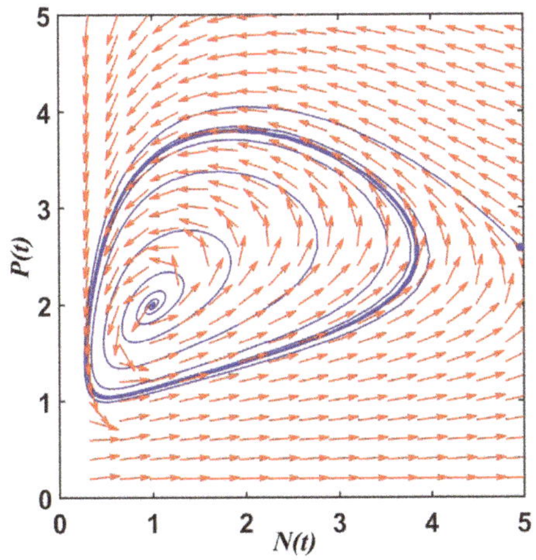

Figure 1.10. Limit cycle for the Holling–Tanner model, i.e. $dP/dt = \mathcal{s}N(1 - \eta p/N)$ for the predator and with parameters given by $\mathcal{r} = 1$, $\mathcal{K} = 7$, $c = 6/7$, $\mathcal{k} = 1$, $\mathcal{s} = 0.2$, $\eta = 0.5$.

model: two species are assumed to diminish each other's per capita growth rate by direct interference. But the model does not specify either the limiting resource(s) the two species are competing for (nutrients, light, shelter, nesting sites, whatever), or the mechanism of competition. Thus it is a *descriptive* or ***phenomenological*** model of competition rather than a *mechanistic* model.

We begin by assuming that two species, with populations densities N_1 and N_2, each grow logistically in the absence of the other. Each species has a per capita growth rate that decreases linearly with population size,

$$\frac{dN_1(t)}{dt} = \mathcal{r}_1 N_1(t)\left(1 - \frac{N_1(t) + \alpha_{12}N_2(t)}{\mathcal{K}_1}\right), \qquad (1.42a)$$

$$\frac{dN_2(t)}{dt} = \mathcal{r}_2 N_2(t)\left(1 - \frac{N_2(t) + \alpha_{21}N_1(t)}{\mathcal{K}_2}\right), \qquad (1.42b)$$

where the competition coefficients α_{ij} are positive parameters measuring the per capita effect of species j on the abundance of species i; all other parameters are as before, except that subscripts indicate the particular species referred to.

1.4.2 Stable equilibrium: competitive exclusion or species coexistence?

To understand the dynamics and equilibria of this model, let us start by finding the nullclines. Both the nullclines and the equilibrium solutions are found by setting either the $\mathcal{r}_i N_i$ terms or the $[1 - (N_i - \alpha_{ij}N_j)/\mathcal{K}_i]$ terms equal 0. The nullclines for N_1 are:

$$N_1 = 0 \qquad (1.43a)$$

and

$$N_1 = \mathcal{K}_1 - \alpha_{12}N_2 \quad (1.43b)$$

Likewise, the nullclines for N_2 are:

$$N_2 = 0 \quad (1.43c)$$

and

$$N_2 = \mathcal{K}_2 - \alpha_{21}N_1 \quad (1.43d)$$

As with predator–prey models, the N_1 and N_2 axes are the 'extinction' nullclines for the other species. The N_1 nullcline (1.43b) intersects the N_1-axis at $\mathcal{K}_1$ and intersects the N_2-axis at $\mathcal{K}_1/\alpha_{12}$. Similarly, the N_2 nullcline (1.43d) intersects the N_1-axis at $\mathcal{K}_2/\alpha_{21}$ and the N_2-axis at $\mathcal{K}_2$. One of these nullclines may lie entirely above the other (figures 1.11(a) and (b)), and it is reasonable to expect that the first will always outcompete the other and drive the other to extinction. Alternatively, these two zero-growth isoclines may cross. Thus, there are four cases, depending on the relative positions of the horizontal and vertical intercepts of these two nullclines in the first quadrant (remember population densities must be positive). Each of the four cases corresponds to a qualitatively different phase portrait and is depicted in a separate panel in figure 1.11. Let us consider each phase portrait in turn, the corresponding equilibria and their stability (in exercise 1.13 we check the qualitative results obtained by this graphical analysis by solving numerically equations (1.42)):

I. If each intercept of the nullcline (1.43d), $N_1 = \mathcal{K}_2/\alpha_{21}$ and $N_2 = \mathcal{K}_2$, is greater than the corresponding intercept of that for (1.43b), $N_1 = \mathcal{K}_1$ and $N_2 = \mathcal{K}_1/\alpha_{12}$, (figure 1.11(a)) i.e. $\mathcal{K}_2/\alpha_{21} > \mathcal{K}_1$ and $\mathcal{K}_2 > \mathcal{K}_1/\alpha_{12}$, which implies that

$$\alpha_{12} > \mathcal{K}_1/\mathcal{K}_2,\ a_{21} < \mathcal{K}_2/\mathcal{K}_1 \quad (1.44a)$$

N_2 excludes N_1. Indeed, if species 2 has a relatively large negative effect on species 1 and species 1 has a relatively small negative effect on species 2, an expected outcome is that species 1 will go extinct while species 2 will approach its carrying capacity $\mathcal{K}_2$. Therefore, it appears that $N_1^* = 0$, $N_2^* = \mathcal{K}_2$ or $(0, K_2)$, indicated as a black circle in figure 1.11(a), is a stable equilibrium point under conditions (1.44a). Likewise, $N_1^* = 0$, $N_2^* = 0$ or (0, 0), indicated as a white circle, appears as an unstable equilibrium point.

II. If each intercept of the nullcline (1.43b), $N_1 = \mathcal{K}_1$ and $N_2 = \mathcal{K}_1/\alpha_{12}$, is greater than the corresponding intercept of that for (1.43d), $N_1 = \mathcal{K}_2/\alpha_{21}$ and $N_2 = \mathcal{K}_2$, (figure 1.11(b)), i.e. $\mathcal{K}_1 > \mathcal{K}_2/\alpha_{21}$ and $\mathcal{K}_1/\alpha_{12} > \mathcal{K}_2$, which implies that

$$\alpha_{12} > \mathcal{K}_1/\mathcal{K}_2,\ a_{21} < \mathcal{K}_2/\mathcal{K}_1 \quad (1.44b)$$

N_1 excludes N_2. That is, species 2 will go extinct while species 1 will approach its carrying capacity $\mathcal{K}_1$. Therefore, it appears that $N_1^* = \mathcal{K}_1$, $N_2^* = 0$ is a stable equilibrium point, indicated as a black circle in figure 1.11(b), is a stable equilibrium point under conditions (1.44b).

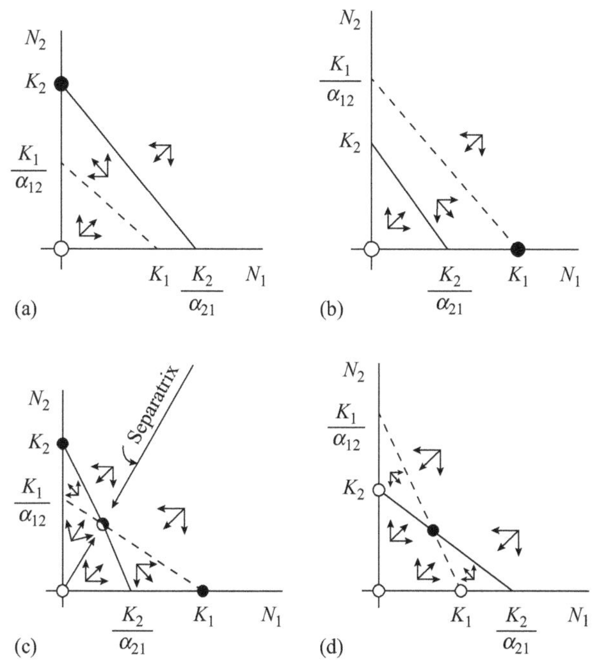

Figure 1.11. Schematic representation of the four possible phase portraits for the competition Lotka–Volterra equations (1.42) depicting vector fields at the different regions delimited by the nullclines of (dashed line for species 1 and filled line for species 2). Stable equilibria are denoted by filled black circles, unstable equilibria by empty circles and saddle points by half filled circles. (a) and (b) When one nullcline lies completely above the other in the first quadrant, the equilibria are stable monocultures. (c) Strong competition results in stable monocultures of either species. (d) Weak competition results in stable coexistence. Adapted from Pastor 2008, with permission from John Wiley & Sons.

Again, $N_1^* = 0$, $N_2^* = 0$ or (0, 0), indicated as a white circle, appears as an unstable equilibrium point. Notice that the reversing of inequalities implies a reversing of the competitive outcome.

III. If the two nullclines cross in such a way that $\mathcal{K}_1 > \mathcal{K}_2/\alpha_{21}$ and $\mathcal{K}_2 > \mathcal{K}_1/\alpha_{12}$, which implies that

$$\alpha_{12} > \mathcal{K}_1/\mathcal{K}_2,\ a_{21} < \mathcal{K}_2/\mathcal{K}_1 \tag{1.44c}$$

i.e. both species are **strong competitors** against the other, then an examination of the vector field of figure 1.11(c) shows that either species can displace the other to extinction, depending on the initial conditions. That is, the two equilibria $(\mathcal{K}_1, 0)$ and $(0, \mathcal{K}_2)$, corresponding to the exclusion of one or the other species, are now both stable nodes (figure 1.11(c)). It is apparent that the species which is initially most abundant will always displace the other. The intersection of the nullclines is therefore a saddle-node: it will be approached only if the system initially lies on the line dividing the two regions of local stability of $(\mathcal{K}_1, 0)$ and $(0, \mathcal{K}_2)$. This line is called a ***separatrix*** (see Glossary) since it separates the two regions. Except for perturbations exactly on the separatrix, any perturbation away from the coexistence equilibrium will be attracted to the nearest monoculture equilibrium.

IV. Finally, if both species are **weak competitors** as defined by inequalities

$$\alpha_{12} > \mathcal{K}_1/\mathcal{K}_2,\ a_{21} < \mathcal{K}_2/\mathcal{K}_1, \tag{1.44d}$$

the vector field of figure 1.11(d) now shows that once trajectories arrive in the triangular regions between the two nullclines, the vectors point to the coexistence equilibrium, which appears to be stable to perturbations away from it in any direction. In other words, the equilibria $(\mathcal{K}_1, 0)$ and $(0, \mathcal{K}_2)$ are now unstable saddle points and trajectories are drawn towards a stable node in the interior of the first quadrant at $\left(\frac{\mathcal{K}_1 - \alpha_{12}\mathcal{K}_2}{1 - \alpha_{12}\alpha_{21}}, \frac{\mathcal{K}_2 - \alpha_{21}\mathcal{K}_1}{1 - \alpha_{12}\alpha_{21}}\right)$. Thus the four possible equilibria are:

$$N_1^* = N_2^* = 0 \tag{1.45a}$$

$$N_1^* = \mathcal{K}_1,\ N_2^* = 0 \tag{1.45b}$$

$$N_1^* = 0,\ N_2^* = \mathcal{K}_2 \tag{1.45c}$$

$$N_1^* = \frac{\mathcal{K}_1 - \alpha_{12}\mathcal{K}_2}{1 - \alpha_{12}\alpha_{21}},\ N_2^* = \frac{\mathcal{K}_2 - \alpha_{21}\mathcal{K}_1}{1 - \alpha_{12}\alpha_{21}} \tag{1.45d}$$

The above four possible competitive situations, defined by the relative positions of the nullclines, are summarized in table 1.1.

It is enlightening to nondimensionalize the Lotka–Volterra competition model by writing, using ***relative yields*** $y_i = N_i/\mathcal{K}_i$ (the species yield in mixture normalized by its yield in monoculture), as

Table 1.1. Summary of effects of relative competitive ability on the outcome of competition between two species, equilibria and their stability.

	Effect of species 2 on species 1	
Effect of species 1 on species 2	Weak: $\alpha_{12} < K_1/K_2$	Strong: $\alpha_{12} > K_1/K_2$
Weak: $\alpha_{21} < K_2/K_1$	**Coexistence** equilibrium at $\left(\frac{\mathcal{K}_1 - \alpha_{12}\mathcal{K}_2}{1 - \alpha_{12}\alpha_{21}}, \frac{\mathcal{K}_2 - \alpha_{21}\mathcal{K}_1}{1 - \alpha_{12}\alpha_{21}}\right)$	**Species 2 always displaces species 1**, equilibrium at $(0, K_2)$
Strong: $\alpha_{21} > K_2/K_1$	**Species 1 always displaces species 2**, equilibrium at $(K_1, 0)$	**Winner depends on initial conditions**, if $N_1(0) > N_2(0)$, equilibrium at $(K_1, 0)$ if $N_1(0) < N_2(0)$, equilibrium at $(0, K_2)$

$$\frac{dy_1(\tau)}{d\tau} = y_1(\tau)(1 - y_1(\tau) + a_{12}y_2(\tau)), \tag{1.46a}$$

$$\frac{dy_2(\tau)}{d\tau} = \rho y_2(\tau)(1 - y_1(\tau) + a_{12}y_2(\tau)), \tag{1.46b}$$

where $\tau = r_1 t$, $\rho = r_1/r_2$, and the a_{ij} denote the interspecific interaction coefficients in terms of these relative yields, defined as $a_{ij} = \alpha_{ij}\ \mathcal{K}_j/\mathcal{K}_i$ (see exercise 1.14).

Notice that in terms of these new interspecific interaction coefficients the condition separating weak from strong competition of species j over species i becomes $a_{ij} = 1$, i.e.

$$\text{if}\begin{cases} a_{ij} < 1 \text{ weak competition of species } j \text{ over species } i \\ a_{ij} > 1 \text{ strong competition of species } j \text{ over species } i \end{cases}$$

Since the intraspecific competition term is taken = 1 (to recover the logistic equation) we can conclude form table 1.1 that whenever at least one of the two interspecific effects is strong, i.e. stronger than intraspecific competition, we end up with competitive exclusion of one species. Only in case IV, where **both interspecific effects were weak relative to intraspecific effects, did the two competing species coexist**. This forms the basis for *Gause's Principle* (Gause 1934) aka the *Principle of Competitive Exclusion* (Hardin 1960). This principle states that, two species competing for the same *ecological niche* cannot coexist. This leads either to the extinction of the weaker competitor or to an evolutionary or behavioral shift toward a different ecological niche. Thus it begs the question: just how similar can the two niches be in order that the two species manage to coexist? We will revisit this question in the second part of this book when dealing with *niche theory*.

1.4.3 Transforming the competition model into a mechanistic model

As we said, the Lotka–Volterra competition equations are a phenomenological approach for the population dynamics of competition since they describe the process of competition by the rather abstract parameters of competition coefficients and carrying capacity. Therefore, these equations have limited ability to explore how the outcomes of competition might change in environments that differ in, for example, the availability of different resources or among species that differ in the use of these resources.

The descriptive Lotka–Volterra model can be transformed into a mechanistic model by explicitly introducing resources as population variables which are consumed or utilized by the two species whose competition we want to model. That is, we explicitly model the dynamics of the resources for which competition occurs and allow species to interact solely through their consumption of shared resources by using the Lotka–Volterra predator–prey equations. In this way we can express the competition coefficients and carrying capacities in terms of rates of utilization and renewal of resources. Hence, imagine two populations N_1 and N_2

(say of herbivores) consuming one resource R (e.g. grass); we can begin by writing the resource equation:

$$\frac{dR}{dt} = r R\left(1 - \frac{R}{k} - b_1 N_1 - b_2 N_2\right), \tag{1.47a}$$

where k is the carrying capacity of the resource and all the other parameters are defined as for equation (1.25a): r denotes, as usual, the growth rate of the prey species in the absence of interaction with the predator species, the parameter $c_i = r.\ b_i$ is a per-capita consumption rate of resource by species i. Now we can write the corresponding dynamical equations for N_1 and N_2 as if they were predators governed by equation (1.25b):

$$\frac{dN_1}{dt} = N_1(a_1 R - d_1), \tag{1.47b}$$

$$\frac{dN_2}{dt} = N_2(a_2 R - d_2). \tag{1.47c}$$

Hence, in essence, the model describes a simple food web. We now assume that the dynamics of the resource is much more rapid compared to that of the consumer, such that we can set the derivative of equation (1.47a) equal to zero, solve for the equilibrium value of R:

$$R^* = (1 - b_1 N_1 - b_2 N_2)k, \tag{1.48}$$

and approximate the consumer dynamics at any particular value of N_1 and N_2 by simply substituting R^* in place of R in equations (1.47b) and (1.47c). Thus we obtain

$$\frac{dN_1}{dt} = N_1(a_1 k - d_1 - a_1 b_1 k N_1 - a_1 b_2 k N_2), \tag{1.49a}$$

$$\frac{dN_2}{dt} = N_2(a_2 k - d_2 - a_2 b_1 k N_1 - a_2 b_2 k N_2). \tag{1.49b}$$

By rearranging we can write equations (1.49a) and (1.49b) as:

$$\frac{dN_1}{dt} = (a_1 k - d_1)N_1\left(1 - \frac{a_1 b_1 k}{a_1 k - d_1}N_1 - \frac{a_1 b_2 k}{a_1 k - d_1}N_2\right), \tag{1.50a}$$

$$\frac{dN_2}{dt} = (a_2 k - d_2)N_2\left(1 - \frac{a_1 b_1 k}{a_2 k - d_2}N_1 - \frac{a_1 b_2 k}{a_2 k - d_2}N_2\right). \tag{1.50b}$$

Hence, identifying coefficients with equations (1.42a) and (1.42b) we can express parameters r_i and $\mathcal{K}_i$ in terms of the mechanistic parameters a_i, b_i, d_i and k as :

$$r_i = a_i k - d_i, \tag{1.51a}$$

$$\mathcal{K}_i = \frac{a_i k - d_i}{a_i b_i k} \tag{1.51b}$$

$$\alpha_{ij} = \frac{b_j}{b_i}. \tag{1.51c}$$

Notice that:

(a) As expected, equation (1.51a) tells us that the net growth rate of a species, r_i, grows with its predation rate, a_i, and with the carrying capacity of the resource, k (and of course decreases with its death rate).

(b) There is a proportionality relationship between the net growth rate of a species, r_i, and its carrying capacity, $\mathcal{K}_i$, since the numerator of equation (1.51b) is r_i. Equation (1.51b) also tells us that, when the death rate of a species, d_i, is negligible with respect to the product $a_i\, k$, its carrying capacity, $\mathcal{K}_i$, is inversely proportional to its per-capita consumption rate of resource, b_i.

(c) The competition coefficients α_{ij}s represent the use of resources by the competing species j relative to the use of resources by the species whose dynamics are being considered i. It is important to emphasize that they do *not* represent the absolute intensity of competition but are rather a ratio of the intensity of interspecific to intraspecific competition.

Remark. Equations (1.47) dealt with a single resource. However, interspecific competition is almost never for a single resource. Indeed, by virtue of the principle of Competitive Exclusion, most ecologists think that two species competing for a single resource cannot persist together forever. MacArthur (1972) proposed a mechanistic model of two species consuming two distinct resources, in terms of predator–prey equations, that overcomes this problem. Results become algebraically more cumbersome.

1.5 The Lotka–Volterra equations for two mutualist species

To complete the description of elementary interactions between species in this section we consider mutualisms, i.e. reciprocally positive interactions between species. Other kinds of beneficial associations of at least one species, like commensalism (0/+ interaction), can be considered as a limiting case of mutualism (likewise, the amensalism −/0 interaction can be thought as a limiting case of prey–predator interaction).

Mutualisms are ubiquitous in nature. Prominent examples include most vascular plants engaged in mutualistic interactions with mycorrhizal fungi, the association of *Rhizobium* bacteria with leguminous plants, and the diverse community of bacteria in all mammalian guts.

There are two broad types of mutualism, ***facultative*** and ***obligate***. Facultative mutualists can live independently of each other, the mutualism is non-essential but the growth of each species is enhanced in the presence of each other. Examples of facultative mutualism are the fish-cleaning mutualism, one fish (the cleaner) removes and eats parasites from the surface of the other (the client) or seed dispersal mutualism, such as the association between the Clark's nutcracker and whitebark pine. Obligate mutualism is essential for both parties, the two partners cannot survive without each other. Ruminants and the bacteria in their digestive tracts that actually digest the cellulose in the animal's food, is an example of obligate mutualists.

A candidate for a mutualism model results from inverting the sign of the interspecific competition coefficients of the Lotka–Volterra competition equations (1.42), from negative to positive. That is:

$$\frac{dN_1(t)}{dt} = r_1 N_1(t)\left(\frac{\mathcal{K}_1 - N_1(t) + \alpha_{12}N_2(t)}{\mathcal{K}_1}\right), \ (\alpha_{12} > 0) \qquad (1.52a)$$

$$\frac{dN_2(t)}{dt} = r_2 N_2(t)\left(\frac{\mathcal{K}_2 - N_2(t) + \alpha_{21}N_1(t)}{\mathcal{K}_2}\right), \ (\alpha_{21} > 0). \qquad (1.52b)$$

We will see in a moment that equations (1.52), with all their quantities **positive**, serve to describe mutualism between two facultative species.

Proceeding in a completely similar way as we did for the competition model, let us calculate the nullclines. The nullclines for N_1 are:

$$N_1 = 0 \qquad (1.53a)$$

and $N_1 = \mathcal{K}_1 + \alpha_{12}N_2$, or equivalently

$$N_2 = (1/a_{12})N_1 - \mathcal{K}_1/a_{12}. \qquad (1.53b)$$

Similarly, the nullclines for N_2 are:

$$N_2 = 0 \qquad (1.53c)$$

$$N_2 = \alpha_{21}N_1 + \mathcal{K}_2. \qquad (1.53d)$$

These are identical to the nullclines of the competition equations, except that they have a positive slope (figure 1.12). Notice first that equations (1.53b) and (1.53d) imply that increasing the density of species j, the mutualist of species i, supports an increase in N_i *above* its carrying capacity (defined as its equilibrium density when alone). And, second that if species j disappears each species converges to its carrying capacity. Therefore, according to our definition, equations (1.52) describe a facultative mutualism (species do not collapse when the mutualist species disappears).

The equilibria are obtained as usual by intersecting the nullclines and are:

$$N_1^* = N_2^* = 0 \qquad (1.54a)$$

$$N_1^* = \mathcal{K}_1, \ N_2^* = 0 \qquad (1.54b)$$

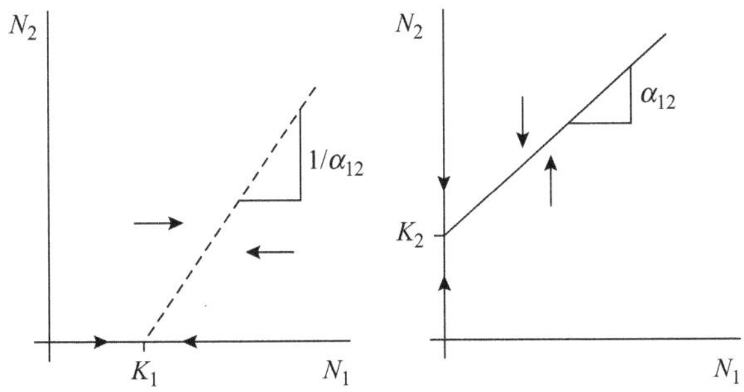

Figure 1.12. Nullclines of the Lotka–Volterra mutualism model equations (1.52) and a schematic representation of the vector field. Left: nullclines for N_1, horizontal axis and dashed line, with slope $1/\alpha_{12}$. Right: nullclines for N_2, vertical axis and filled line with their slope α_{21}. Adapted from Pastor 2008, with permission from John Wiley & Sons.

$$N_1^* = 0,\ N_2^* = \mathcal{K}_2 \tag{1.54c}$$

$$N_1^* = \frac{\mathcal{K}_1 + \alpha_{12}\mathcal{K}_2}{1 - \alpha_{12}\alpha_{21}},\ N_2^* = \frac{\mathcal{K}_2 + \alpha_{21}\mathcal{K}_1}{1 - \alpha_{12}\alpha_{21}} \tag{1.54d}$$

Notice they become identical to the equilibria of the competition model, (equations (1.45)), by replacing α_{ij} by $-\alpha_{ij}$ in the coexistence equilibrium (equation (1.54d)).

In the same way as we defined weak and strong competition, let us now define weak and strong mutualism as:

$$\begin{cases} \alpha_{12}\alpha_{21} < 1: & \text{weak mutualism} \\ \alpha_{12}\alpha_{21} > 1: & \text{strong mutualism} \end{cases}$$

Since all parameters, the carrying capacities and the interaction coefficients, are positive, it is straightforward to show analytically that the coexistence equilibrium (1.54d) is only possible for weak mutualism. (The strong mutualism condition implies that the equilibrium densities would be negative!) The same can be shown geometrically. The weak mutualism condition implies that the slope of the nullcline (1.53b) for N_1, equal to $1/\alpha_{12}$, is always larger than the slope of the nullcline (1.53d) for N_2, equal to $1/\alpha_{21}$. Consequently, both nullclines will intersect in the first quadrant (figure 1.13(a)). Furthermore, the vector field tells us that for weak mutualism this coexistence equilibrium is stable. Remember that the same happened for two competing species: the coexistence equilibrium was stable only for weak competition. The difference is that populations are larger, rather than smaller, than their respective carrying capacities (figure 1.13(a)). In contrast, the strong mutualism condition implies that the slope of the nullcline (1.53b) for N_1 is always smaller than the slope of the nullcline (1.53d) for N_2 and thus both nullclines will not intersect in the first quadrant (figure 1.13(b)). Rather, under strong mutualism, the vector field indicates that both populations increase indefinitely (figure 1.13(b)).

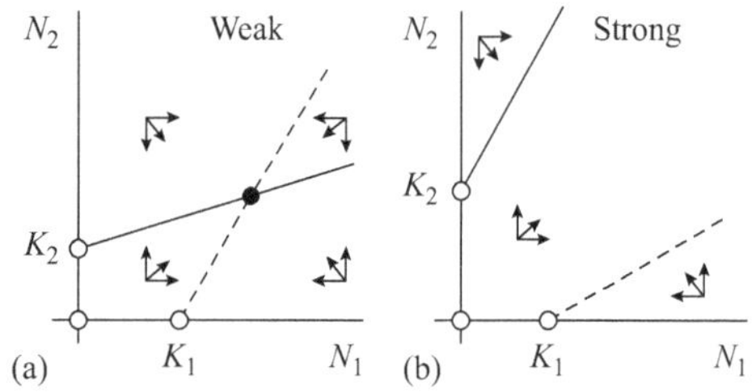

Figure 1.13. Nullclines, equilibria with their stability (a stable equilibrium is denoted by a filled black circle and an unstable one by an open white circle) and a schematic representation of the vector field for **facultative mutualism**. (a) Weak mutualism. (b) Strong mutualism. Adapted from Pastor 2008, with permission from John Wiley & Sons.

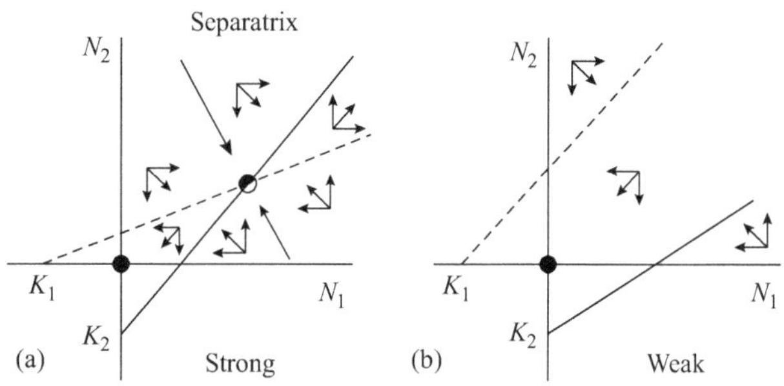

Figure 1.14. Nullclines, equilibria with their stability (stable equilibria denoted by a filled black circle and an unstable one by an open white circle) and a schematic representation of the vector field for **obligate mutualism**. (a) Strong mutualism. (b) Weak mutualism. Adapted from Pastor 2008, with permission from John Wiley & Sons.

In other words, for coexistence to be possible, the equilibrium solution must lie in the first quadrant. And, this is only possible for facultative mutualism if $\alpha_{12}\alpha_{21} < 1$, which is weak mutualism. This means that the positive feedback between the two species is not strong enough to result in destabilizing positive feedback. On the other hand, in the case of strong mutualism $\alpha_{12}\alpha_{21} > 1$ the nullclines in the first quadrant diverge from each other and the coexistence equilibrium is in the third quadrant (both N_1^* and N_2^* are negative). So, we conclude that **for two facultative mutualists to coexist, mutualism must be weak.**

What about equations describing obligate mutualism? Indeed, figure 1.13(b) gives us a clue about how to modify equations (1.52) to obtain a coexistence equilibrium for obligate mutualism. Notice that if we move the nullcline of species 1 parallel to itself toward the left until the intersection point with the horizontal axis becomes $-\mathcal{K}_1$ and the nullcline of species 2 parallel to itself downward until the intersection point with the vertical axis becomes $-\mathcal{K}_2$, the two nullclines would intersect in the first quadrant (figure 1.14(a)). This is equivalent to making the two carrying capacities negative. Obviously one cannot have a negative carrying capacity. But a negative carrying capacity in the equations can be interpreted simply as requiring that there is a minimal number of individuals of the mutualist partner that can actually initiate the mutualism process and result in increasing population numbers. Replacing $\mathcal{K}_1$ and $\mathcal{K}_2$ by $-\mathcal{K}_1$ and $-\mathcal{K}_2$ equations (1.52) transform into:

$$\frac{dN_i(t)}{dt} = r_i N_i(t)\left(\frac{-\mathcal{K}_i - N_i(t) + \alpha_{ij}N_j(t)}{-\mathcal{K}_i}\right)$$

$$= -r_i N_i(t)\left(\frac{-\mathcal{K}_i - N_i(t) + \alpha_{ij}N_j(t)}{\mathcal{K}_i}\right).$$

We are almost done. As we just mentioned, we want that species i grows if the density of its mutualist species j is large enough. The above equation would describe this provided r_i is also negative (this is because the minus sign in front of r_i and the fact that the bracketed factor is positive for large values of N_j). Thus, in the same way that facultative mutualism has parameters $r_1 > 0$, $r_2 > 0$, $\mathcal{K}_1 > 0$, $\mathcal{K}_2 > 0$, let us define obligate mutualism as having parameters $r_1' < 0$, $r_2' < 0$, $\mathcal{K}_1' < 0$, $\mathcal{K}_2' < 0$. Therefore the equations for obligate mutualism can be written as:

$$\frac{dN_1(t)}{dt} = r'_1 N_1(t)\left(\frac{\mathcal{K}_1' - N_1(t) + \alpha_{12}N_2(t)}{\mathcal{K}_1'}\right), \; (\alpha_{12} > 0, \, r'_1 < 0, \, \mathcal{K}_1' < 0) \quad (1.55a)$$

$$\frac{dN_2(t)}{dt} = r'_2 N_2(t)\left(\frac{\mathcal{K}_2' - N_2(t) + \alpha_{21}N_1(t)}{\mathcal{K}_2'}\right), \; (\alpha_{21} > 0, \, r'_2 < 0, \, \mathcal{K}_2' < 0). \quad (1.55b)$$

They have precisely the same form as the original mutualism equations, with the intrinsic rates of increase and with the carrying capacities both negative (we use the primes for stressing this). And, since the quotient $r_i'/\mathcal{K}_i'$ is positive, they tell us that in the obligate case, the mutualism coefficients must be large enough to make the growth of each species positive in **the presence of** the other species, but each species on its own will not have positive equilibria.

Could two obligate mutualists coexist? As we have observed, for a strong obligate mutualism the nullclines intersect at one point in the first quadrant. But, is this coexistence equilibrium stable? The vector field for the strong facultative mutualism indicates that the coexistence equilibrium is a semistable saddle-point (figure 1.14(a)). In turn, the weak obligate mutualism has an equilibrium solution in the third quadrant because its nullclines diverge in the first quadrant.

We can verify analytically the stability of the coexistence equilibrium both for two weak facultative mutualists and for two strong obligate mutualists (exercise 1.15) by examining the stability of the Jacobian matrix evaluated at each of these coexistence points (see appendix I).

So far we have discussed symmetrical relationships, i.e. either facultative–facultative or obligate–obligate mutualisms. What about asymmetrical facultative–obligate associations? For example, if N_1 is the facultative species

Table 1.2. Summary of the possibility and stability of coexistence for different mutualistic associations, both weak and strong, and either symmetrical or not.

Intensity	Species 1	Species 2: Facultative	Species 2: Obligate
Weak $\alpha_{12}\alpha_{21} < 1$	Facultative	**Stable node**	**Stable node if** $\mathcal{K}_1 > -\alpha_{12}\mathcal{K}_2'$, $\alpha_{12}\mathcal{K}_1 > -\mathcal{K}_2'$
	Obligate	**Stable node if** $\alpha_{12}\mathcal{K}_2 > -\mathcal{K}_1'$, $\mathcal{K}_2 > -\alpha_{12}\mathcal{K}_1'$	**Impossible**
Strong $\alpha_{12}\alpha_{21} > 1$	Facultative	**Impossible**	**Unstable**
	Obligate	**Unstable**	**Saddle node**

($r_1 > 0$, $\mathcal{K}_1 > 0$) and N_2 the obligate species ($r_2' < 0$, $\mathcal{K}_2' < 0$) the coexistence equilibrium becomes:

$$N_1^* = \frac{\mathcal{K}_1 + \alpha_{12}\mathcal{K}_2'}{1 - \alpha_{12}\alpha_{21}}, \quad N_2^* = \frac{\mathcal{K}_2' + \alpha_{21}\mathcal{K}_1}{1 - \alpha_{12}\alpha_{21}}. \tag{1.56}$$

For weak mutualism the denominator of equation (1.56) is positive and therefore a coexistence equilibrium exists in the first quadrant provided both numerators are also positive, i.e. $\mathcal{K}_1 > -\alpha_{12}\mathcal{K}_2'$, $\alpha_{12}\mathcal{K}_1 > -\mathcal{K}_2'$. The stability of non symmetrical cases can be easily derived by an analysis similar to the one we performed for the symmetrical cases[9]. The outcome for all cases of mutualism, both symmetrical and asymmetrical, are summarized in table 1.2.

Exercises

Exercise 1.1

Fibonacci numbers and the Golden Ratio

(A) A certain man had one pair of rabbits together in a certain enclosed place. How many pairs of rabbits can be created from the pair in one year if we suppose that each month each pair begets a new pair, which on the second month on becomes productive?

(B) Show that if we call F_n the number of rabbits at month n, the following relation holds: $F_{n+2} = F_{n+1} + F_n$, for $n = 0, 1, 2, \ldots$.

(C) To solve the above recurrence equation try $F_n = \lambda^n$, where λ is a constant you have to compute.

(D) Plot F_n, versus n.

[9] We refer the reader interested in these non symmetrical mutualistic associations to Vandermeer and Boucher (1978) or Pastor (2008).

Exercise 1.2

Obtaining the logistic parameters by least square regression method.

Using the least square regression method obtain from the data below (Gause 1932) the carrying capacity $\mathcal{K}$, and the growth rate $\mathscr{r}$ as well as the initial volume V_0.

Time (h)	Yeast vol (ml)
6	0.37
7.5	1.63
15	6.2
16	8.87
24	10.66
29	12.5
31.5	12.6
33	12.9
40	13.27
44	12.77
48	12.87
51.5	12.9

Exercise 1.3

(from Hale and McCarthy 2005).

Let us consider the population of white-tailed deer in the state of Kentucky. The Kentucky Department of Fish and Wildlife Resources (KDFWR) sets guidelines for hunting and fishing in the state, and it reported an estimate of 900 000 deer prior to the hunting season of 2004. Johnson (2003) notes: 'A deer population that has plenty to eat and is not hunted by humans or other predators will double every three years.' This corresponds to a rate of increase $r = \ln(2)/3 = 0.2311$. (This assumes—with plentiful food supply and no predation—that the population grows exponentially, which is reasonable, at least in the short term.) The KDFWR also reports deer densities for 32 counties in Kentucky, the average of which is approximately 27 deer per square mile. Suppose this is the deer density for the whole state (39 732 square miles). The carrying capacity $\mathcal{K}$ is 39 732 sq. mi. × 27 deer/sq. mi. or 1 072 764 deer.

Investigate if the above data are consistent with the logistic growth model.

Exercise 1.4

Consumption by fixed quota for a population with logistic growth.

Consider the logistic model with fixed consumption per share:

$$\frac{dX}{dt} = \mathscr{r}X\left(1 - \frac{X}{\mathcal{K}}\right) - C,$$

(A) Calculate the equilibria and analyze their stability using pencil and paper.
(B) Write a script (in MATLAB, Fortran, R, etc) to solve this differential equation (for example using the Runge–Kutta second order method) and compute $X/\mathcal{K}$ for $\mathscr{r} = 1$ and $C = \mathscr{r}\mathcal{K}/2$, $C = \mathscr{r}\mathcal{K}/4$ and $C = \mathscr{r}\mathcal{K}/8$ with the initial condition $X(0) = \mathcal{K}/2$, and $X(0) = 2\mathcal{K}$.
(C) Solve the equation using some ODE integrator package.

Exercise 1.5

Calculation of H_{min} for Lotka–Volterra predator–prey phase trajectories.

Equation (1.30) is obeyed by the family of phase trajectories generated by the Lotka–Volterra predator–prey equations (1.27a) and (1.27b), with one phase trajectory for each value of H. Show that the minimum possible value for H is $H_{min} = 1 + \alpha$ and it occurs at the singular point $x = y = 1$.

Exercise 1.6

Graphical solution of equation (1.30) for the phase trajectories to conclude the singular point is a center.

We can solve equation (1.30) graphically by considering the two functions

$$z_1 = ye^{-y} \text{ and } z_2 = \frac{e^{-\alpha x}}{x^{\alpha}}$$

(A) Plot z_1 versus y and z_2 versus x.
(B) Verify that for a given value of x, there are either two, one, or zero values of y such that $z_1(y) = z_2(x)$ and vice versa, for a given value of y, there are either two, one, or zero values of x such that $z_1(y) = z_2(x)$. The only one of the three scenarios depicted in figure 1.5 that has no more than two points of intersection between each orbit and each vertical line is the center scenario. We conclude that the nontrivial equilibrium is a center surrounded by a family of periodic orbits, and not a stable or unstable focus.

Exercise 1.7

Numerical solution of the dimensional Lotka–Volterra predator–prey.

A couple of hints to explore numerically the Lotka–Volterra predator–prey dimensional model:

1. There are four parameters to choose $\mathscr{a}$, $\mathscr{b}$, $\mathscr{c}$, $\mathscr{d}$. The linear coefficients for the prey and the predator, $\mathscr{a}$ and $\mathscr{d}$, respectively, should be chosen between 1 and 10 and $\mathscr{a} < \mathscr{d}$ (try different values and see which one convinces you most).
2. You can use some ODE solver package, for example the MATLAB ODE functions.

Exercise 1.8

Numerical solution of the Lotka–Volterra predator–prey non dimensional model.

By a change of variables $x = (c/d)N$ and $y = (b/a)P$, the Lotka–Volterra predator–prey equations can be rewritten in terms of a single parameter $\alpha = d/a$.

(A) Solve this system of equations for different initial conditions. (You can use some ODE solver package or write your own script implementing Runge–Kutta method.)

(B) Graph the vector diagram (the vector field at each point has by components dx/dt and dy/dt) in the phase plane x–y.

Exercise 1.9

Lynx and hares.

Assume initial populations of 100 lynx and 5000 hares and you want to predict what will happen to them in the next 10 years. Write a program in MATLAB to solve the corresponding Lotka–Volterra differential equations.

Exercise 1.10

Realistic Lotka–Volterra predator–prey model.

(a) Show that a simple inclusion of density-dependence in the prey population replacing the prey equation by $\frac{dN}{dt} = rN(1 - \frac{N}{\mathcal{K}}) - cNP$ is enough to shift the behavior of the model from neutral stability to local stability about the equilibrium point.

(b) Now modify both equations by

$$\frac{dN}{dt} = rN\left(1 - \frac{N}{\mathcal{K}}\right) - c\frac{N^2}{N^2 + k^2}P$$

$$\frac{dP}{dt} = \beta c\frac{N}{N^2 + k^2}P - dP$$

Solve this system of equations for $r = 0.02$, $c = 6.2 \times 10^{-4}$, $k = 2.3$ and $\beta = 81$ (in some appropriate system of units) and show that there appear stable ***limit cycle*** oscillations.

Exercise 1.11

Holling–Tanner model.

Solve the Holling–Tanner model (Tanner 1975)

$$\frac{dN}{dt} = rN\left(1 - \frac{N}{\mathcal{K}}\right) - \frac{cN}{N + k}P$$

$$\frac{dP}{dt} = sP\left(1 - \eta\frac{P}{N}\right).$$

with parameters given by $r = 1$, $\mathcal{K} = 7$, $c = 6/7$, $k = 1$, $s = 0.2$, $\eta = 0.5$ and reproduce the figure 1.10.

Exercise 1.12

From one second order differential equation to two first order differential equations.

The equation governing the dynamics of a mass m attached to a linear spring of constant k is the second order equation $md^2x/dt^2 = -kx$ for small displacements x with respect to the equilibrium position (when the spring is neither compressed nor stretched). Introducing the new variable $y = dx/dt$ write the two-dimensional system of first order equations for the variables x and y.

Exercise 1.13

Lotka–Volterra competition (LVC) model for two species

The Lotka–Volterra model for competition between 2 species is given by equation (1.42).

(A) Check that this system has the four equilibria of equations (1.45a)–(1.45d).
(B) Write a script in MATLAB or Fortran or R, etc, to solve the system of two LVC equations.
(C) Graph for $\mathcal{K}_1 = \mathcal{K}_2 = 10$, $\alpha_{12} = \alpha_{21} = 0.5$ the trajectories of the system in the phase plane X_1–X_2 starting from different initial conditions along with the velocity field.
(D) Repeat (C) for $\mathcal{K}_1 = \mathcal{K}_2 = 10$, $\alpha_{12} = \alpha_{21} = 2$. What difference do you find with respect to (C) and how do you explain it?

Exercise 1.14

Non dimensional Lotka–Volterra competition (LVC) model for two species

Check that the LVC system can be re-written in terms of non dimensional variables as:

$$\frac{dy_1(\tau)}{d\tau} = y_1(\tau)(1 - y_1(\tau) + a_{12}y_2(\tau)), \tag{1.57a}$$

$$\frac{dy_2(\tau)}{d\tau} = \rho y_2(\tau)(1 - y_1(\tau) + a_{12}y_2(\tau)), \tag{1.57b}$$

where $y_i = N_i/\mathcal{K}_i$, $\tau = r_1 t$, $\rho = r_1/r_2$, and the a_{ij} denote the interspecific interaction coefficients in terms of these relative yields, defined as $a_{ij} = \alpha_{ij}\ \mathcal{K}_j/\mathcal{K}_i$.

Exercise 1.15

Stability of the coexistence equilibria for two mutualist species.

(A) Evaluate the Jacobian matrix at the coexistence equilibrium of two mutualist species $\mathbf{J}_{\text{coex}}$.
(B) The sign of the trace tr($\mathbf{J}_{\text{coex}}$) and of the determinant det ($\mathbf{J}_{\text{coex}}$) of this matrix allow to determine the stability of the equilibrium (see appendix I): for stability, the trace must be negative and the determinant positive. Compute tr($\mathbf{J}_{\text{coex}}$) and det ($\mathbf{J}_{\text{coex}}$).

(C) Show that for the weak facultative–facultative mutualism the coexistence equilibrium is a stable node and that for the strong obligate–obligate mutualism the coexistence equilibrium is a semistable saddle-node.

References

Gilpin M E 1973 Do hares eat lynx? *Am. Nat.* **107** 727–30

Hale B M and McCarthy M L 2005 *An Introduction to Population Ecology* (Washington, DC: MAA)

Hardin G 1960 The competitive exclusion principle *Science* **131** 1292–7

Holling C S 1959 The components of predation as revealed by a study of small-mammal predation of the European pine sawfly *Can. Entomol.* **91** 293–320

Gause G F 1932 Experimental studies on the struggle for existence *J. Experimental Biol.* **9** 389–402

Gause G F 1934 *The Struggle for Existence* (New York: Hafner)

Johnson G 2003 *The Problem of Exploding Deer Populations has No Attractive Solutions* (accessed January 29, 2020) http://biologywriter.com/on-science/articles/deerpops/

Lotka A J 1925 *Elements of Physical Biology* (Baltimore, New York: Williams & Wilkins)

MacArthur R 1972 *Geographical Ecology* (New York: Harper and Row)

Malthus T R 1798 *An Essay on the Principle of Population As It Affects the Future Improvement of Society, with Remarks on the Speculations of Mr. Goodwin, M. Condorcet and Other Writers* 1st edn (London: J. Johnson in St Paul's Church-yard)

May R M 1976 Simple mathematical models with very complicated dynamics *Nature* **261** 459–67

Michaelis L and Menten M L 1913 Die Kinetik der Invertinwirkung *Biochem. Z.* **49** 333–69

Monod J 1949 The growth of bacterial cultures *Annu. Rev. Microbiol.* **3** 371–94

Morozov A and Poggiale J C 2012 From spatially explicit ecological models to mean-field dynamics: The state of the art and perspectives *Ecol. Comp.* **10** 1–11

Odum E P 1953 *Fundamentals of Ecology* (Philadelphia, PA: Saunders)

Pastor J 2008 *Mathematical Ecology of Populations and Ecosystems* (Singapore: Wiley)

Pearl R and Reed L 1920 On the rate of growth of the population of the United States *Proc. Natl. Acad. Sci.* **6** 275–88

Real L A 1977 The kinetics of functional response *Am. Nat.* **111** 289–300

Tanner J T 1975 The stability and the intrinsic growth rates of prey and predator populations *Ecology* **56** 855–67

Vandermeer J H and Boucher D 1978 Varieties of mutualistic interaction in population models *J. Theor. Biol.* **74** 549–58

van der Vaart H R 1973 A comparative investigation of certain difference equations and related differential equations *Bull. Math. Biol.* **35** 195–211

Verhulst P F 1838 Notice sur la loi que la population poursuit dans son accroissement *Corresp. Math. Phys.* **10** 113–21

Volterra V 1926 Fluctuations in the abundance of a species considered mathematically *Nature* **118** 558–60

Chapter A1

Extensive livestock farming: a quantitative management model in terms of a predator–prey dynamical system

1. Extensive livestock farming, based on natural grasslands, increasingly needs quantitative tools to evaluate the biophysical and economic impacts to manage risk and optimize revenues. For example, precision livestock production (PLP), the manipulation of livestock activity taking into account the different components of agroecosystems to improve production, has acquired growing importance in recent years in agronomy.
2. We discuss here an ecological predator–prey model, the so-called ***predator–prey grassland livestock*** (**PPGL**) model, in terms of two variables, the grass height and the individual liveweight of animals, to simulate livestock weight gain under different management conditions as well as the effects of forage deficiency on the farm's economic performance. Both variables are linked by a nonlinear interaction between them: animal performance (liveweight) is related with grass consumption, which depends on forage availability, which in turn is affected by the grazing pressure.
3. To check the PPGL model internal coherence: (A) we study the long-term evolution for its two variables and found oscillations which capture the general observed dynamics both for grass and animal liveweight. From a mathematical point of view this behavior is robust since we show that it corresponds to a frequency locked limit cycle (i.e. forced oscillations). (B) Varying the stocking rate we show that the PPGL model, based on a mechanistic animal–grass interaction, leads to the well-known Mott's curves with optimal stocking rates for native pastures which is are good agreement with values assessed by empirical means.
4. Next, we show how the PPGL model can be used by farmers to quantify gross margins per hectare in different scenarios of animal stocking rate, grass allowance and climate.

doi:10.1088/978-0-7503-2432-8cha1

5. Finally, we discuss possible improvements to the PPGL model. For example, including grass digestibility. Digestibility is a measure of the proportion of a grass (on a dry matter basis) that can be utilized by an animal.

A1.1 Background information: the growing demand for quantitative livestock models

Grassland production systems occupy about 25% of the world surface and contribute to the livelihoods of over 800 million people (Steinfield *et al* 2006). They play an important role in world food supply and in the economy of many countries. This is particularly the native grasslands at the Pampa biome, which extends from the south of Brazil to Argentina and Uruguay. These natural pastures are indeed mega diverse ecosystems comprising over 400 species (Berretta 2006). The average low stocking rate $\mathcal{S}$ (cattle head per hectare) reflects the extensive condition for grassland systems; $\mathcal{S}$ is typically 0.8 livestock units (LU) per ha (the LU is defined for such extensive production systems as a breeding cow of 380 kg liveweight).

Livestock production based on direct cattle grazing the whole year entails complexities like (a) nonlinearity, i.e. the interaction between herbivores and grass is nonlinear; (b) feedback, i.e. animal performance (e.g. liveweight variations) is linked to grass consumption, which depends on forage availability, which in turn is affected by the grazing pressure from animals (Olson 2005). On top of this there are several other complications. For example, meat production on grassland is highly dependent on voluntary forage intake (Allison 1985). This animal grass intake responds to many factors, like forage availability and quality, dry matter (DM) content and chemical composition of the sward, species selection by cattle, stocking rate, among others (Allison 1985). An important external factor that contributes to the complexity of grassland livestock in the Pampa biome is the climate variability, mainly through frequent drought stress (Morales *et al* 2012).

Getting insights about such complex systems is crucial to enhance resilience of livestock farming for sustainable animal protein production (Steiner *et al* 2014). Improving modelling capabilities for native grasslands agroecosystems seems to be important to discuss with stakeholders key management or control variables (e.g. season stocking rate adjustment) as well as to mitigate climate risks, aiming to construct an adaptive management. However, there is a tradeoff between management practices oriented to enhancing yield and conservation issues like keeping the native grass biodiversity. Important lists of agricultural models and their classification as decision-making tools for farmers are provided by Janssen and van Ittersum (2007) and Dury *et al* (2011). Hart (1993) reviewed the 'stocking rate theory' with the general accepted influence of this system control parameter (a major management key that farmers can control to adjust grass intake pressure) being to obtain a maximal meat yield per land use units. Additional examples in which models were used as a simulation tool and/or for optimization are: GRAZE (Loewer *et al* 1987), GRAZPLAN (Freer *et al* 1997), GRASP (McKeon *et al* 2000), TGM (Kothmann and Hinnant 1997),

INFORM (Rendel *et al* 2013) and FARMAX (http://www.farmax.co.nz). In particular, for extensive livestock systems on Pampa Biome, bio-economic models as decision support for farmers can be found in Romera *et al* (2008), Machado (2010) and Tanure *et al* (2013).

Such recent proliferation of models, and particularly of quantitative models, is related to *precision livestock production* (PLP) (Laca 2009, Soca 2015), i.e. the manipulation of livestock activity taking into account the different components of agroecosystems to improve production. PLP has acquired growing importance in recent years within the new paradigm of 'ecological intensification' in agronomy (Temple *et al* 2011, Tittonell 2014). The objective of quantitative livestock modelling is to identify and quantify resources, requirements and constraints to the achievement of specified levels of livestock product demand. Quantitative livestock modelling also provides a means of *ex ante* assessment of the effects of development programmes, like management or feed, aimed at changing selected parameters of the system in which attempts are being made to raise productivity (Dieguez and Fort 2017).

A1.2 A predator–prey model for grassland livestock or PPGL

To develop the quantitative model for grassland livestock we will follow the recipes of section 0.2.4 formulated as questions in box 0.1. So we will start with the **why?** question, i.e. which is our **goal** and conclude with **how can we improve the model?** But previously we will devote section A1.1 to providing some context about the growing demand for quantitative models in grassland livestock.

A1.2.1 What is our goal?

Good management of rangeland resources is crucial for providing forage for livestock grazing and other ecosystem services such as soil fertility, environmental benefits and carbon storage. Particularly important is determining appropriate stocking rates since it has both economic and environmental consequences (Ritten *et al* 2010). Focusing on either biologically or economically optimal stocking rates may not adequately address interactions between financial and environmental consequences of stocking decisions (Wilson and MacLeod 1991). Indeed often the biological optimum stocking decisions will not coincide with economic optimal stocking rates (Izac *et al* 1990).

How to reconcile the seemingly contradictory goals of achieving a high level of food production and simultaneously providing ecosystem services is a core problem for agroecosystems like grassland livestock. The Food and Agriculture Organization (FAO 2009) recently defined 'ecological intensification' as 'maximization of primary production per unit area without compromising the ability of the system to sustain its productive capacity'.

So our goal with this *predator–prey grassland livestock* (PPGL) model is to contribute to a quantitative management tool to analyze different managements, depending on climate events, to support decision-making of the stakeholders.

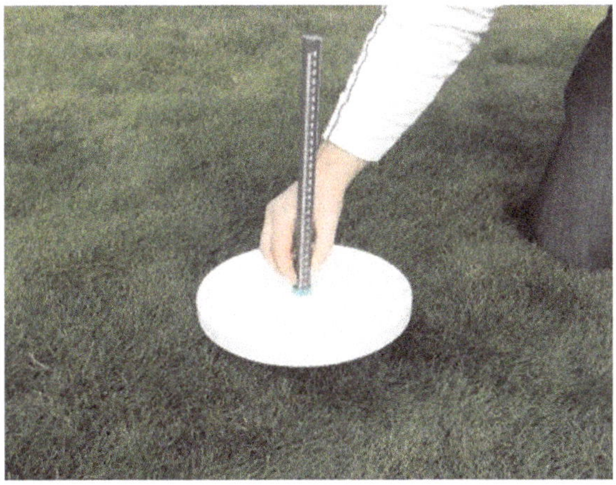

Figure A1.1. A device to measure the grass height. Courtesy of New Zealand Sports Turf Institute.

A1.2.2 What do we know? and what do we assume?: identifying measurable relevant variables for grass and animals

Here we will develop an integral ecological approach to grassland livestock production by modelling the dynamics of combined animals–grass system as a predator–prey dynamical system. Cattle and grass will play, respectively, the role of predator and prey. Cattle and grass will be described each one by an aggregated variable[1]. That is, we consider (a) a single representative or **'average' grass species** describing the mixture of hundreds of coexisting species in natural Pampa's grasslands and (b) an **'average' animal**—i.e. homogeneity of breed, age, sex, etc.

The variable for the grass will be its height x (mean over the entire paddock). The rationale for choosing the height is that it is a good proxy of the available dry matter (DM) content which is relatively simple to estimate by the proper farmers. Depending on the type of soil, 1 cm of grass height corresponds to 180–400 kg of dry matter (DM) for basaltic natural and natural improved fields (Dieguez *et al* 2012, Duarte *et al* 2018). The grass height can be measured with devices like the one shown in figure A1.1 or with the so-called 'green ruler' (figure A1.2). This green ruler, introduced by the *Instituto Nacional de Investigación Agropecuaria* (INIA, Uruguay) to support livestock farmers also provides warnings and recommendations for farmers (Jaurena *et al* 2018).

The variable for cattle is the individual liveweight of animals w (mean over the population). Both variables will be governed by a pair of coupled differential equations describing the nonlinear interaction between them: animal performance (liveweight variations) is linked with grass consumption, which depends on

[1] Remember the ***Aggregation and compartmental modelling*** tip of the previous chapter.

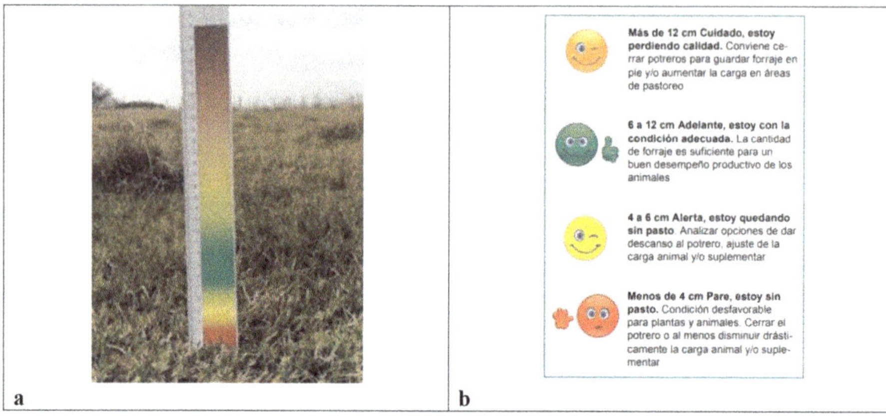

Figure A1.2. The green ruler from INIA (Jaurena *et al* 2018). Panel (b) explains the color code. Red: '**Less than 4 cm Stop, I am out of grass**. Unfavorable condition for grass and animals. Close the paddock or at least drastically decrease the stocking rate.' Yellow: '**4 to 6 cm Warning, I am running out of grass.** Analyze the options of either let the paddock recover or reduce the stocking rate.' Green: '**6 to 12 cm Go ahead, I am in the right condition**. The amount of grass is enough for good animal performance.' Orange: '**Above 12 cm Warning, I am losing quality**. Increase the stocking rate.' Reprinted from Jaurena 2018 with permission of INIA.

forage availability (grass height), which in turn is affected by the grazing pressure.

A1.2.3 How? Adapting a predator–prey model

As we have seen in section 1.3.3, a realistic as well as a robust dynamical system to model Predator–prey dynamics in the case of vertebrate predators is given by the pair of equations:

$$\frac{dN}{dt} = r N\left(1 - \frac{N}{\mathcal{K}}\right) - cP\frac{N^2}{N^2 + \mathcal{H}^2}, \tag{A1.1a}$$

$$\frac{dP}{dt} = \beta c P\frac{N^2}{N^2 + \mathcal{H}^2} - dP. \tag{A1.1b}$$

Furthermore, there is evidence that the intake for large herbivores is well described by this Holling type III functional response because these animals hardly graze if the vegetation is too poor (de Boer 2012). The density variables N and P as well as parameters $\mathcal{K}$ and $\mathcal{H}$ are measured in kg ha^{-1}.

Let us see how to modify the above equations for approaching our problem:

We start by connecting our variables x and w with prey and predator yields (biomass densities) N and P. This is straightforward because N and P are proportional to the grass height x (measured in cm) and the individual animal weight w (measured in kg), respectively. We have seen that 1 cm of height corresponds to between 180–400 kg of DM per hectare. Let us take this conversion constant as 300 kg cm^{-1} ha^{-1} of DM, i.e. $x = N/300$ cm. In the case of P we have $P = \mathcal{S}w$, where the proportionality constant $\mathcal{S}$ is the ***stocking rate*** i.e. the number of animals or heads on a given amount of land, generally measured in units of animals per hectare (heads ha^{-1}).

We also know that the empirical Kleiber's law stipulates that an animal's metabolic rate scales to the 3/4 power of the animal's weight (Kleiber 1932). Hence, the predator per capita functional responses will be multiplied by $w^{3/4}$ (rather than w). Therefore, we can rewrite equations (A1.1) as:

$$\frac{dx}{dt} = \mathcal{r}x\left(1 - \frac{x}{K}\right) - \mathcal{c}\mathcal{S}w^{3/4}\frac{x^2}{x^2 + \mathcal{h}^2}, \qquad (A1.2a)$$

$$\frac{dw}{dt} = \beta \mathcal{c} w^{3/4}\frac{x^2}{x^2 + \mathcal{h}^2} - \mathcal{d}w. \qquad (A1.2b)$$

where parameters K and $\mathcal{h}$ have dimension of length (measured in cm) and are, respectively, proportional to $\mathcal{K}$ and $\mathcal{H}$: $K = \mathcal{K}/300$ cm and $\mathcal{h} = \mathcal{H}/300$ cm.

Actually, the carrying capacity K is not a constant parameter but rather it exhibits an annual cyclic variation that can be described by a sinusoidal function of time $K(t)$ (figure A1.1), associated with the daylight duration:

$$K(t) = (K_{max} - K_{min})/2 * \cos[(2\pi t)/365] + (K_{max} + K_{min})/2, \qquad (A1.3)$$

where $K_{\rm min}$ and $K_{\rm max}$ are two extreme values and t denotes the time, measured in days (d).

The simplest way to include climate variation is by multiplying $K(t)$ by a nondimensional climate coefficient $\mathcal{CC}$. That is, $\mathcal{CC} = 1$ is a normal or average year and separating 'bad' years ($\mathcal{CC} < 1$) from 'good' years ($\mathcal{CC} > 1$). E.g. a $\mathcal{CC} = 0.5$ implies a reduction of 50% of the grass offer, whereas a $\mathcal{CC} = 1.5$ has the effect of enhancing 50% the grass offer.

Finally, we know empirically that if x is below 5 cm the average daily gain (ADG) for bovines, independently of their weight w, becomes negative (Pereira 2011). This is because the ADG can be conceived as the remnant nutrients that can be used for liveweight gain, i.e. the difference between the total intake rate and a minimum intake rate I_m for maintenance requirements.

So we rewrite equations (A1.2) as:

$$\frac{dx}{dt} = \mathcal{r}x\left(1 - \frac{x}{\mathcal{CC}K(t)}\right) - \mathcal{c}\mathcal{S}w^{3/4}\frac{x^2}{x^2 + \mathcal{h}^2}, \qquad (A1.4a)$$

$$\frac{dw}{dt} = \beta \mathcal{c} w^{3/4}\left(\frac{x^2}{x^2 + \mathcal{h}^2} - \frac{5^2}{5^2 + \mathcal{h}^2}\right) - \mathcal{d}w. \qquad (A1.4b)$$

Parameter estimation

We have to estimate a total of six parameters from empirical data:

- three parameters for grass growth function: $\mathscr{r}$, K_{max} and K_{min};
- two parameters for the bovine consumption function: $\mathscr{c}$ and $\mathscr{h}$;
- one parameter for the conversion of grass into body weight: β.

$\mathcal{CC}$ is an external parameter while $\mathcal{S}$ is an internal control parameter that the farmer can vary depending on grass availability and the external climate conditions, measured by $\mathcal{CC}$, to optimize productivity.

$\mathscr{r} = 0.02$ value was estimated by field experts and subsequently empirically validated in a workshop with stakeholders (Bommel *et al* 2014).

We calculate K_{max} and K_{min} from the net primary production (NPP) of grass using satellite remote sensing data for the basaltic region of Uruguay, for a series from March 2000 to January 2014 and taking monthly averages for the whole series (Dieguez and Fort 2017). To convert NPP into cm we used a conversion coefficient ($kgDMcm$) of 300 kg DM cm^{-1} (Dieguez *et al* 2012). Thus, $K_{min} = 5.45$ and $K_{max} = 27.70$ are the couple of values that provide the best fit to data. Figure A1.3 shows that the monthly variation of K is well described by a sinusoidal function with these K_{min} and K_{max}.

To estimate c let us consider the energy maintenance requirements leading to a daily energy requirement (DER) in Mcal per kg of metabolic liveweight per day. There is an empirical relationship connecting DER with ADG (NRC 1996):

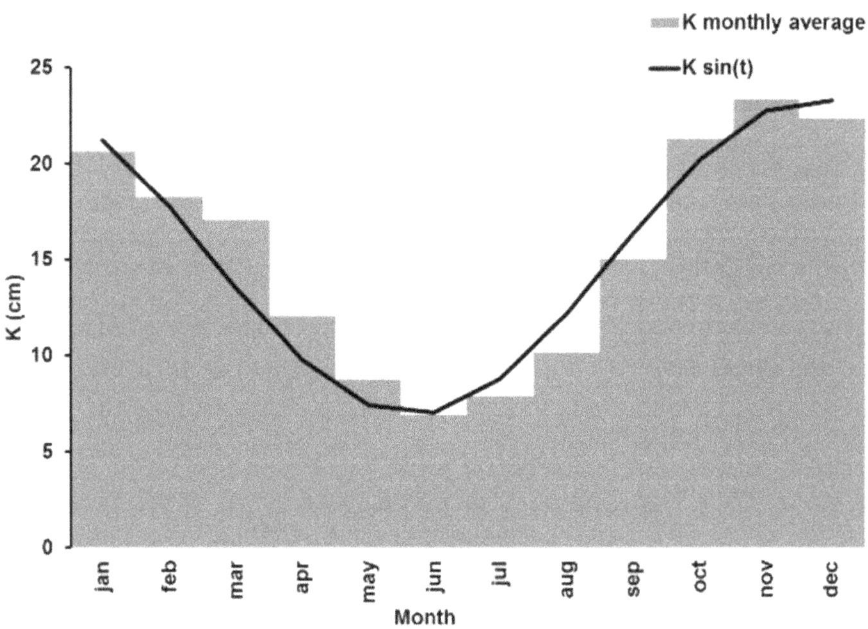

Figure A1.3. Monthly variation of the carrying capacity *K*. Reprinted from Dieguez and Fort 2017. Copyright 2017 with permission from Elsevier.

$$\text{DER} = 0.107 + 0.141\text{ADG}. \tag{A1.5}$$

To obtain c we have to consider the maximum ADG (ADG_{max}) an animal can achieve, substitute this value into equation (A1.5) and divide it by the pasture DM content and its energy concentration. Hence, c can be written as:

$$c = (0.107 + 0.141\text{ADG}_{\text{max}})/(\textit{pastureEnergy} \cdot \textit{kgDMcm}), \tag{A1.6}$$

where ADG_{max} was empirically estimated as 0.660 kg animal^{-1} day^{-1}, i.e. 60 kg animal^{-1} season^{-1} divided by 91 days per season (Dieguez *et al* 2012), *pastureEnergy* is the grass energetic density, 1.08 Mcal kg^{-1} of metabolizable energy per kg of DM (Mieres *et al* 2004) and *kgDMcm* is the conversion factor from kg of DM per ha to cm, defined as 300 kg DM cm^{-1} ha^{-1} (Dieguez *et al* 2012). Therefore we get: $c = 6.2 \times 10^{-4}$ cm ha kg metabolic liveweight^{-1} day^{-1}.

Estimating h from empirical data is less straightforward. We calculated this parameter by fitting the Holling III response function to empirical data (Dieguez and Fort 2017) producing $h = 2.3$ cm.

Finally, β can be obtained as follows. The efficiency of metabolic conversion from pasture intake to animal biomass is the product of two main components: the conversion of grass intake (in cm) to energy, *pastureEnergy* · *kgDMcm*, and the efficiency of metabolizable energy to liveweight gain (kg), which varies from 16% for feeds with high fiber content (like straw or low quality hay) to 60% for grass new shoots or first regrowth stage of pasture (NRC 1996). We take an annual average value of 25% due to a high fiber concentration of native pastures (Mieres *et al* 2004). Hence we obtain:

$$\beta = 0.25 \times \textit{kgDMcm} \cdot k_g \cdot \textit{pastureEnergy} = 81 \text{ kg metabolic liveweight cm}^{-1} \tag{A1.7}$$

All model parameters appear in table A1.1. The reader can check dimensional homogeneity and consistency.

Table A1.1. Value of parameters of PPGL model.

Symbol	Meaning	Type	Value	Units
r	Logistic grass growth rate	Constant	0.02	day^{-1}
c	Maximal grass intake	Constant	6.2×10^{-4}	cm ha kg metabolic liveweight^{-1} day^{-1}
h	Half saturation grass intake	Constant	2.3	cm
β	Grass conversion efficiency	Constant	81	kg metabolic liveweight cm^{-1} ha^{-1}
K_{max}	Maximal grass carrying capacity	Constant	27.7	cm
K_{min}	Minimal grass carrying capacity	Constant	5.45	cm

A1.2.4 What will our model predict?

The input of the PPGL model is the initial grass height x_0 and the initial animal weight w_0. The output will be the grass height and animal weight after a discrete time t (measured in days) x_t and w_t. Since the time variable is discrete we rewrite equations (A1.4) as finite difference equations:

$$x_t = x_{t-1} + r x\left(1 - \frac{x_{t-1}}{\mathcal{CC}K_{t-1}}\right) - c\mathcal{S}w_{t-1}^{3/4}\frac{x_{t-1}^2}{x_{t-1}^2 + h^2}, \tag{A1.8a}$$

$$w_t = w_{t-1} + \beta c w_{t-1}^{3/4}\left(\frac{x_{t-1}^2}{x_{t-1}^2 + h^2} - \frac{5^2}{5^2 + h^2}\right) - d w_{t-1}. \tag{A1.8b}$$

Then for different choices of the initial conditions and parameters $\mathcal{CC}$ and $\mathcal{S}$ equations (A1.8) produce the height of grass and the live animal weight after t days (from a season to a year). Of particular economic interest for the farmer is w_t. Subtracting production costs from the product of w_t times $\mathcal{S}$ times the price of a kg of meat the farmer can estimate gross margins per ha.

A1.3 Model validation

A1.3.1 Are predictions valid?

To test the PPGL model, first we will study the system short and long-term dynamics for x and w. Even though short term periods of time (e.g. one season) are the most important for management of natural resources at farm level, the long-term simulations are helpful to test the internal coherence of the PPGL model. Second, we will analyze the response when varying the stocking rate in order to check if we recover the zootechnical Mott's curve (Mott 1960) for the relationship between animal output (per animal and per unit of land area) and stocking rate. Third, we will perform a sensitivity analysis respect to parameters either estimated with less certainty or which can vary strongly from year to year.

Short and long-term system dynamics

To check PPGL produces a reasonable behavior for short and long times we will take the stock rate fixed to $\mathcal{S} = 1$ and the climate coefficient fixed to $\mathcal{CC} = 1$. For short term dynamics we check the ADG along a season of 91 days. We consider 21 different initial conditions for the grass height x_0, from 5 to 10 cm in steps of 0.25 cm (y_0 was always fixed to 380 kg, i.e. a livestock unit). The ADG produced by equations (A1.8) depends on the initial conditions and varies from 0.239 to 0.599 kg day^{-1} animal^{-1} (mean = 0.472 kg day^{-1} animal^{-1}), a range of values entirely compatible with empirical data for the basaltic region (Berreta 2006).

Results for long-term simulations are shown in figure A1.4. Panel (a) shows the variation of x and w variables for two grass height initial conditions ($x_0 = 4$ and 9 cm) and three animal initial liveweight (y_0= 300, 400 and 500 kg). After a transient, independent of the initial conditions, both variables reach a forced annual cycle

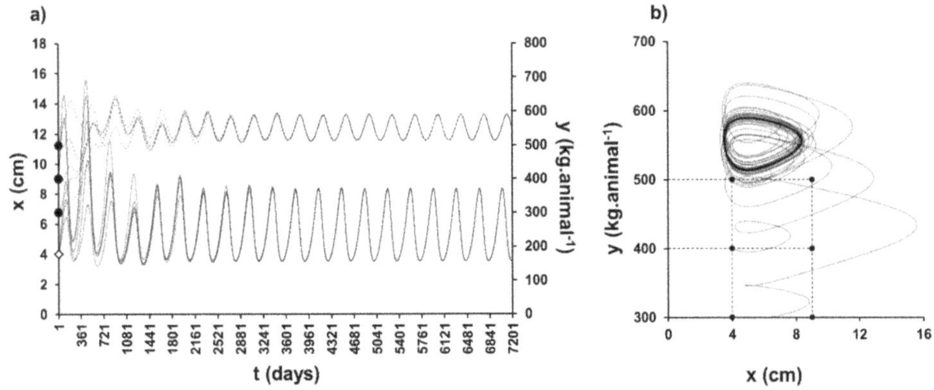

Figure A1.4. a) Variation of x and w variables for two grass height initial conditions (x_0 = 4 and 9 cm) and three animal initial liveweight (y_0 = 300, 400 and 500 kg). b) Convergence to a stable cycle around the x^*= x_m = 5 cm and w^*= 546 kg when starting from different initial conditions in the phase space (see text). Reprinted from Dieguez and Fort 2017. Copyright 2017 with permission from Elsevier.

(see remark below) around equilibrium values $x^* \cong 5$ cm, i.e. coinciding with the maintenance grass height x_m, and $w^* \cong 546$ kg, a liveweight which is near the adult bovine weight for breeder systems in Uruguayan extensive livestock grassland farming (we show how to obtain analytically this value from equation (A1.4a) in box I). That is, we get periodic oscillations typical from predator–prey population models. Likewise, panel (b) shows the convergence to a stable cycle around the $x^*= x_m = 5$ cm and $w^*= 546$ kg when starting from different initial conditions in the phase space, i.e. by plotting w versus x.

However, an important remark regarding these oscillations is in order. The seasonal variation in the carrying capacity operates as an **external periodic forcing** which stabilizes the oscillations, after a transient lasting around three to four years, with a periodicity of one year. Thus, the stable cycle towards trajectories in the phase converge is a **forced cycle** rather than a true limit cycle. Had we taken a constant carrying capacity K this limit cycle would have appeared, but its period would be different from a year. In other words, when the constant K is replaced by a sinusoidal seasonal one the phenomenon of frequency locking (Hilborn 2000) transforms this limit cycle into the observed forced cycle.

At any rate, from the point of view of livestock management, the relevant time scales are shorter (months or sometimes seasons) and **what matters is the transient regime**.

Until now the stocking rate was fixed to $\mathcal{S}$ =1. In fact $\mathcal{S}$ is a main control parameter that farmers can vary to adjust the nutrient demand on extensive production systems. Thus, it is interesting to simulate different scenarios to explore the system behavior to variation on grazing pressure by total grass intake. To explore the effect we consider three values of $\mathcal{S}$, 0.8, 1.0 and 1.2 animals ha^{-1}. We keep x_0 = 7 cm and w_0 = 380 kg fixed. Figure A1.5 depicts the resulting phase diagrams. In all three the system converges to a forced cycle whose center has the same abscission ($x_c \approx 5$ cm) and ordinate w_c which decreases with S. This forced cycle varies as expected with the stocking rate, i.e. the resulting liveweight varies inversely proportional with the grazing pressure.

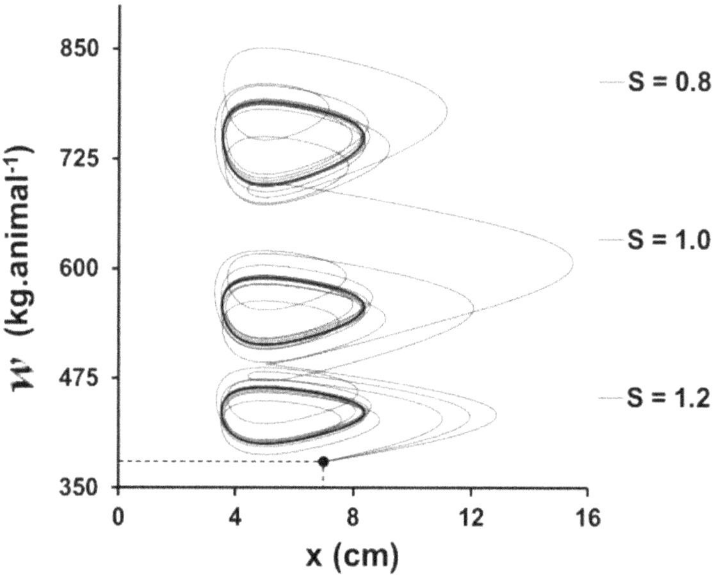

Figure A1.5. Phase diagrams for different stocking rates $\mathcal{S}$ (x_0 = 7 cm and w_0 = 380 kg are kept fixed). Reprinted from Dieguez and Fort 2017. Copyright 2017 with permission from Elsevier.

The optimal stocking rate

Mott (1960) proposed that there exists an optimal stocking rate $\mathcal{S}_{\text{opt}}$ for grassland livestock systems which is achieved when the maximal yield (in kg ha^{-1}) is reached. The effect of increasing $\mathcal{S}$ beyond this optimal value is a reduction of the cattle yield (in kg animal^{-1} ha^{-1}), as reviewed by Hart (1993).

To test the effect of $\mathcal{S}$ on individual liveweight gain and yield (calculated as individual liveweight gain multiplied by $\mathcal{S}$) we vary $\mathcal{S}$ from 0.1 to 2 animal ha^{-1}, with w_0 = 380 kg for one year. Four initial grass height scenarios are considered: x_0 = 9, 7, 5 and 3 cm. The corresponding results produced by equations (A1.8) show that PPGL can reproduce the Mott's relationship (figure A1.6).

Hence, the PPGL model, based on a mechanistic animal–grass interaction, leads to an optimal stocking rate for native pastures, which is frequently assessed in an empirical way.

Depending on the initial grass allowance, x_0, the model yields values of maximal yield for $\mathcal{S}$ between 0.6 and 1.0 responds to general assumptions for native grasslands average conditions stated on table A1.2. This is in agreement with the recommended stocking rate for a secure livestock management in native pastures in Uruguay (0.8 LU ha^{-1}).

A1.3.2 Sensitivity analysis

An important component of a model validation is regarding the sensitivity of the model with respect to variation of its parameters, particularly with respect to parameters estimated with large uncertainty. So, let us perform a sensitivity analysis for two parameters: $\mathcal{k}$ and $\mathcal{C}$. The value of parameter $\mathcal{k}$ (2.3 cm) was calculated

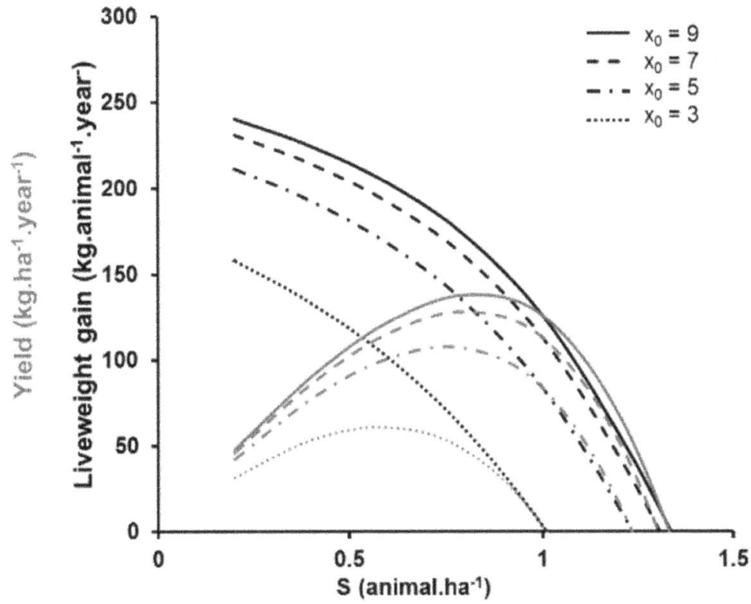

Figure A1.6. Liveweight gain (black) and yield (gray) = liveweight gain multiplied by the stocking rate $\mathcal{S}$ produced by equations (A1.8) for different values of $\mathcal{S}$ for one year with $w_0 = 380$ kg. Four initial grass height scenarios are considered: $x_0 = 9$, 7, 5 and 3 cm. Reprinted from Dieguez and Fort 2017. Copyright 2017 with permission from Elsevier.

Table A1.2. Maximum gross margins (USD ha^{-1}) for the 12 different productive scenarios considered. $\mathcal{S}_{\text{opt}}$ is the stocking rate that produces the maximum gross margin.

x_0 (cm)	$\mathcal{CC}$	Maximum GM per ha (USD ha^{-1})	$\mathcal{S}_{\text{opt}}$ (heads ha^{-1})
3	1	31.35	0.82
3	0.5	12.46	0.72
3	0.25	−29.86	0.52
3	0.125	−79.75	0.24
5	1	107.78	1.03
5	0.5	71.58	0.9
5	0.25	1.9	0.64
5	0.125	−68.62	0.32
7	1	146.27	1.13
7	0.5	99.82	0.97
7	0.25	16.3	0.69
7	0.125	−63.85	0.35

indirectly by fitting a Holling III functional response with empirical data. Given the difficulty of empirically validating its value, it is important to analyze the sensitivity of the PPGL model with respect to this parameter. To do this we repeat calculations for different values of this parameter approximately 20% below and 20% above i.e. $\mathcal{k} = 1.8$ cm and 2.8 cm (and $\mathcal{S} = 1$ animal ha^{-1}, $w_0 = 380$ kg and $x_0 = 5$ cm) and measure the variation of w^* with h (x^* doesn't change, remaining approximately at 5 cm). A useful way to quantify the sensitivity of data to variations of a parameter is in terms of relative changes. That is, if w^*changes in $\Delta w^*$when h changes in Δh then the sensitivity S_h is defined by:

$$S_h = (\Delta w^*/w^*)/(\Delta h/h) \tag{A1.9}$$

In our case for $\Delta h = \pm 0.5$ cm we get a $\Delta w^* = 70$ kg animal^{-1} and substituting in equation (A1.9) we get $S_H = 0.56$. This means that an error of 20% in the estimation of parameter h would introduce in the mean liveweight per animal w^* an error of 11%. For model parameters whose values are not known with high certainty, like h, a relatively low sensitivity ($S_h < 1$) is a desirable property since it provides robustness to model uncertainties. Figure A1.7 presents the sensitivity of liveweight and grass height to parameter h.

In a similar way, we can perform a sensitivity analysis with respect to variations of $\mathcal{CC}$ by varying it from values of 0.2–2.0 (in steps of 0.2). Given that animal performance is affected by $\mathcal{S}$, this analysis includes simultaneously different values of $\mathcal{S}$ (we take $x_0 = 5$ cm fixed). The system response considered was its yield (in kg ha^{-1} year^{-1}) and is shown in figure A1.8.

The effect of two varying parameters ($\mathcal{S}$ and $\mathcal{CC}$) shown in figure A1.8(a) reproduces the Mott's curve in different climate situations. As $\mathcal{CC}$ increases, the maximum yield increases for all $\mathcal{S}$ values tested. It is worth mentioning that whole year $\mathcal{CC}$ values do not appear constant in Nature, but the aim of this analysis is to make explicit the effect of this parameter in a systematic way.

Maximal yield results suggest a linear increase when $\mathcal{CC}$ increases (figure A1.8(b)). Results of simulations reinforces the concept that an optimal $\mathcal{S}$ must be linked with the climatic conditions, where the recommended value for natural grassland ($\mathcal{S} = 0.8$ LU ha^{-1}) depends on climatic conditions that affect pasture growth rate and then grass availability.

A1.3.3 Verdict: model validated

Since the outputs of the PPGL model are in quite good agreement with empirical results for the different tests we considered, we can conclude that its predictions are valid.

A1.4 Uses of PPGL by farmers: estimating gross margins in different productive scenarios

Simulation scenarios can be built by varying two aspects: management and climate conditions. Management involves two key parameters: the stocking rate $\mathcal{S}$—as an internal regulator of the system dynamics by farmers—and initial values of grass allowance, described by the initial grass height x_0. The external climate forcing factor, in turn, is modelled through $\mathcal{CC}$.

i. Animal management: we consider variations of $\mathcal{S}$ from 0.43 to 1.33 heads ha^{-1} in 100 steps of 0.01 heads ha^{-1}. For simulations here we will set the initial cattle liveweight w_0 at 1 LU = 380 kg animal^{-1}.

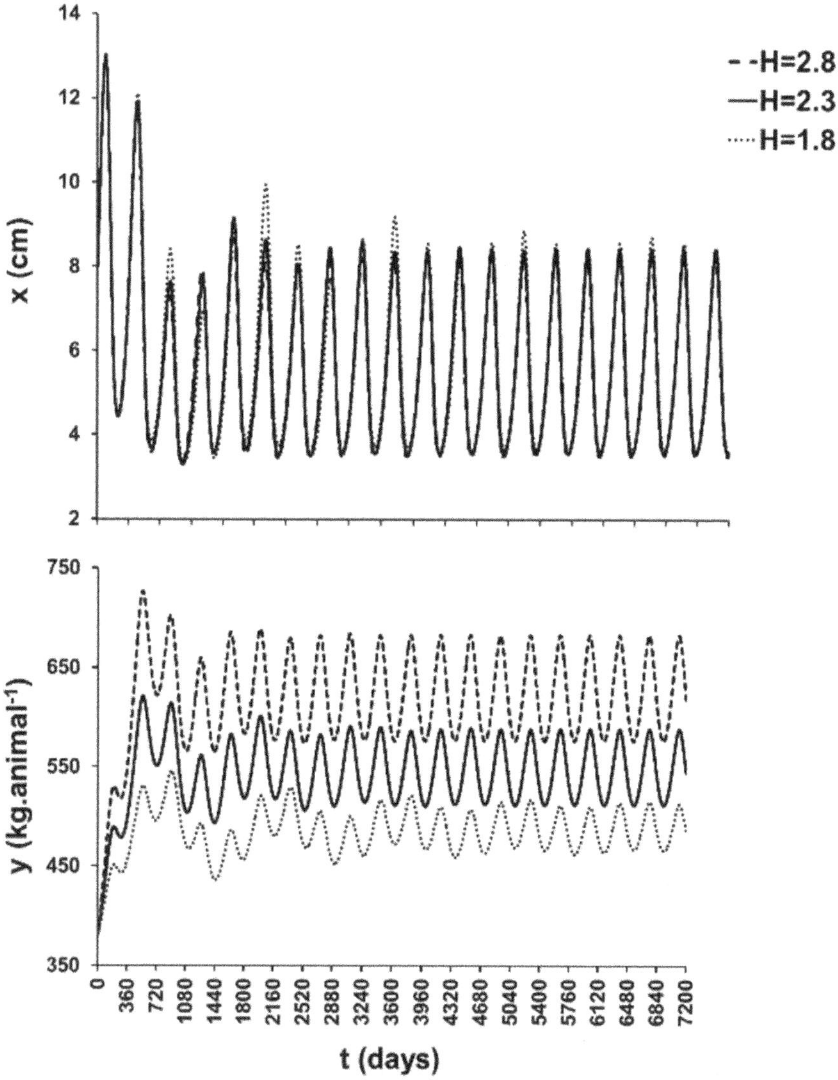

Figure A1.7. Sensitivity of liveweight and grass height to parameter h. Reprinted from Dieguez and Fort 2017. Copyright 2017 with permission from Elsevier.

ii. Initial forage allowance: in the scenarios we consider, three values of initial grass height are used: x_0 = 3, 5 or 7 cm (corresponding to 900, 1500 and 2100 kg of dry matter/ha, respectively).

iii. Environmental conditions: we consider a scenario of grass shortage making grass below the required threshold of animal food needs. This forage stress is applied in spring (September, October and November in the Southern Hemisphere). Spring is particularly important because it is the season

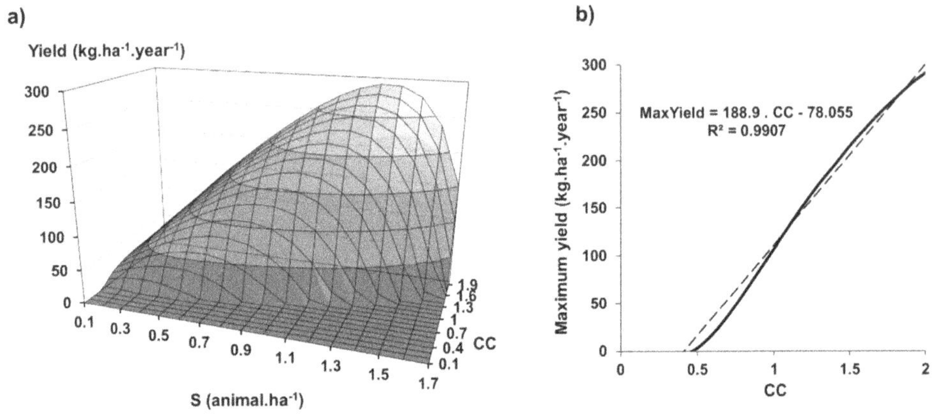

Figure A1.8. Sensitivity analysis with respect to variations of the parameter $\mathcal{CC}$ for different values of the stocking rate $\mathcal{S}$ ($x_0 = 5$ cm is fixed). Reprinted from Dieguez and Fort 2017. Copyright 2017 with permission from Elsevier.

expected to exhibit the highest grass growth rate in native grasslands and simultaneously is crucial for animal fattening (it is expected that cattle recover body condition and gain liveweight in spring, after a normal weight loss during the winter season, and before the summer mating season). Then, the standard scenarios we consider are: a 'control' value of $\mathcal{CC} = 1$, and three scenarios with values of 0.5, 0.25 and 0.125 to force a forage shortage for the three spring months.

Considering the above combinations ii and iii, we have $3 \times 4 = 12$ different scenarios.

So we use equations (A1.8) for these 12 scenarios, always starting at the beginning of autumn, on March 21 (Julian day = 80), which is a date recommended for technical herd management reasons (MGAP 2011), and ending on March 20 of the next year.

The economic data we consider are for Uruguayan livestock farms for the economical year 2016–2017 (IPA 2018). Farmers received on average an amount of 1.44 USD kg^{-1} of liveweight meat and total production costs, taking averages on main items of livestock farming from the IPA (2018) report, amounting to 95.75 USD per hectare (Dieguez and Fort 2019).

The 12 curves of gross production per hectare and per animal (GP/ha and GP/animal, respectively) as well as the 12 curves of gross margin per ha (GM/ha) versus $\mathcal{S}$ are shown in figure A1.9. Each column corresponds to a different value of x_0.

Table A1.2 shows maximum GM ha^{-1} values (in USD ha^{-1} for the 12 productive scenarios considered). This table also includes the corresponding optimal stocking rate producing such maximum GM.

From table A1.1 we can see that:

1. The GM curve is asymmetric around $\mathcal{S}_{opt}$: If $\mathcal{CC} \leqslant 0.5$ it decreases faster for larger values of $\mathcal{S}$ than for smaller ones.
2. The initial grass height of pasture is then a crucial factor for economic performance. If $x_0 \geqslant 5$ cm:

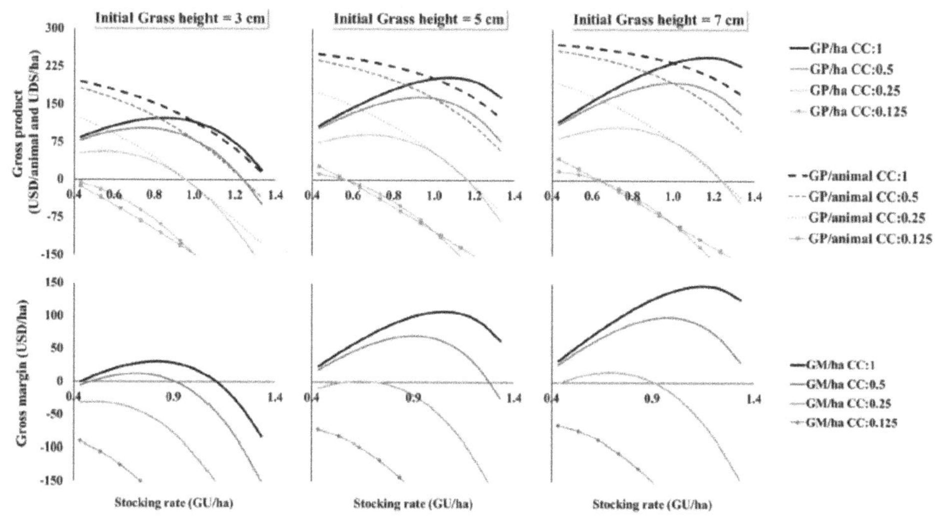

Figure A1.9. The 12 curves of GP and GM versus $\mathscr{S}$ curves generated by the PPGL model for the 12 different productive scenarios. Reprinted from Dieguez and Fort 2019, copyright AIMS 2020.

- Even for severe climate situations ($\mathcal{CC}$ between 0.25 and 0.5) a positive gross margin can be attained provided the livestock farmer reduces the stocking rate from its recommended value for secure livestock management (0.8 heads ha^{-1}) around 20%.
- On the other hand, under normal climate conditions ($\mathcal{CC}$ = 1), a stocking rate close to one animal/ha is the optimum and it allows gross margins above 100 USD ha^{-1}.
- The safer stocking rate of 0.8 ensures around an 80% of the maximum GM.

Therefore, depending on her/his tolerance to risk a livestock farmer should set the stocking rate between 0.8 and 1 and start the productive cycle with a grass height not below 5 cm.

When comparing simulations with actual data (IPA 2018) for the economical year 2016–2017, we can see that predictions of the PPGL model are in line with the actual average results of 58 USD/ha of gross margin with an average stocking rate = 0.79 animals/ha (unfortunately no information on the initial grass height is specified).

A1.5 How can we improve our model?

Adding grass digestibility

One way to improve the PPGL model is by including grass digestibility. Digestibility is a measure of the proportion of a grass (on a dry matter basis) that can be utilized by an animal. For example, if an animal eats 10 kg of dry matter and 3 kg is expelled as dung, then the grass is 70% digestible. Digestibility influences the speed with which a grass passes through the digestive system. Grasses with higher digestibility will be processed more rapidly, allowing stock to eat more and so have higher weight gain. Digestibility is also a useful measure of the quality of a grass as it is directly related to the energy content of a feed.

Roughly: high livestock production requires feed of 70%–80% digestibility; moderate production requires 60%–70% digestibility; and 55%–60% digestibility is required to maintain dry stock. Below 55% digestibility, dry stock will lose weight. Digestibility is strongly influenced by the plant's stage of growth. Grasses that are green, leafy and actively growing will have a higher digestibility than those that are in head or have hayed off.

For example, digestibility can be modelled by an empirical linear function of the grass height $D(x)$ (Dieguez and Fort 2017):

$$D(x) = \begin{cases} 1 & \text{if} x \leqslant 5\text{cm} \\ -0.8\dfrac{(x-5)}{K_{\max}-5} + 1 & \text{if} x > 5\text{cm} \end{cases} \tag{A1.10}$$

MATLAB codes

Main code: LVPPGL_Ap1%

```
%LVPPGL_Ap1: LVPP is for Lotka-Volterra Pre-predator, GL for Grass-Livestock
%uses FINITE DIFFERENCES eqs. (A1.8) rather than differential equations

clear all
% Input data:
Nyears=input(' enter number of years = (default = 1) = ')
Tf = Nyears*365;
x0 =input(' enter initial grass height in cm = (default = 8) = ' )
y0 =input(' enter initial animal weight in kg = (default = 380) = ' )
S =input(' enter stockong rate (heads/ha) = (default = 0.8) = ' )
CC =input(' enter Climate coefficient = (default = 1) = ')
Kmax=27.7; %Maximum and minimum grass height (cm).
Kmin=5.45;


% Fixed parameters:
r = 0.02;                  % Maximum growth rate for logistic equation.
c = 6.185 32* 10^(-4);     % Maximum consumption rate of grass per animal
                           % metabolic weight per unit of time.
H = 2.2955;                % Half-saturation constant.
cMant = c*5^2/(5^2+H^2);   % Consumption rate required for maintaining
                           % animal weight (5.0985* 10^(-4)).
B = 81;                    % Metabolic conversion constant (kg of grass into
                           % kg of animal weight).
```

```
xf=zeros(1,4);
cumulated=zeros(1,4);
for iT0=1:4     % Dividing the year in four seasons
 if iT0==1
  T0=60;
  K=0.5*(Kmax-Kmin)*cos(2*pi*([T0:T0+Tf-1])/365)+0.5*(Kmax+Kmin);
 elseif iT0==2
  T0=152;
  K=0.5*(Kmax-Kmin)*cos(2*pi*([T0:T0+Tf-1])/365)+0.5*(Kmax+Kmin);
 elseif iT0==3
  T0=244;
  K=0.5*(Kmax-Kmin)*cos(2*pi*([T0:T0+Tf-1])/365)+0.5*(Kmax+Kmin);
 elseif iT0==4
  T0=335;
  K=0.5*(Kmax-Kmin)*cos(2*pi*([T0:T0+Tf-1])/365)+0.5*(Kmax+Kmin);
 end

 % Including grass digestibility via the function Digest(x,var_relat)
 % for x le. 5 cm the digestibility is maximum
 rel_var= 0.8/(Kmax-5) ;%(from 1 for x < 5 cm to 0.2 for x = Kmax)
 T = 1:Tf;              % Time vectors.
 X = zeros(Tf,1);
 Y = zeros(Tf,1);
 X(1) = x0;
 Y(1) = y0;
 Xd = zeros(Tf,1);
 Yd = zeros(Tf,1);
 Xd(1) = x0;
 Yd(1) = y0;
 for i=2:Tf

  % (A) Without taking into account grass digestibility
  X(i) = X(i-1) + r* X(i-1)* (1 - X(i-1)/(CC*K(i-1))) - (c*S*...
        (X(i-1))^2 / (H^2 + (X(i-1))^2))* Y(i-1)^(3/4)* S;
  Y(i) = Y(i-1) + ((c* (X(i-1))^2 /(H^2 + (X(i-1))^2) - cMant))...
         * B* (Y(i-1))^(3/4);

  % (B) Taking into account grass digestibility
  Xd(i) = Xd(i-1) + r* Xd(i-1)* (1 - Xd(i-1)/(CC*K(i-1))) -(c*S*...
         (Xd(i-1))^2 / (H^2 + (Xd(i-1))^2))* Yd(i-1)^(3/4)* S;
  Yd(i) = Yd(i-1) + Digest(Xd(i-1),rel_var)*(c* Xd(i-1)^2/(H^2 +...
          Xd(i-1)^2) - cMant)* B* (Yd(i-1))^(3/4);

 end
 xf(iT0)=X(91);
end
```

```
xf
cumulated=(xf-x0)*180
figure()
subplot(2,1,1)
hold on
plot(T,X,'b')
plot(T,Xd,'r')
xlabel('Time (days)')
ylabel('Grass height (cm)')
subplot(2,1,2)
hold on
plot(T,Y,'b')
plot(T,Yd,'r')
xlabel('Time (days)')
ylabel('Animal eight (kg)')
legend('No digestibility','With digestibility')
figure
plot(T,K,'--',T,X,'b')
xlabel('Time (days)')
ylabel('Grass height (cm)')
title('Starting day')
```

Function 'Digest'

```
function [dd] = Digest(x,rel_var)
  if x<=5
   dd=1;
  else
   dd=-rel_var*(x-5)+1;
  end
end
```

References

Allison C D 1985 Factors affecting forage intake by range ruminants: a review *J. Range Manag.* **38** 305–11

Berretta E 2006 *Uruguay-Country Pasture and Forage Resource Profile* (Rome: FAO) Available at: http://fao.org/ag/agp/agpc/doc/counprof/PDF%20files/Uruguay_English.pdf

Bommel P, Dieguez F, Bartaburu D, Duarte E, Montes E, Pereira M, Corral J, Pereira C J and Morales H 2014 A further step towards participatory modelling. Fostering stakeholder involvement in designing models by using executable UML *J. Artif. Soc. Soc. Simulat.* **17** 6 de Boer R 2012 Modeling population dynamics: a graphical approach *Theoretical Biology and Bioinformatics* (Utrecht: Utrecht University)

Dieguez F, Bommel P, Corral J, Bartaburu D, Pereira M, Montes E, Duarte E and Morales Grosskopf H 2012 Modelización de una explotación ganadera extensiva criadora en basalto *Agrociencia* **16** 120–30

Dieguez F and Fort H 2017 Towards scientifically based management of extensive livestock farming in terms of ecological Predator–prey modeling *Agric. Syst.* **153** 127–37

Dieguez F and Fort H 2019 An application of a dynamical model with ecological Predator–prey approach to extensive livestock farming in Uruguay: Economical assessment on forage deficiency *J. Dyn. Games* **6** 119–29

Duarte E *et al* 2018 Elaboración participativa de metodologías de extensión que contribuyan a aumentar la producción en sistemas ganaderos sobre campo natural mediante el control de la asignación de forraje *Revista INIA* **vol 55** 48–51 http://inia.uy/Publicaciones/Documentos%20compartidos/Revista-INIA-55-diciembre-2018.pdf

Dury J, Schaller N, Garica F, Reynaud A and Bergez J 2011 Models to support cropping plan and crop rotation decisions. A review *Agron. Sustain. Dev.* **32** 567–80

FAO 2009 http://fao.org/3/Y4176E/y4176e0a.htm

Freer M, Moore A D and Donnelly J R 1997 GRAZPLAN: decision support systems in Australian grazing enterprises—II *Agric. Syst.* **54** 77–126

Hart R 1993 Viewpoint: 'Invisible colleges' and citations clusters in stocking rate research *J. Range Manag.* **46** 378–82

Hilborn R 2000 *Chaos and Nonlinear Dynamics: An Introduction for Scientists and Engineers* 2nd edn (Oxford: Oxford University Press)

IPA 2018 *Resultados del ejercicio 2016-2017, Instituto Plan Agropecuario Montevideo, Uruguay*

Izac A-M N, Anaman K A and Jones R 1990 Biological and economic optima in a tropical grazing ecosystem in Australia *Agric. Ecosyst. Environ.* **30** 265–79

Janssen S and van Ittersum M 2007 Assessing farm innovations and responses to policies: a review of bio-economic farm models *Agric. Syst.* **94** 622–36

Jaurena M *et al* 2018 La regla verde: Una herramienta para el manejo del campo natural *Revista INIA* **54** 24–7 http://inia.uy/Documentos/P%C3%BAblicos/INIA%20Salto%20Grande/2018/2018.09.25-Gira_INIA_Norte/Articulo_La_Regla_Verde.pdf

Kleiber M 1932 Body size and metabolism *Hilgardia* **6** 315–51

Kothmann M M and Hinnant R T 1997 The Grazing Manager: a new application of the carrying capacity concept *Proc. of the XVIII Int. Grassland Assoc. (Winnepeg, Manitoba)* Available at: http://www.internationalgrasslands.org/files/igc/publications/1997/2-26-009.pdf

Laca E A 2009 Precision livestock production: tools and concepts *R. Bras. Zootec.* **38** 123–32

Loewer O J, Taul K L, Turner L W, Gay N and Muntifering R 1987 Graze: amodel of selective grazing by beef animals *Agric. Syst.* **25** 297–309

Machado C F, Morris S T, Hodgson J, Arroqui M and Mangudo P A 2010 A web-based model for simulating whole-farm beef cattle systems *Comput. Electron. Agric.* **74** 129–36

McKeon G M, Ash A J, Hall W B and Stafford-Smith M 2000 Simulation of grazing strategies for beef production in north-east Queensland *Applications of Seasonal Climate Forecasting in Agricultural and Natural Ecosystems—The Australian Experience* ed G Hammer, N Nicholls and C Mitchell (Dordrecht: Kluwer) 227–52

MGAP 2011 *Ministerio de Ganaderia, Agricultura y Pesca, Pautas para el manejo del campo natural, Montevideo, Uruguay* Available at: https://planagropecuario.org.uy/web/24/librillos/pautas-para-el-manejo-del-campo-natural.html.

Mieres J, Assandri L and Cúneo M 2004 Tablas de valor nutritivo de alimentos *Guía para la alimentación de rumiantes* vol 142 (Montevideo: INIA) pp 13–68

Morales H, Bartaburu D, Dieguez F, Bommel P and Tourrand J F 2012 Local knowledge, agents and models for the adaptation to climatic variability of livestock farmers in Uruguay *Proc. of the X Int. Farming Assoc. (Aarhus)* Available at: http://ifsa.boku.ac.at/cms/fileadmin/Proceeding2012/IFSA2012_WS3.1_Grosskopf.pdf

Mott G O 1960 Grazing pressure and the measurement of pasture production *Proc. 8th Int. Grassland Congress* ed C L Skidmore, P J Boyle and L W Raymond (Oxford: Alden) 606–11

NRC-National Research Council 1996 *Nutrient Requirements of Beef Cattle* (Washington DC: National Academy)

Olson K C 2005 Range management for efficient reproduction *J. Anim. Sci.* **8** (E. Suppl.) E107–16

Pereira M 2011 *Manejo y conservación de las pasturas naturales del Basalto* (Montevideo: Instituto Plan Agropecuario)

Rendel J M, MacKay A D, Manderson A and O'Neill K 2013 Optimising farm resource allocation to maximise profit using a new generation integrated whole farm planning model *Proc. N. Z. Grassl. Assoc.* 75 85–90

Ritten J, Bastian C and Fraiser M 2010 Economically optimal stocking rates: A bioeconomic grazing model *Rangeland Ecol. Manag.* **63** 407–14

Romera A J, Burges J C, Morris S T, Hodgson J and Woodwards S J R 2008 Modelling Spring and Autumn calving systems in beef herds of the Salado region of Argentina *Livestock Sci.* **115** 62–72

Soca P 2015 Fundamentos y propuestas del cambio técnico para mejorar la sostenibilidad de la ganadería familiar en Uruguay *Seminario la ganadería familiar en un escenario dinámico del siglo XXI, INIA* Available at: http://cnfr.org.uy/uploads/files/Presentacin_Ing_Pablo_Soca___FAGRO.pdf

Steiner J, Engel D, Xiao X, Saleh A, Tomlinson P and Rice C 2014 Knowledge and tools to enhance resilience of beef grazing systems for sustainable animal protein production *Ann. N. Y. Acad. Sci.* **1328** 10–7

Steinfield H, Gerber P, Wassenaar T, Castel V, Rosales M and De Haan C 2006 *Livestock's Long Shadow: Environmental Issues and Options* (Rome: FAO) Available at: http://ftp.fao.org/docrep/fao/010/a0701e/a0701e.pdf

Tanure S, Nabinger C and Becker J L 2013 Bioeconomic model of decision support system for farm management. Part I: systemic conceptual modeling *Agric. Syst.* **115** 104–16

Temple L, Kwa M, Tetang J and Bikoi A 2011 Organizational determinant of technological innovation in food agriculture and impacts on sustainable development *Agron. Sustain. Dev.* **31** 745–55

Tittonell P 2014 Ecological intensification of agriculture—sustainable by nature *Curr. Opin. Environ. Sustain.* **8** 53–61

Wilson A D and MacLeod N D 1991 Overgrazing: present or absent? *J. Range Manag.* **44** 475–82

Chapter 2

Lotka–Volterra models for multispecies communities and their usefulness as quantitative predicting tools

'The question is, I suppose, how exact do biologists want the correspondence between theory and fact before they admit that the theory tells us something about a biological problem?'

B G Murray 1992

Summary

The linear Lotka–Volterra equations are considered by many ecologists as exploratory models with limited direct relationship to real ecosystems because of their many simplifying assumptions, their popularity being partly attributable to sentimental attachment (Keddy 2001).

This chapter has three main goals: firstly, to present the linear Lotka–Volterra generalized equations (LLVGE) for describing the dynamics of a community of species connected by different kinds of interspecific interactions: mutual competition, amensalism, predation, commensalism and mutualism. Secondly, to test the LLVGE as a *quantitative* tool for describing/explaining/predicting the equilibrium species abundances. That is, how accurately does the LLVGE predict the yields of S interacting species? Thirdly, to analyze which quantitative predictions are possible with an incomplete knowledge of the LLVGE parameters.

Next, we show that, by estimating the LLVGE parameters from the yields in monoculture and biculture experiments, the LLVGE produce quite accurate predictions for species yields in single trophic level communities of $S > 2$ species, either artificial or natural.

doi:10.1088/978-0-7503-2432-8ch2

Furthermore, we can obtain reasonable reliable predictions with an incomplete knowledge of the LLVGE parameters. That is, depending on the level of knowledge we have about the model parameters, these LLVGE can be used as a quantitative tool at two different levels. First, when we only know a fraction of the model parameters, the mean field approximation allows making predictions on aggregate or average quantities. Second, for cases in which all the interaction parameters involving a particular species are available, we have the focal species approximation for predicting the yield of this focal species.

2.1 Many interacting species: the Lotka–Volterra generalized linear model

The most popular mathematical models for describing the species abundances in terms of their interactions are based on the *Lotka–Volterra equations* (Lotka 1925, Volterra 1926). Lotka and Volterra derived two different sets of equations: one set applies to situations involving competition for food or space and the other set to the *predator–prey* situation. In the previous chapter we have introduced both sets of equations for two interacting species. Now we will present a unified framework than can accommodate all kinds of pair interactions between pairs of species, namely:

- mutual competition –/–, each species has an inhibiting effect on the growth of the other;
- amensalism –/0, the growth of one species is negatively affected while the growth of the other is unaffected;
- predation –/+, the 'predator' ('prey'), has an inhibiting (accelerating) effect on the growth of the prey (predator);
- commensalism 0/+, the growth of one species accelerates while the other is unaffected,
- mutual cooperation or mutualism +/+, each species has an accelerating effect on the growth of the other.

(Notice that according to these definitions, the host-parasite and the plant-herbivore interactions would be classified as 'predation'.)

This theory in its simplest formulation consists in a set of linear equations for the per capita growth rates $\frac{1}{n_i}\frac{dn_i}{dt}$ where n_i is a variable denoting the *yield* of species i (its density = abundance per unit of area, or its biomass density). A straightforward way to write these *linear* Lotka–Volterra generalized equations (LLVGE) is:

$$\frac{1}{n_i}\frac{dn_i}{dt} = \mathfrak{r}_i\left(1 + \sum_{j=1}^{S}\mathfrak{a}_{ij}n_j\right) \quad i = 1, \ldots, S, \tag{2.1}$$

where $\mathfrak{r}_i$ is the intrinsic growth rate of species i, with dimension of time^{-1}, and $\mathfrak{a}_{ij}$ is an interaction coefficient quantifying the *per capita* effect of species j on species i through their pairwise interaction, which can be either negative (competition), positive (facilitation) or zero, with dimension of density^{-1}. Notice that when the number of species, or *species richness*, S is equal to 2 equation (2.1) reduces to:

- the prey equation, equation (1.25a), by taking

$$\begin{cases} \mathscr{a}_{11} = 0 \\ \mathscr{r}_1\mathscr{a}_{12} = \mathscr{c} \end{cases}, \tag{2.2a}$$

and to the predator equation, equation (1.25b), by taking

$$\begin{cases} \mathscr{a}_{22} = 0 \\ \mathscr{r}_2\mathscr{a}_{21} = \mathscr{a} \end{cases}; \tag{2.2b}$$

- to competition equations (1.29) by taking

$$\begin{cases} \mathscr{a}_{ii} = -1 \\ \mathscr{a}_{ij} = \alpha_{ij}/\mathcal{K}_i \text{ for } i \neq j \end{cases}. \tag{2.3}$$

These LLVGE can be thought as the first order or linear approximation in a Taylor series expansion of the per-capita growth rates of species about the equilibrium points of a more complex and general theory (Volterra 1926, 1931):

$$\frac{1}{n_i}\frac{dn_i}{dt} = f_i(n_1, n_2, \cdots, n_S) \qquad i = 1, \ldots, S, \tag{2.4}$$

where $f_i(n_1, n_2\ldots, n_S)$ are arbitrary functions of the S species making up the community, which we denote more compactly in terms of the S-dimensional vector of yields $\mathbf{n} = [n_1, n_2, \ldots n_S]$ as $f_i(\mathbf{n})$. Therefore, let $\delta\mathbf{n} = \mathbf{n} - \mathbf{n}^*$, be a small displacement from an equilibrium point $\mathbf{n}^*$, which by definition satisfies $f_i(\mathbf{n}^*) = 0$, and if we perform a Taylor expansion of $f_i(\mathbf{n})$ around $\mathbf{n}^*$ we get:

$$\begin{aligned} f_i(\mathbf{n}) &= f_i(\mathbf{n}^*) + \sum_{j=1}^{S} \frac{\partial f_i(\mathbf{n})}{\partial n_j}\bigg|_{\mathbf{n}^*} \delta n_i + \sum_{j=1}^{S}\sum_{k=1}^{S} \frac{\partial^2 f_i(\mathbf{n})}{\partial n_j \partial n_k}\bigg|_{\mathbf{n}^*} \delta n_j \delta n_k \\ &\quad + \sum_{j=1}^{S}\sum_{k=1}^{S}\sum_{l=1}^{S} \frac{\partial^3 f_i(\mathbf{n})}{\partial n_j \partial n_k \partial n_l}\bigg|_{\mathbf{n}^*} \delta n_j \delta n_k \delta n_l + \cdots \\ &= 0 + \sum_{j=1}^{S} \frac{\partial f_i(\mathbf{n})}{\partial n_j}\bigg|_{\mathbf{n}^*} \delta n_j + O(\delta n^2) \\ &= -\sum_{j=1}^{S} \frac{\partial f_i(\mathbf{n})}{\partial n_j}\bigg|_{\mathbf{n}^*} n_j^* + \sum_{j=1}^{S} \frac{\partial f_i(\mathbf{n})}{\partial n_j}\bigg|_{\mathbf{n}^*} n_j + O(\delta n^2), \end{aligned}$$

where $O(\delta n^2)$ involve quadratic terms given by $\sum_{j=1}^{S}\sum_{k=1}^{S}\left.\frac{\partial^2 f_i(\mathbf{n})}{\partial n_j \partial n_k}\right|_{\mathbf{n}^*}\delta n_j \delta n_k$, or terms of order greater than two, i.e. cubic terms given by $\sum_{j=1}^{S}\sum_{k=1}^{S}\sum_{l=1}^{S}\left.\frac{\partial^3 f_i(\mathbf{n})}{\partial n_j \partial n_k \partial n_l}\right|_{\mathbf{n}^*}\delta n_j \delta n_k \delta n_l$, in such a way that we obtain equation (2.1) by identifying

$$\mathfrak{r}_i = -\sum_{j=1}^{S}\left.\frac{\partial f_i(\mathbf{n})}{\partial n_j}\right|_{\mathbf{n}^*} n_j^*, \tag{2.5a}$$

$$\mathfrak{r}_i \mathfrak{a}_{ij} = \left.\frac{\partial f_i(\mathbf{n})}{\partial n_j}\right|_{\mathbf{n}^*}. \tag{2.5b}$$

Let us introduce at this point a widely used matrix in community ecology, which is closely related with the matrix $[\mathfrak{a}_{ij}]$ but different and that often has been used interchangeably, the *community matrix* J_{ij} (Levins 1968). The community matrix, is nothing but the mathematical Jacobian matrix (see appendix I), defined by:

$$J_{ij} \equiv \left.\frac{\partial\left(\frac{dn_i}{dt}\right)}{\partial n_j}\right|_{\mathbf{n}^*}. \tag{2.6}$$

Thus, equation (2.5b) can be rewritten as connecting the matrix $[\mathfrak{a}_{ij}]$ with the community (i.e. Jacobian) matrix (see appendix I):

$$\mathfrak{a}_{ij} = \frac{J_{ij}}{\mathfrak{r}_i n_i^*}. \tag{2.5b'}$$

The linear Lotka–Volterra equations have been often criticized as being too simple for quantitative modelling real systems involving many interacting species. For example, for $S > 2$ species the model would be valid only to the extent that higher-order interactions are null or negligible compared to pairwise interactions (Simberloff 1982). When a pair of species is entangled within a large and complex web of interacting species, like in food webs (see section 2.5), it is possible to generate interactions with third or fourth parties that create mutually negative effects for the pair under consideration. This phenomenon is called 'apparent competition' (Keddy 2001). As a consequence, these equations have been mostly used to approach qualitative general ecological issues, like community stability and species coexistence, rather than on making quantitative predictions to be tested against experimental data, like the species abundances in equilibrium (Hubbel 2002).

However, hereafter we will adopt a pragmatic stance, regarding the Lotka–Volterra models as ***descriptive*** or ***phenomenological*** models, i.e. they describe how the abundance of one species affects the abundance of another, without specifically including a particular mechanism for such interaction (Morin 2011). Neither the nature of the particular competition mechanism nor if it is direct or mediated by other(s) species, is specified. Rather, ***effective*** interaction coefficients summarize the per capita effects of one species on another.

2.2 The Lotka–Volterra linear model for single trophic communities

All life forms can be broadly lumped into one of two categories: the autotrophs, who produce organic matter from inorganic substances (like plants, algae) and the heterotrophs, organisms that cannot produce their own food, relying instead on the consumption of other organisms (either autotrophs or heterotrophs).

Indeed, the heterotrophs can be arranged in different levels according to who eats whom, called trophic levels. So, for example, the green plants (producers) are the first trophic level, the organisms that feed on plants are the second level (primary consumers), carnivores are the third level and carnivore predators are the fourth level (secondary and tertiary consumers). This is an example of a general **trophic chain** or **food chain**.

The choice (2.3) for $S > 2$ allows specializing the generalized Lotka–Volterra linear equation (2.1) for a set of many species competing for resources in one of such trophic levels, i.e. in an ecosystem, each of several hierarchical levels comprising organisms that share the same function in the food chain and the same nutritional relationship to the primary sources of energy.

In this section we will consider different versions of single-trophic equation: firstly, the purely competitive case, and secondly the more general case in which species, despite belonging to a single trophic level, can exert besides competition facilitative interactions one on another.

2.2.1 Purely competitive communities

If the effect of all interactions between species has an inhibiting effect on the growth of the other species, i.e. interspecific coefficients obey equation (2.3), we can write the Lotka–Volterra competition equations (LVCE) as:

$$\frac{dn_i}{dt} = r_i n_i \left(1 - \frac{\sum_{j=1}^{S} \alpha_{ij} n_j}{\mathcal{K}_i} \right) \quad i = 1, \dots, S \ \ (\alpha_{ij} > 0 \text{ for all } i \text{ and } j, \ \alpha_{ii} = 1). \tag{2.7}$$

Notice that equation (2.7), alternatively, can be obtained straightforwardly as an extension of the equations for two species (1.29) to an arbitrary number of species S.

2.2.2 Single trophic communities with interspecific interactions of different signs

As we have said, the Lotka–Volterra generalized equations (Hofbauer and Sigmund 1998, Pastor 2008) are a set of equations which are more general than either the competitive or predator–prey examples of Lotka–Volterra equations and thus can accommodate other interactions beyond mutually negative –/– interaction between species, like −/+ or +/+. Indeed, in general, these other interspecific interactions have been considered in the past to occur only between different trophic levels (Morin 2011). For instance, the typical example of −/+ is predation and of +/+ is mutualism, like the plant–pollinator relationship. However, several examples of these types of interspecific interactions have been also found for species sharing the same trophic level; from viruses (Turner and Chao 1999, Arbiza *et al* 2010) to natural plant communities (Holmgren *et al* 1997) or artificial plant polycultures (Halty *et al* 2017), just to mention a few examples.

The LLVGE for S interacting species within a single trophic level can be written as:

$$\frac{dn_i}{dt} = r_i n_i \left(1 + \frac{\sum_{j=1}^{S} \alpha_{ij} n_j}{\mathcal{K}_i} \right) \quad i = 1, \ldots, S \quad (\alpha_{ii} = -1), \tag{2.8}$$

i.e. identical to the competition equation (2.1) but without the restrictive condition that the interspecific α_{ij} coefficients be negative.

At this point it is worth summarizing the definitions and interpretations of the three alternative interaction strength matrices we have introduced (table 2.1, adapted from Novak *et al* 2016).

Table 2.1. Summary of the definitions and interpretations of the three alternative interaction strength matrices.

	Community matrix	a_{ij} matrix	LV interaction matrix
Mathematical definition	$J_{ij} \equiv \left. \frac{\partial\left(\frac{dn_i}{dt}\right)}{\partial n_j} \right\|_{\mathbf{n}^*}$	$a_{ij} \equiv \left. \frac{\partial\left(\frac{1}{r_i n_i}\frac{dn_i}{dt}\right)}{\partial n_j} \right\|_{\mathbf{n}^*}$	$\alpha_{ij} \equiv \left. \frac{\partial\left(\frac{\mathcal{K}_i}{r_i n_i}\frac{dn_i}{dt}\right)}{\partial n_j} \right\|_{\mathbf{n}^*}$
Defined for	Any model of differentiable functions $f_i(n_1, n_2 \ldots, n_S)$	Any model of differentiable functions $f_i(n_1, n_2 \ldots, n_S)$	The linear generalized Lotka–Volterra model
Dimensions	time^{-1}	time^{-1} N^{-1} L^2 or time^{-1} M^{-1} L^2	Non-dimensional
Interpretation	Direct effect of the average species individual j on species i's		
	Population growth rate	Per-capita growth rate relative to its intrinsic growth rate	Per-capita growth rate relative to its intrinsic growth rate

2.2.3 Obtaining the parameters of the linear Lotka–Volterra generalized model from monoculture and biculture experiments

In order to obtain the model parameters our starting point is the equilibrium abundances predicted by equations (2.7) or (2.8) which verify:

$$n_i^*\left(\mathcal{K}_i + \sum_{j=1}^{S}\alpha_{ij}n_j^*\right) = 0 \quad i = 1, \ldots, S, \tag{2.9}$$

where the asterisks denote quantities at equilibrium. A main difficulty for obtaining from the set of equations (2.8) the equilibrium species yields n_i^* is how to get the set of parameters $\{\mathcal{K}_i, \alpha_{ij}\}$. A straightforward procedure is to perform, during sufficiently long enough periods (in order that the equilibrium state is reached): (a) the S single species or monoculture experiments, and from each of them to estimate the carrying capacities as the yield of the species i in monoculture $m_i^{\text{ex}*}$ (we use m to emphasize that they are yields in monoculture, while the superscript 'ex' is for denoting experimentally measured quantities to distinguish them from the theoretical ones); (b) the $S \times (S-1)/2$ pairwise experiments and for each of them, obtain the pair of the *biculture* (pairwise experiments) yields, $n_{i(j)}^{\text{ex}*}$ and $n_{j(i)}^{\text{ex}*}$ (the subscripts $i(j)$ and $j(i)$ stand for the relative yield of species i in presence of species j and vice versa). Using (a) we obtain $\mathcal{K}_i$, and then from (b) we obtain α_{ij} and α_{ji} by solving equation (2.9) for $S = 2$, as:

$$\mathcal{K}_i = m_i^{\text{ex}*}, \tag{2.10a}$$

$$\alpha_{ij} = \left(n_{i(j)}^{\text{ex}*} - m_i^{\text{ex}*}\right)/n_{j(i)}^{\text{ex}*}, \quad \alpha_{ji} = \left(n_{j(i)}^{\text{ex}*} - m_j^{\text{ex}*}\right)/n_{i(j)}^{\text{ex}*}. \tag{2.10b}$$

Therefore, if the yield of species i in biculture with species j is $> \mathcal{K}_i$ ($<\mathcal{K}_i$) then $\alpha_{ij} > 0$ (<0) and the interaction of j on i is facilitative (competitive). This is the kind of approach followed by Vandermeer (1969) in a pioneering experimental study with protozoa.

In box 2.1 we summarize the different yields we are considering.

Box 2.1. Summary of the different yields we are considering.

Symbol	Equilibrium experimental yield	Necessary for obtaining parameters	Equations
$m_i^{\text{ex}*}$	Of species i in monoculture	$\mathcal{K}_i$; α_{ij}, α_{ji}	(2.10a), (2.10b)
$n_{i(j)}^{\text{ex}*}$	Of species i in biculture with species j	α_{ij}, α_{ji}	(2.10b)
$n_i^{\text{ex}*}$	Of species i in polyculture with $S-1$ species		

Remark. A criticism is that model predictions in general assume that the community is in equilibrium. In the strict mathematical sense, a community is at equilibrium only when the rate of change for all species, i.e. the left-hand side of equation (2.8) is zero. But this theoretical ideal is rarely achieved in natural communities (Wiens 1984, Roxburgh and Wilson 2000). Moreover, in nature the intensity of interactions may vary with a range of environmental factors including climate, kind of resources, spatial distribution of resources and temporal variation in all of the foregoing. Thus, the coefficients themselves become variables.

2.3 Food webs and trophic chains

As we have said, species in a community can be arranged in different trophic levels. For instance, the first and lowest trophic level contains the producers, e.g. green plants. The plants or their products are consumed by the second trophic level organisms—the herbivores, or plant eaters. Carnivores, in turn, constitute the third level; carnivore predators are the fourth level and so on.

A **food web** can be thought as a diagram depicting which species in a community interact (basically who eats whom). A food chain differs from a food web, because the complex network of different animals' feeding relations are aggregated and the chain only follows a direct, linear pathway. Food chains intertwine making a food web because most organisms consume more than one type of animal or primary producer.

In trophic chains pairs of adjacent trophic levels are customarily modelled by predator–prey equations. And the first or lowest trophic level, corresponding to producers, is commonly modelled by means of self-regulating equations involving carrying capacities. Therefore, if we have a trophic chain of L trophic levels with S_1 producer species, S_2 primary consumer species,..., S_L top consumer species, the corresponding Lotka–Volterra linear equations can be written as:

$$\frac{dn_i^{(1)}}{dt} = r_i n_i^{(1)} \left(1 + \frac{\sum_{j=1}^{S_1} \alpha_{ij}^{(1)} n_j^{(1)}}{\mathcal{K}_i} - \sum_{k=1}^{S_1} \alpha_{ik}^{(2)} n_k^{(2)} \right) \tag{2.11a}$$

$$i = 1, \dots, S_1, k = 1, \dots, S_2 \ (\alpha_{ii} = -1),$$

$$\frac{dn_k^{(\mu)}}{dt} = a_k n_k^{(\mu)} \left(\sum_{j=1}^{S_{\mu-1}} \alpha_{kj}^{(\mu-1)} n_j^{(\mu-1)} - \sum_{l=1}^{S_{\mu+1}} \alpha_{kl}^{(\mu+1)} n_l^{(\mu+1)} - e_k \right) \tag{2.11b}$$

$$\mu = 2, \dots, L-1, \ k = 1, \dots, S_\mu,$$

$$\frac{dn_k^{(\mu)}}{dt} = a_k n_k^{(\mu)} \left(\sum_{j=1}^{S_{\mu-1}} \alpha_{kj}^{(\mu-1)} n_j^{(\mu-1)} - e_k \right) \qquad \mu = L, \ k = 1, \dots, S_\mu, \tag{2.11c}$$

where we use Latin sub-indices for denoting species and Greek supra-indices (between parentheses) for the trophic levels, so that $n_k^{(\mu)}$ corresponds to the density of individuals of species k belonging to the trophic level μ and $\alpha_{ij}^{(\nu)}$ to the interaction coefficient measuring the effect of species j belonging to the trophic level ν over species i belonging to the trophic level μ (the possibilities for ν being: $\mu - 1$, μ and $\mu + 1$). The parameter a_k denotes the net per capita rate of growth of the species k, and the parameter e_k is the per-capita loss rate d_k divided by a_k.

2.4 Quantifying the accuracy of the linear model for predicting species yields in single trophic communities[1]

The linear generalized Lotka–Volterra equations constitute the simplest mathematical model for a community of S interacting species. As we have seen, the minimality of the LLVGE often raises doubts on their ability to make quantitative predictions, like species abundances. Rather, Lotka–Volterra models are regarded more as a qualitative than a quantitative tool in population or community ecology (Brown *et al* 2001). More 'realistic' and complex theories, e.g. equations involving nonlinear response functions (Vandermeer and Goldberg 2013) and higher order interactions (Abrams 1983) implying non-additive effects (Morin 2011), are often preferred because they are perceived as more reliable (although they can be as intractable as the real systems they aim to model). However, this is done at the price of including additional parameters which are very hard to measure. Actually, in many natural communities—like tropical forests, plankton or mutualistic networks—the species richness S is of the order of hundreds. Hence, estimating all the parameters of the LLVGE from empirical data is an unfeasible task, let alone estimating additional parameters of more complex models.

Our viewpoint is that models should be mainly evaluated not on the basis of the realism of their assumptions but on the basis of the accuracy of their predictions. So here we will analyze how well the LLVGE work as a *quantitative* tool for describing/explaining/predicting the outcome of experiments for single trophic species belonging to the same taxonomic group (plants, algae, etc) assuming that a state of equilibrium was reached. And the way we will discuss it is by asking questions. Therefore, our first question will be:

(1) Can LLVGE accurately predict the species yields in a community of interacting species?

As we will see, overall the LLVGE do a pretty good job. This is what is shown by different metrics to assess global accuracy of models. Nevertheless, in some cases they fail at predicting with reasonable accuracy at least one of the species yields. So our next question will be:

(2) In those cases in which the LLVGE fail at predicting with reasonable accuracy at least one of the species yields; with relatively slight

[1] This section is based on Fort (2018a).

modifications of few interaction coefficients within their confidence intervals can they accurately reproduce, *ex post*, all the species yields[2]?

However, a main limitation of the experimental procedure of performing all the monoculture and biculture treatments, summarized in box 2.1, is that it is only feasible provided S is not too large. This is because the number of required experiments grows as S^2. Thus, for large values of S, only a fraction of these experiments is commonly carried out and consequently we have an incomplete knowledge of the S^2 parameters required to compute the equilibrium yields from the LLVGE. It turns out that with incomplete information on the parameters, perhaps we still are able to predict some valuable quantities related to species abundances. Hence, a third pertinent question is:

(3) Are there global or aggregate quantities that can still be quantitatively predicted when our knowledge on the set of LLVGE parameters is incomplete?

Another common situation is when we are interested in predicting the abundance of a *particular* species embedded in a community with other species belonging to the same trophic level (imagine for example a crop coexisting with different weeds). Suppose that for this species we have quite detailed data. Or fourth question is:

(4) Is it possible to predict the performance of a given particular species when our knowledge on the set of LLVGE parameters is incomplete?

We will analyze questions (1) and (2) in the next section while questions (3) and (4) will be approached in section 2.4 on working with imperfect information.

To answer questions (1) and (2) we will use a dataset of experiments designed to measure the effects of intra and interspecific interactions in single-trophic communities with S > 2 species. The approach followed was firstly used by John Vandermeer (1969) for protozoa communities. The idea is to perform experiments that measure *all* the yields for the treatments listed below:

(i) the yields of the S coexisting species (one treatment),
(ii) the yields of species in monoculture (S treatments),
(iii) the yields of species in bicultures ($S \times (S - 1)/2$ treatments).

In addition, such experiments must be carried out until all the species yields (biomass density in most of the cases but also biovolume and number of individuals in one experiment) versus time seem to stabilize at equilibrium constant values.

The polyculture experimental yields at equilibrium, $n_i^{ex^*}$, and the monoculture yields $m_i^{ex^*} = \mathcal{K}_i$ (as well as their SE) are obtained, respectively, by taking averages over replicas of (i) and (ii). Similarly, the biculture yields $n_{i(j)}^{ex^*}$ and $n_{j(i)}^{ex^*}$ and their SE

[2] Notice that this is different than the *ex ante* prediction (1) since it uses the information of the empirical values of $n_i^{ex^*}$ to modify some of the interaction parameters α_{ij} in order to improve the accuracy of the corresponding theoretical relative yields.

are obtained from replicas of (iii). The interaction coefficient between species i and j can in turn be computed from $n_{i(j)}^{\mathrm{ex}*}$ and $n_{j(i)}^{\mathrm{ex}*}$ by using equation (2.10).

So we will use a total of 33 experiments verifying the above requirements compiled from the literature (Fort 2018a). Some of these experiments were completed in the laboratory and others on the field under natural conditions; they included mostly plants but also algae, crustacean and protozoa.

We compare the empirical $n_i^{\mathrm{ex}*}$ of all species for each experiment against the values obtained from the LLVGE[3]. In the next two subsections we will analyze firstly how to obtain the theoretical yields and secondly how to quantify their accuracy.

2.4.1 Obtaining the theoretical yields: linear algebra solutions and simulations

The straightforward way of obtaining the n_i^* is by solving equation (2.9) for the given set of empirically determined parameters $\{\mathcal{K}_i, \alpha_{ij}\}$. The equilibrium state in which all the S species coexist reduces equation (2.9) to this simpler equation:

$$\sum_{j=1}^{S} \alpha_{ij} n_j^* = -\mathcal{K}_i \qquad i = 1, \ldots, S, \tag{2.12}$$

or in matrix form:

$$\mathbf{A}\mathbf{n}^* = -\mathbf{k},$$

where $\mathbf{A}$ denotes the $S \times S$ interaction matrix $[\alpha_{ij}]$, $\mathbf{n}^*$ and $\mathbf{k}$ are column vectors of S entries with, respectively, the yields n_i and the carrying capacities. And then, by inverting this matrix relationship we obtain the column vector $\mathbf{n}^*$ with the S yields as:

$$\mathbf{n}^* = -\mathbf{A}^{-1}\mathbf{k}. \tag{2.13}$$

Thus, a first required condition to do this inversion is that the matrix A is invertible i.e. det $\mathbf{A} \neq 0$. All the 33 interaction matrices obtained from biculture yields are invertible. Let us see an example of obtaining the theoretical equilibrium state using equation (2.13). We will use the experiment involving four species of winter annuals plants of Rees *et al* (1996)—*Erophila verna* (E), *Cerastiums emidecandrum* (C), *Miyosotis ramosissima* (M), and *Valerianella locusta* (V)[4]. For the area A and year 1979 community, the carrying capacity vector $\mathbf{k}$ and the interaction matrix $\mathbf{A}$ are given by:

[3] Actually in a few experiments there were temporal series that would allow going beyond the single equilibrium point. For example, in Vandermeer (1969) the abundances of species were measured at 32 different times and this permitted him to draw the entire dynamical curve by eye through the observed data points. Nevertheless, this is not the case for the vast majority of the experiments considered here; rather, the species yields were measured for a few different times and this was not enough to obtain the time evolution of such yields and fitting an additional set of parameters (the species intrinsic growth rates).

[4] In this experiment all the yields were measured as the number of flowering individuals.

$$\mathbf{k} = \begin{bmatrix} 331 \\ 499 \\ 145 \\ 335 \end{bmatrix} \tag{2.14}$$

$$\mathbf{A} = \begin{bmatrix} -1 & 0 & 0 & -0.0806 \\ -0.2257 & -1 & 0 & 0 \\ -0.0135 & -0.003 & -1 & -0.147 \\ 0 & -0.0282 & -0.2262 & -1 \end{bmatrix} \tag{2.15}$$

$\det(\mathbf{A}^{A79}) = 0.9675$, and thus the matrix is invertible. Substituting (2.14) and (2.15) into (2.13) produces the equilibrium:

$$\mathbf{n}^* = \begin{bmatrix} 306.7 \\ 429.8 \\ 95.3 \\ 301.3 \end{bmatrix} \tag{2.16}$$

consistent with the experimental one (i.e. within the error bars):

$$\mathbf{n}^{ex^*} = \begin{bmatrix} 248.2 \pm 139.7 \\ 680.8 \pm 279.2 \\ 135.9 \pm 84.2 \\ 259.5 \pm 91.5 \end{bmatrix}, \tag{2.16$'$}$$

that is, the percentage of predictions which fall within the error bars, *P*95, is equal to 100%. What about the stability of this equilibrium? It turns out that an equilibrium $\mathbf{n}^*$ is locally stable if all the real parts of the eigenvalues of the *Jacobian* or *community* matrix, whose elements are given by $J_{ij} = \alpha_{ij} n_j^*$, are negative (appendix I). Since the four eigenvalues λ_i of the Jacobian matrix have negative real parts: $\lambda_1 = -0.908$, $\lambda_2 = -4.311$, $\lambda_3 = -3.056 + 0.123i$ and $\lambda_4 = -3.056 - 0.123i$ we conclude that this equilibrium is locally stable. Moreover, simulating the dynamical equation (2.8) we see that the system converges towards this equilibrium state for different initial conditions (figure 2.1) (exercise 2.1).

So far so good. However, for several pairs of **A** and **k** the matrix equation (2.13) produces at least one negative value. This means that the pair {**A**, **k**}—obtained by equations (2.10)—is not fully consistent with the equilibrium in which all the species coexist, i.e. this equilibrium is unfeasible and thus $n_l = 0$ for some species l. Therefore, rather than equation (2.12), which is valid for an equilibrium in which all species coexist, we have to consider the full equilibrium equation (2.9). The problem is that instead of a single equilibrium the system of equation (2.9) now has multiple equilibria (in which at least one species extinguishes). This theoretical equilibrium with $n_l^* = 0$ would still be consistent with the experimental one provided that the corresponding empirically measured $n_l^{ex^*}$ is small enough (so that $n_l^* = 0$ falls within the 95% confidence interval around $n_l^{ex^*}$). To find among this set of possible equilibria the one towards which the system converges we can simulate the

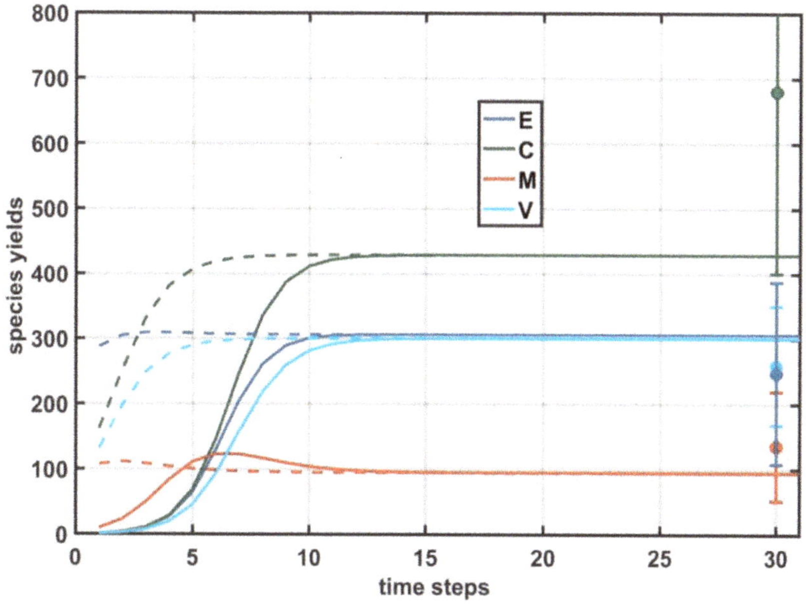

Figure 2.1. Theoretical yields produced by two different simulations, i.e. different initial conditions (filled and dashed lines), converging to experimental values (discs with error bars) for the Rees *et al* (1996) experiments.

dynamical equation (2.9), starting from a random set of initial values n_i^0 (for more details see exercise 2.2).

2.4.2 Accuracy metrics to quantitatively evaluate the performance of the LLVGE

A commonly used metric to assess how well a model fits observed data is the familiar Pearson's correlation coefficient (r) or its square, the coefficient of determination (R^2). Nevertheless, a problem with these two statistics is that they actually describe the degree of *collinearity* between the observed and model-predicted values rather than their numerical agreement (Willmott 1984). In fact, by their very definition, both indices are insensitive to additive and proportional differences between the model predictions and observations (Willmott 1984). Thus, both suffer from limitations that make them poor measures of model performance. For further discussion on why r and R^2 are incorrect measures of predictive accuracy we refer the reader to Li (2017) and references therein. Therefore, to quantitatively assess the degree to which the LLVGE match the observations, we will resort to different indices. We summarize the four metrics of goodness-of-fit we will use in table 2.2 (some of them are commonly used in atmospheric and hydrologic sciences). **Thereafter in this chapter, since we always assume an equilibrium or quasi-equilibrium state, we will omit the * for all yields in the understanding that they are yields in equilibrium.**

Notice that all the above metrics are in terms of the absolute value instead of the square of differences. This is because absolute values are preferable over squares since by using absolute values, o errors and differences are given a more appropriate

Table 2.2. Mathematical definitions of error/accuracy measures used in this study. $\overline{n^{ex}}$ and SE denote, respectively, the mean and standard error for the experimental yields (from Fort 2018a).

Error/accuracy measure	Definition
Relative mean absolute error (*RMAE*)	$RMAE = \frac{\sum_{i=1}^{S} \left\| n_i^{ex} - n_i \right\| / S}{\overline{n^{ex}}} 100\ (\%)$
Predictions within 95% confidence intervals (*P*95)	$P95 = \frac{\sum_{i=1}^{S} \delta(\left\| n_i^{ex} - n_i \right\|)}{S} 100\ (\%),\quad \delta(x) = \begin{cases} 1 \text{ if } x < 1.96\ \text{SE} \\ 0 \text{ if } x > 1.96\ \text{SE} \end{cases},$
Modified coefficient of efficiency	$E_1 = 1 - \frac{\sum_{i=1}^{S} \left\| n_i^{ex} - n_i \right\|}{\sum_{j=1}^{S} \left\| n_j^{ex} - \overline{n^{ex}} \right\|}$
Modified index of agreement	$d_1 = 1 - \frac{\sum_{i=1}^{S} \left\| n_i^{ex} - n_i \right\|}{\sum_{j=1}^{S} \left(\left\| n_j - \overline{n^{ex}} \right\| + \left\| n_j^{ex} - \overline{n^{ex}} \right\| \right)}$

weighting, not inflated by their squared values (Willmott 1981). Squaring in statistics is useful because squares are easier to manipulate mathematically than are absolute values, but use of squares forces an arbitrarily greater influence on the statistic by way of the larger values (Legates and McCabe 1999).

Let us briefly comment on these indices:

The relative mean absolute error (*RMAE*) is obtained from dividing the mean absolute error (*MAE*) between the mean of the species yields. *MAE* and the similar root mean square error (*RMSE*) are two commonly used measures for assessing the predictive accuracy in the environmental sciences (Li and Heap 2008). To avoid any dependence of *MAE* on *S* we will use the relative metric *RMAE*. In order to quantify the accuracy we need to introduce some reference point. Actually error measures, like *RMAE*, are not accuracy measures, so they can only tell which model produces less error but they are unable to tell how accurate a model is (Li 2017). At any rate, a very tolerant measure of model goodness would be $RMAE < 100\%$, i.e. every quantity is measured with an error smaller than the size of the quantity itself. We will consider here the more stringent threshold of $RMAE < 50\%$. A reference point of 50% might seem too high. Still, it is comparable with the typical SE of the experimental yields (as we shall see in the companion Application 2 chapter).

*P*95 measures the percentage of predictions which fall within the confidence intervals of 1.96σ (within the error bars shown in figure 2.2). $P95 = 100(0)\%$ means that all (none of) the yields predicted by the model fall within the error bars around the corresponding experimental values. For *P*95 we will consider two thresholds: its maximum possible value of 100% and the (arbitrary) 66.7%, so that $P95 \geqslant 66.7\%$ indicates the model does a decent job.

E_1 (Legates and McCabe 1999) is a modified version of the coefficient of efficiency (Nash and Sutcliffe 1970) defined by $E = 1 - MSE/\sigma^2$ (*MSE* = mean

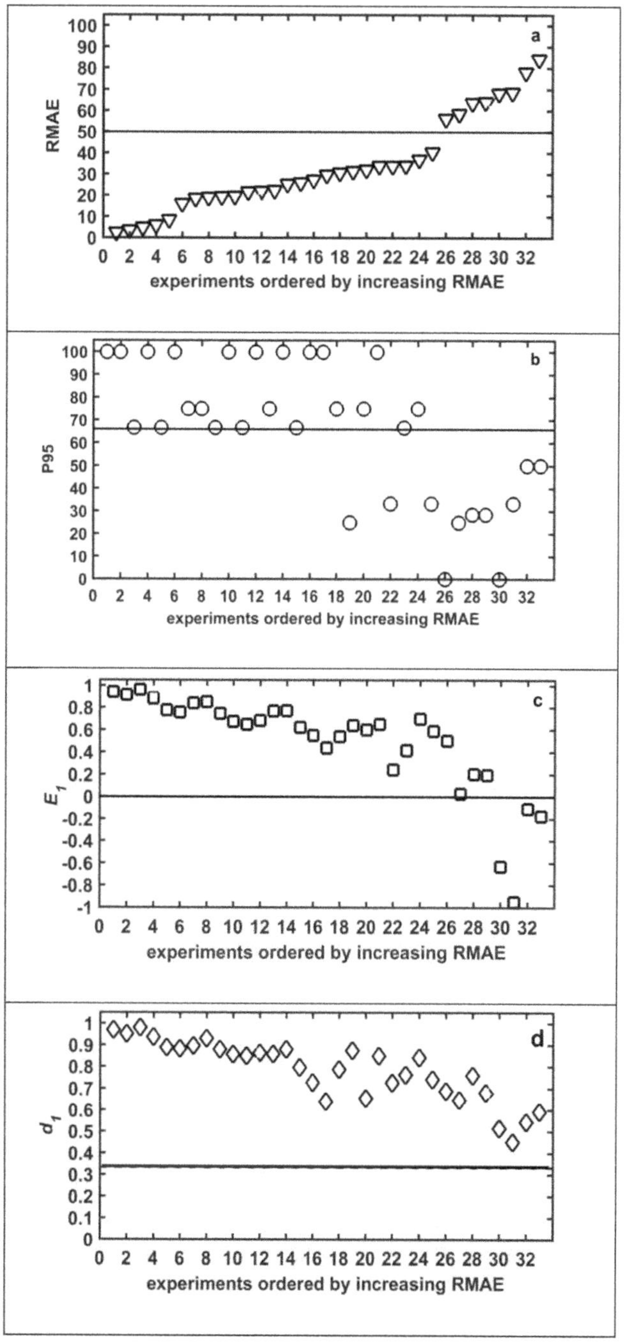

Figure 2.2. The four error/accuracy indices for the 33 experiments of table 2.3 ordered by increasing *RMAE*. Filled lines correspond to the reference values for *RMAE*, *P*95, E_1, and d_1, distinguishing good from bad performance, respectively: 50%, 66.7%, 0, and 1/3. Reprinted from Fort 2018a.

square error), but in terms of absolute differences rather than square differences. It ranges from minus infinity to 1, the larger its value the better the agreement. In particular, $E_1 = 1$ indicates perfect match between model predictions and measures. For example, if the absolute differences between the model and the observation is as large as the variability in the observed data (measured by $\sum_{i=1}^{S} | n_i^{\text{ex}} - \overline{n^{\text{ex}}} |$), then $E_1 = 0.0$, and if it exceeds it, then $E_1 < 0.0$ (i.e. the observed mean is a better predictor than n_i). In other words, a value of zero for E_1 indicates that the observed mean $\overline{n^{\text{ex}}}$ is as good a predictor as the model, while negative values indicate that the observed mean is a better predictor than the model (Wilcox *et al* 1990).

d_1 (Legates and McCabe 1999) is an index similar to E_1. An advantage over E_1 is that it is bounded, i.e. it varies from 0.0 to 1.0 (again the higher its value the better the agreement between model and observations). By the same token it loses the meaningful reference point of 0.0 for the coefficient of efficiency, which serves to assess when the model is a better predictor than the observed mean. However, it is possible to introduce a reference point for d_1 by observing that for two completely uncorrelated random vectors drawn from a uniform distribution, d_1 is on average = 1/3 (see exercise 2.5). Therefore, if $d_1 \leqslant 1/3$ we conclude that the model is a poor predictor.

The values of the above four error/accuracy metrics for each of the 33 experimental studies are listed in table 2.3[5]. We also provide the corresponding taxon.

Figure 2.2 shows the *RMAE, P*95, E_1, and d_1 for these 33 experiments. Filled lines correspond to the reference values for distinguishing good from bad performance, respectively: 50%, 66.7%, 0, and 1/3.

2.4.3 The linear Lotka–Volterra generalized equations can accurately predict species yields in many cases

In figure 2.2 these 33 experiments were ordered from smaller to larger *RMAE.* As expected, when the *RMAE* increases the three accuracy metrics decrease. Thus qualitatively we see that the LLVGE do a good job in predicting yields up to the 25th experiment.

More quantitatively: all the 33 experiments have *RMAE* < 100% (25 of them with *RMAE* < 50%) and simultaneously $d_1 > 1/3$; 22 have *P*95 ⩾ 66.67% and 28 have $E_1 > 0$. Furthermore, two thirds of the 33 experiments simultaneously verify the four above inequalities with *RMAE* < 50% and *P*95 ⩾ 66.7% (this number reduces to 10 experiments when the maximum *P*95 = 100% is required).

Therefore, now we are in a position to answer question (1): Can LLVGE accurately predict species yields?

[5] The interested reader can find the data for these experimental studies—including interaction matrices **A**, n_i^{ex} and n_i as well as the contribution of each species to the four metrics used for estimating error/accuracy—in the supplementary table S1 of Fort (2018a).

Table 2.3. The 33 experiments considered in this study with their references. *S* is the number of species. Columns from *RMAE* % to d_1 are the four metrics (defined in table 2.2) employed to assess the error/accuracy used of the models. For all the experiments, except those with numbers in italics in the *P*95% column, by varying the interaction coefficients within their confidence intervals it is possible to achieve *P*95 = 100% (adapted from Fort 2018a).

Paper	Taxon	Community	S	*RMAE* %	*P*95%	E_1	d_1
Aarssen (1988)	Plants	1939 pasture	4	78.1	50	−0.108	0.544
Hooper (2004)	Plants[a]	1993	3	19.5	100	0.675	0.858
		1998	3	68.7	33.3	−0.948	0.452
		1999	3	68.2	0	−0.63	0.515
Huisman *et al* (1999)	Algae		4	19.0	75	0.852	0.931
Kastendiek (1982)	Algae		3	33.9	100	0.651	0.850
Neill (1974)	Crustacean		4	3.5	100	0.915	0.952
Picasso *et al* (2008)	Plants[b]	A–WC–IBF	3	4.8	66.7	0.962	0.981
		A–OG–IW	3	33.9	33.3	0.244	0.726
		A–OG–SW	3	21.6	66.7	0.648	0.850
		A–SW–EG	3	21.9	100	0.684	0.864
		WC–OG–IW	3	34.2	66.7	0.418	0.763
		IBF–IW–EG	3	56.3	0	0.502	0.688
		IBF–SW–EG	3	40.4	33.3	0.590	0.742
		IBF–OG–IW	3	25.9	66.7	0.622	0.796
		IBF–OG–SW	3	19.3	66.7	0.747	0.881
		A–IW–EG	3	16.1	100	0.758	0.884
		WC–SW–EG	3	27.3	100	0.550	0.727
		WC–OG–SW	3	25.2	100	0.772	0.880
		A–WC–OG–IW	4	58.6	25	0.029	0.646
		A–WC–SW–EG	4	36.9	75	0.699	0.842
		A–IBF–SW–EG	4	18.5	75	0.841	0.897
		A–IBF–OG–IW	4	30.6	75	0.540	0.788
		WC–IBF–OG–IW	4	31.4	25	0.641	0.875
		WC–IBF–SW–EG	4	84.2	50	−0.173	0.593
Rees *et al* (1996)	Plants	area A, 1979	4	29.6	100	0.440	0.640
		area B, 1979	4	32.1	75	0.602	0.654
Roxburgh and Wilson (2000)	Plants		7	64.2	28.6	0.196	0.681
Vandermeer (1969)	Protozoa		4	22.3	75	0.769	0.859
Wilson and Keddy (1986)	Plants[c]		7	63.7	28.6	0.206	0.760
Zarnetske *et al* (2013)	Plants	LOW SAND	3	2.4	100	0.942	0.970
		MID SAND	3	8.4	66.7	0.779	0.889
		HIGH SAND	3	6.0	100	0.883	0.938
Average				**33.5**	**65.4**	**0.494**	**0.785**

[a] Relative yields (yields divided by carrying capacities) for functional groups rather than yields for species.
[b] Averages of biomass over the three annual harvests (2004, 2005 and 2006), A = alfalfa, W = white clover, IBF = Illinois bundleflower, O = orchardgrass, IW = intermediate wheatgrass, SW = switchgrass, EG = eastern gamagrass.
[c] Relative yields (yields divided by carrying capacities) rather than yields.

Answer to question (1): We found that by obtaining the generalized Lotka–Volterra parameters as mean values over samples of biculture and monoculture experiments, the LLVGE can accurately *predict* (*ex ante*) not all but the majority of the species yields for most of the experiments.

2.4.4 Often a correction of measured parameters, within their experimental error bars, can greatly improve accuracy

What's the usefulness of tuning the parameters of a model in order to reproduce experimental results? In fact, this provides assurance that this model actually correctly describes a phenomenon provided we feed it with the right parameters.

In the case of the LLVGE, due to the variability of yields among replicas in experiments the Y_i have in general confidence intervals or error bars which may be quite large. It turns out that in all but five of the 23 experiments with $P95 < 100$, by modifying the interaction coefficients—only a few of them in most of the cases (as can be seen in table S1 of Fort 2018a)—within their confidence intervals (see box 2.2)[6], allow for accommodating all the theoretically predicted yields within the error bars of their corresponding experimental values. For instance, figure 2.3 shows four examples in which a modification of one or two of the interaction coefficients

Box 2.2. Procedure for varying the interaction coefficients within their confidence intervals.

It turns out that in many cases in which the LGLVE fail at predicting at least one of the species yields within their 95% confidence intervals, with relatively slight modifications of few interspecific interaction coefficients α_{ij} within their own confidence intervals allow for accommodating all the theoretically predicted yields within the error bars of their corresponding experimental values.

The confidence intervals for the interspecific interaction coefficients α_{ij} can be easily computed from equation (2.10b) used for obtaining them:

$$\alpha_{ij} = \left(n_{i(j)}^{\text{ex}} - m_i^{\text{ex}}\right)/n_{j(i)}^{\text{ex}}, \quad \alpha_{ji} = \left(n_{j(i)}^{\text{ex}} - m_j^{\text{ex}}\right)/n_{i(j)}^{\text{ex}}.$$

By means of the error propagation formula we get:

$$\varepsilon_{ij} = \frac{\sqrt{\varepsilon_i^2 + \varepsilon_{i(j)}^2 + \alpha_{ij}\varepsilon_{j(i)}^2}}{n_{j(i)}^{\text{ex}*}},$$

where $\pm\varepsilon_{ij}$, $\pm\varepsilon_i$, $\pm\varepsilon_{i(j)}$, and $\pm\ \varepsilon_{j(i)}$ denote the confidence intervals of, respectively, α_{ij}, m_i^{ex}, $n_{i(j)}^{\text{ex}}$, $n_{j(i)}^{\text{ex}}$.

[6] The confidence intervals for the interaction coefficients are obtained from the confidence intervals of the biculture yields B_i by using standard error propagation formulas. This as well as the procedure used for varying these coefficients is described in box 2.2.

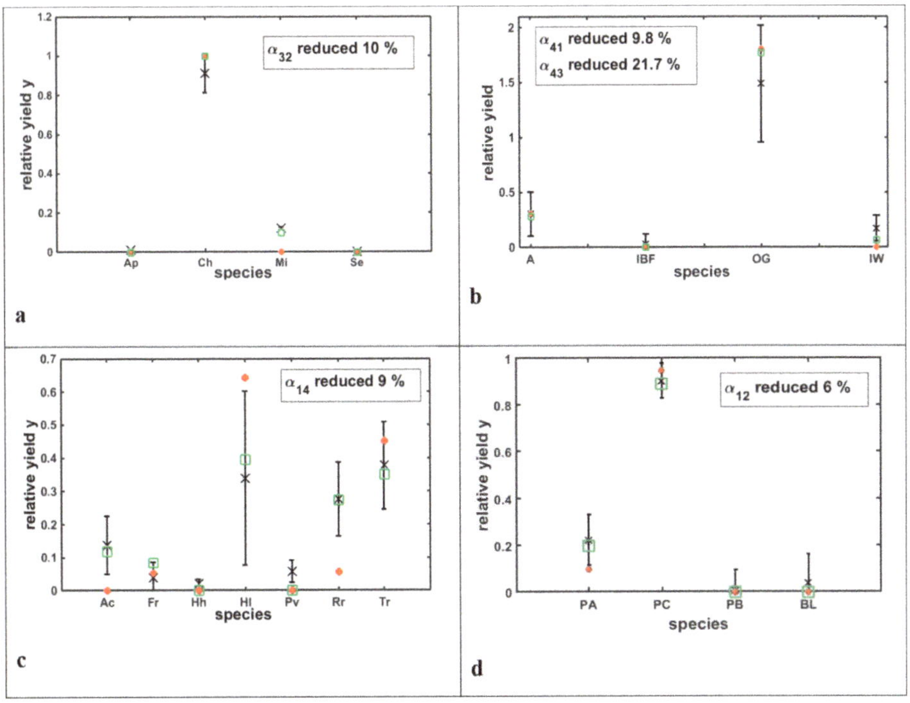

Figure 2.3. Modifying some few interaction coefficients allows ex post reproduction of species yields. Experimental relative yields $y_i = Y_i/K_i$ with their error bars (crosses), predicted values *ex ante* (filled circles, red online) and the corresponding ex post reproduced values by modifying one or two interaction coefficients (open squares, green online): (a) Algae (Huisman *et al* 1999). (b) Plants A–IBF–OG–IW mixture (Picasso *et al* 2008). (c) Plants (Roxburgh and Wilson 2000). (d) Protozoa (Vandermeer 1969). The boxes in the different panels indicate which coefficients were changed and in which percentage. Reprinted from Fort 2018a.

allows all the theoretical yields to fall within the error bars for the corresponding experimentally measured yields. Since the considered experiments involve quite distinct taxonomic groups—e.g. plants, algae, protozoa—different measures of the yield were used: biomass for plants, biovolume for algae and the number of individuals for protozoa. Therefore, just to make easier the comparison among these experiments, we will use the non-dimensional relative yields $y_i = n_i/\mathcal{K}_i$, which additionally in general are <1 and rarely >2.

In some cases, like the data from the experiment by Roxburgh and Wilson (2000), the improvement in the accuracy of predicted yields is particularly remarkable. Notice that just by reducing the a_{14} coefficient by 9% changes *P*95 from 28.6%—two out of seven of the predicted species yields within the error bars around the experimental values—to 85.7%—all but one of the predicted species yields within the error bars around the experimental values.

Therefore,

Answer to question (2): LLVGE can strikingly reproduce *ex post* with accuracy all the species yields for the great majority of the experiments (18 out of 23).

2.5 Working with imperfect information

A common situation we face when trying to predict yields and final diversity of multi-crops or in agriculture or in natural communities is that we don't know all the parameters appearing in the LLVGE. However, we can still predict some relevant quantities about species yields when we have an incomplete knowledge of the LLVGE parameters. In cases in which we know some but not all of the interaction coefficients α_{ij}, to answer question (3), we will show that a recently proposed *mean field* approximation allows us to predict aggregate or mean variables computed in terms of relative yields (Fort 2018b). Additionally, to answer question (4), we will discuss another more refined approximation. We call this approximation, that works in such cases in which we know all the parameters corresponding to a *single* species of particular interest, the *focal species approximation.*

As the reader can check performing a bibliographic search, the number of published experimental studies that measured all the yields (i)–(iii) mentioned at the beginning of section 2.3 decreases quickly as S grows. For $S > 6$ it is very hard to find an experiment in which not the totality of the $S(S+1)/2$ treatments were carried out but only a percentage f_e of the them. Nevertheless, in such cases with $f_e < 100\%$, it is still possible to make quantitative predictions provided we work with relative yields $y_i = n_i/\mathcal{K}_i$ (i.e. the species yield in mixture normalized by its yield in monoculture) rather than yields n_i and use an approximate expression for the interaction matrix.

It is not difficult to show that the interaction matrix can be written in terms of these relative yields, y_i,

$$y_i = n_i/\mathcal{K}_i \tag{2.17a}$$

as:

$$[a_{ij}] = [\alpha_{ij}\mathcal{K}_j/\mathcal{K}_i]. \tag{2.17b}$$

Notice thus that the diagonal elements of the matrix $[a_{ij}]$, corresponding to intra-specific competition, remain equal to -1 since $a_{ii} = \alpha_{ii}K_i/K_i = \alpha_{ii} = -1$. Thus, equation (2.12) for the equilibrium in which all species coexist can be written as (exercise 2.6):

$$\sum_{j=1}^{S} a_{ij}y_j = -1, \qquad i = 1, \ldots, S, \quad \text{or in matrix form: } \hat{\mathbf{A}}\mathbf{y} = -\mathbf{1}, \tag{2.18}$$

where we use $\hat{\mathbf{A}}$ to denote the matrix $[a_{ij}]$ (to distinguish it from $\mathbf{A} = [\alpha_{ij}]$), while $\mathbf{y}$ and $\mathbf{1}$ are column vectors of S entries with, respectively, relative yields y_i and ones.

So now we are in position to consider the two above mentioned approximations for $\hat{\mathbf{A}} = [a_{ij}]$ which allow making predictions about the yields in cases with incomplete knowledge of the LLVGE parameters.

2.5.1 The *'Mean Field Matrix'* (MFM) approximation for predicting global or aggregate quantities

Here we will introduce the so-called Mean Field Matrix (MFM) approximation (Fort 2018b, Fort and Segura 2018), capable of predicting aggregate or mean quantities, expressed in terms of relative yields, with reasonable accuracy. The name alludes to the resemblance of this approximation with the **Mean Field approximation**, commonly used in physics, consisting in replacing spatially dependent variables by a constant equal to their mean value. Here the average will be taken over the values of the interspecific interaction coefficients, i.e. the off-diagonal part of the interaction matrix $[a_{ij}]$ is replaced by a 'mean field' competition coefficient.

The aggregate or mean quantities we will consider are the relative yield total (RYT) (de Wit 1970),

$$RYT = \sum_{i=1}^{S} y_i, \tag{2.19}$$

and the mean relative yield (MRY),

$$MRY = RYT/S = \sum_{i=1}^{S} y_i/S. \tag{2.20}$$

Both these indices allow comparing community productivity on a relative basis. For instance, in agriculture science, the RYT is often used to quantify the overyielding of diverse plant mixtures relative to plant monocultures in studies of biodiversity effects on ecosystem function. A $RYT > 1$ implies that the yield performance will be better in polyculture than in monoculture, a phenomenon termed as *overyielding* (Vandermeer 1989).

The idea then is to replace the off-diagonal elements of the interaction matrix $[a_{ij}]$, corresponding to interspecific interactions, by their average value over the sample of available interspecific competition coefficients:

$$a = \overline{a_{i \neq j}}, \tag{2.21}$$

which thereafter will be called the *mean interspecific interaction strength parameter*. In such a way we get a mean field matrix (MFM):

$$\hat{\mathbf{A}}^{MF} \equiv [a_{ij}]^{MF} = \begin{pmatrix} -1 & a & \cdots & a & a \\ a & -1 & \cdots & a & a \\ & & \ddots & & \\ a & a & \cdots & -1 & a \\ a & a & \cdots & a & -1 \end{pmatrix}. \tag{2.22}$$

For the MFM equation (2.18) reduces to

$$-y_i + a \sum_{j \neq i}^{S-1} y_j = -1, \qquad i = 1, \ldots, S, \tag{2.23}$$

that by summing and subtracting ay_i can be written as:

$$-y_i(1 + a) + a\sum_{j=1}^{S} y_j = -1, \qquad i = 1, \ldots, S,$$

that is,

$$-y_i(1 + a) + aRYT = -1, \qquad i = 1, \ldots, S,$$

which we can solve for y_i as:

$$y_i = \frac{1 + aRYT}{1 + a}, \quad \text{for all } i. \tag{2.24}$$

And, if we sum from 1 to S both sides of equation (2.24), we get

$$RYT = S\frac{1 + aRYT}{1 + a},$$

which allows expressing the RYT and MRY as simple functions of a and S (Fort 2018b, Fort and Segura 2018):

$$RYT^{\mathrm{MF}}(a, S) = \frac{S}{1 - (S - 1)a}, \tag{2.25}$$

$$MRY^{\mathrm{MF}}(a, S) \simeq \frac{1}{1 - (S - 1)a}, \tag{2.26}$$

where the superscript 'MF' is to emphasize that these expressions hold for the MFM approximation. The pair of equations (2.25) and (2.26) tell us that the MFM approximation is particularly suited to addressing how the total or mean yield depends with the species richness and the mean intensity of competition. This is an interesting question in ecology.

A caveat of the method is that if facilitation is the dominant interspecific interaction, so that $a > 0$, the denominator of equations (2.25) and (2.26) could become negative for a or S sufficiently large. Since the RYT and MRY must be positive quantities, a requirement for this approximation to work properly is that facilitation is not too large.

When knowledge of model parameters is incomplete, let us denote by $\mathcal{M}$ the subset of the S monoculture experiments and by $\mathcal{B}$ the subset of the $S \times (S - 1)/2$ biculture or pairwise experiments that were carried out. A recipe we can use for estimating the experimental average interspecific competition coefficient a (to feed equations (2.25) and (2.26)) and the experimental RYT, which has been demonstrated to work quite well (Fort 2018b), is as follows:

First, the average equilibrium relative yields in biculture are computed from the measured yields of subsets $\mathcal{M} + \mathcal{B}$; let us denote it as $\overline{y_{i(j)}^{\mathrm{ex}}}^{\mathcal{M}+\mathcal{B}} = \overline{n_{i(j)}^{\mathrm{ex}}}^{\mathcal{B}} / \overline{m_i^{\mathrm{ex}}}^{\mathcal{M}}$ (that is, the mean of m_i^{ex} is performed over the subset $\mathcal{M}$ and the mean of $n_{i(j)}^{\text{ex}}$ is performed over the subset $\mathcal{B}$). Second, equation (2.10b) can be rewritten as (exercise 2.6):

$$a_{ij} = \left(y_{i(j)}^{\text{ex}} - 1\right)/y_{j(i)}^{\text{ex}}, \quad a_{ji} = \left(y_{j(i)}^{\text{ex}} - 1\right)/y_{i(j)}^{\text{ex}}, \tag{2.27}$$

and therefore we get a mean interaction parameter given by:

$$\bar{a}^{\mathcal{M}+\mathcal{B}} = \left(\overline{y_{i(j)}^{\text{ex}}}^{\mathcal{M}+\mathcal{B}} - 1\right)/\overline{y_{i(j)}^{\text{ex}}}^{\mathcal{M}+\mathcal{B}}. \tag{2.28}$$

Inserting $\bar{a}^{\mathcal{M}+\mathcal{B}}$ into equations (2.25) and (2.26) we get the corresponding theoretical relative yield total and mean relative yield, to compare versus the experimental $RYT^{\mathcal{M}}$ (the superscript $\mathcal{M}$ is to emphasize that this quantity is estimated from the subset of available monoculture experiments, i.e. that the sum for the RYT is over the relative yields which are possible to compute). There are at least three different ways to estimate $RYT^{\mathcal{M}}$. The first and most straightforward procedure is by summing the available empirical relative yields $\{y_i\}^{\mathcal{M}}$. A second one is by approximating $RYT^{\mathcal{M}} = S\bar{n}/\bar{\mathcal{K}}^{\mathcal{M}}$. A third estimation, that only requires knowing the total yield of the S-species polyculture $\sum_i n_i$, is $RYT^{\mathcal{M}} = \sum_i Y_i/\bar{\mathcal{K}}^{\mathcal{M}}$. Box 2.3 illustrates this calculation of $a = \bar{a}^{\mathcal{M}+\mathcal{B}}$ and how to estimate $RYT^{\mathcal{M}}$ (see also exercise 2.7).

Indeed, the example analyzed in box 2.3, corresponding to experiments involving polycultures of 32-species in Switzerland (Hector *et al* 2010), serves to highlight that the MFM approximation can do a good job even in cases in which only a small fraction f_e of the required experimental treatments was carried out. Namely, 528 experimental treatments (32 monocultures and 32 × 31/2 = 496 bicultures) would be required to estimate the totality of LLVGE parameters.

In figure 2.4 we compare for a dataset of 77 experiments, such that for all of the them the mean interspecific interaction strength parameter is negative and the number of species S varies from 2 to 32 (Fort 2018b), the experimental RYT and the corresponding theoretical predictions generated by equation (2.25) as a function of both S and a.

As we mentioned, the MFM approximation is useful to analyze how the yield depends with the species richness and the mean intensity of competition. We observe that:

First, regarding the dependence of the RYT as a function of S, figure 2.4 shows that the RYT increases with S, a phenomenon which has been observed in many empirical studies and interpreted as **niche partitioning** (Cardinale *et al* 2011, 2012). That is, different species can use resources in a complementary way and therefore more diverse communities would lead to a higher RYT. The $RYT > 1$ for the great majority of the experiments considered here, especially for those with $S > 2$ interacting species (24 out of 25 as shown in figure 2.4), is clear *quantitative* evidence of such species' complementarity (Loreau and Hector 2001, de Wit 1970). Notice that for a fixed value of the mean interaction parameter a equation (2.25) becomes a nonlinear saturating relationship between diversity and yield. Non-linear saturating relationships have also been obtained in models by simulation or in other more

Box 2.3. An example showing both the calculation of theoretical RYT (equation (2.25)) and how the corresponding experimental RYT can be estimated with incomplete data.

To illustrate how the MF works we use here an experiment corresponding to the BIODEPTH project (Loreau and Hector 2001), a network of field experiments that examined the functioning of European grassland ecosystems in relation to the direct manipulation of plant diversity.

Specifically we consider the annual measures of yield for experiments involving polycultures of 32-species conducted in Switzerland (Hector *et al.* 2010).

As we have seen, to completely estimate the Lotka–Volterra parameters 32 monoculture and 32 × 31/2 = 496 biculture experiments would be required. For practical reasons, only 17 of these experiments were carried out: 10 monoculture and 7 biculture experiments, or a fraction of the total = (10 + 7)/(32 + 496), i.e. f_e = 3.2%. Since f_e was so low we consider the three available years rather than only the last year to have more statistics.

The $\overline{y_{i(j)}^{\mathrm{ex}*}}^{\mathcal{M}+\mathcal{B}}$ that emerges from the 17 experiments is 0.74 (exercise 2.6), and therefore, by equation (2.28), $\bar{a}^{\mathcal{M}+\mathcal{B}} = 0.26/0.74 = 0.35$. Substituting this value of a in equation (2.25) produces a theoretical $RYT = 32/(31 \times 0.35 + 1) = 2.70$.

As mentioned, there are three different ways to estimate $RYT^{\mathcal{M}}$. A first one is by summing the 10 relative yields obtained by dividing each one of the yields for these 10 species obtained in the 32-species polyculture experiment by the corresponding monoculture yield, and we obtain $RYT^{\mathcal{M}} = 3.05$ (exercise 2.6). A second estimation is by taking the mean of monoculture and polyculture yields we obtain, respectively: $\bar{\mathcal{K}}^{\mathcal{M}} = 275$ g/m^2 and $\bar{n} = 27$ g/m^2 [32], which produces $RYT^{\mathcal{M}} = S\bar{n}/\bar{K}^{\mathcal{M}} = 32 \times 27/275 = 3.14$. A third estimate for $RYT^{\mathcal{M}}$ is by taking the ratio between the total yield in the 32-species polyculture $\sum_i n_i = 661$ g/m^2 [32] and $\bar{\mathcal{K}}^{\mathcal{M}}$: $RYT^{\mathcal{M}} = \sum_i n_i/\bar{\mathcal{K}}^{\mathcal{M}} = 661/275 = 2.40$. Notice that these three estimates are not dramatically different and the relative errors between the theoretical and experimental RYT they yield are, respectively, ε_{TE} = 11.5%, 14.0% and 12.2% which are all smaller than the experimental standard error = 15.2%. Therefore the error calculation seems to be reliable.

complex models (Loreau 1998, Schnitzer *et al* 2011, Downing *et al* 2012). In fact, it is worth remarking that the shape of the relationship between productivity and species diversity measured in nature is highly variable: In many cases, productivity shows a saturating relationship with diversity, nevertheless 'hump-shaped' relationships, where productivity peaks at intermediate diversity, are also quite common (Chase and Leibold 2002).

Second, as expected, for fixed S the RYT and MRY (figure 2.5) both decrease with the absolute value of a. That is, the more intense the average interspecific competition the smaller the yield.

The relative absolute errors of the RYT and MRY obtained when using formulas provided by equations (2.25) and (2.26), in the case of experiments with incomplete knowledge of model parameters, range between 0.8% and 24% (mean 12.4% ± 4.6%) and are not significantly greater than those obtained by just solving the entire set of LLVGE, which was possible for experiments in which all the model parameters were

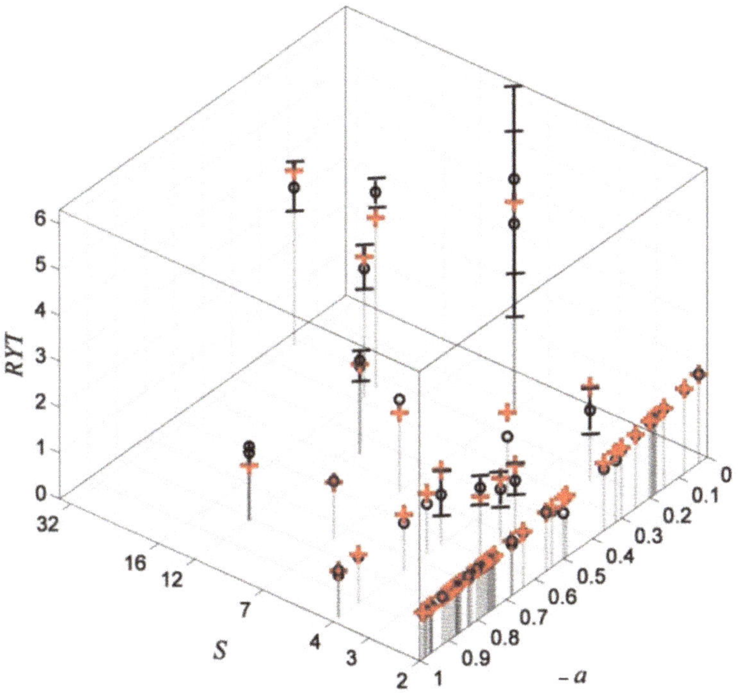

Figure 2.4. Empirical (o) and theoretical RYT, predicted by the MFM approximation equation (2.25) (red crosses), for a set of 77 experiments (see text) as a function of S (log scale) and the mean competition parameter a. The experimental error bars correspond to ± the standard error (SE). Reprinted from Fort 2018a.

available as the ones considered in section 2.4. Furthermore, in general the $RMAE$ is proportional to the size of the experimental error bar and thus the failure of equations (2.25) and (2.26) can be mostly explained by a large variance among experimental replicas or by a low experimental precision.

So, what global or aggregate quantities can be quantitatively predicted when our knowledge of the set of LLVGE parameters is incomplete?

Answer to question (3):

Provided only competition interactions among species exist or at least dominate, we still can predict *aggregate* (RYT) or *mean* (MRY) relative competition indices with an accuracy inversely proportional to the experimental variance of the experimental measurements.

2.5.2 The *'focal species'* approximation for predicting the performance of a given species when our knowledge on the set of parameters is incomplete

Now let us consider a different situation and suppose we are interested in the performance of a particular or *focal* species that coexists with other species. Imagine for instance a crop that coexists with several species of weeds. Since we are not interested in the yield of each individual weed, we can treat all of them as a single species. Thus in order to make more accurate specific predictions for this focal

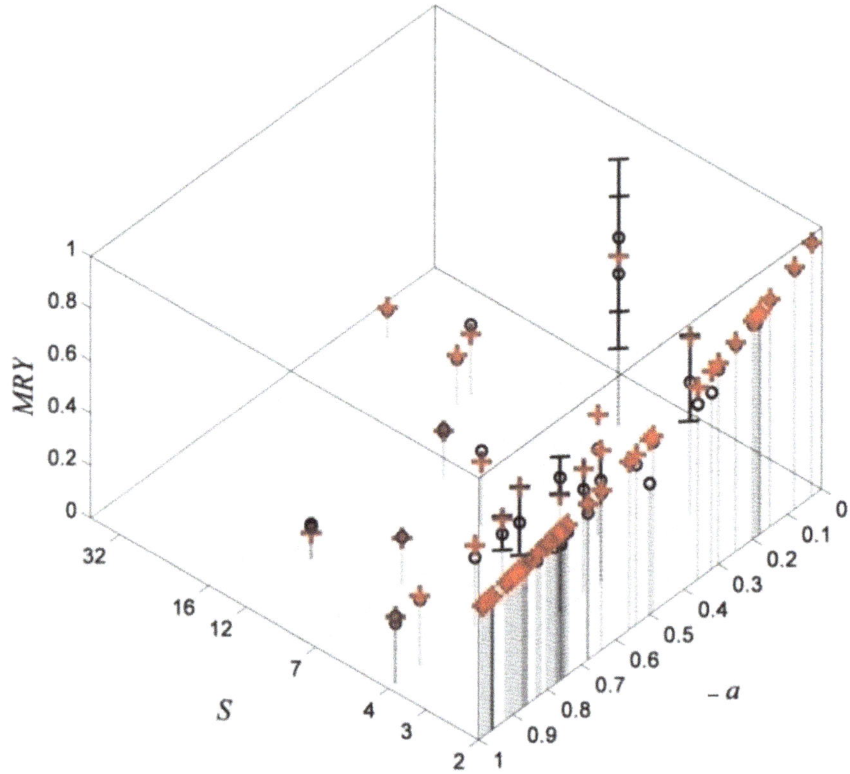

Figure 2.5. Theoretical *MRY* predicted by the MFM approximation equation (2.26) (red crosses) versus empirical *MRY* (o) with error bars corresponding to ± SE for a set of 77 experiments as a function of *S* (log scale) and *a*. Reprinted from Fort 2018a.

species than just the *MRY*, one possibility is to measure the interaction coefficients for this species *k* with all the others (or at least with the most important ones) and use an interaction matrix, in terms of relative yields, of the form:

$$\hat{\mathbf{A}}^{(k)} \equiv [a_{ij}]^{(k)} = \begin{pmatrix} -1 & -1 & \cdots & a_{1k} & \cdots & -1 & -1 \\ -1 & -1 & \cdots & a_{2k} & \cdots & -1 & -1 \\ \vdots & \vdots & \ddots & \vdots & \cdots & \vdots & \vdots \\ a_{k1} & a_{k2} & \cdots & -1 & \cdots & a_{kS-1} & a_{kS} \\ \vdots & \vdots & \cdots & \vdots & \ddots & \vdots & \vdots \\ -1 & -1 & \cdots & a_{S-1k} & \cdots & -1 & -1 \\ -1 & -1 & \cdots & a_{Sk} & \cdots & -1 & -1 \end{pmatrix}, \quad (2.29)$$

i.e. all the interspecific coefficients between species different from the focal species *k* are taken equal to −1 (the rationale for this is we are considering all these species as one and the same, and therefore the interaction coefficient takes the value −1 corresponding to intraspecific interactions), while a_{ik} denotes the interaction coefficient of the focal species over species *i* and a_{ik} the interaction coefficient of species *i* over the focal one.

The recipe is then to use matrix (2.29) to solve equation (2.18), **by simulation of the dynamical equations until the system reaches its equilibrium state**, and only take into account the relative yield for the focal species $y_k^{(k)}$ (while neglecting the values $y_j^{(k)}$ for $j \neq k$). That is, from a matrix of the form (2.29) we obtain just one relative yield for the focal species ($y_k^{(k)}$). Let us call this approximation, which represents an improvement of the mean field one, the *focal species* approximation.

To illustrate how this focal approximation works we will consider the experiment of four algal species of Huisman *et al* (1999): *Aphanizomenon flos-aquae* (*Ap*, species #1), *Chlorella vulgaris* (*Ch*, species #2) a *Microcystis* strain (*Mi*, species #3) and *Scenedesmus protuberans* (*Se*, species #3) and. The population densities (measured as millions per mL) of the four species in polyculture were:

$$\mathbf{n} = \begin{bmatrix} 0.002 \\ 18.600 \\ 2.150 \\ 0.001 \end{bmatrix} \tag{2.30}$$

The carrying capacity vector $\mathbf{k}$ (millions per mL) and the interaction matrix $\mathbf{A}$ are given by (Fort 2018a):

$$\mathbf{k} = \begin{bmatrix} 0.180 \\ 20.40 \\ 17.90 \\ 0.531 \end{bmatrix} \tag{2.31}$$

$$\mathbf{A} = \begin{bmatrix} -1 & -0.0086 & -0.0060 & -0.4000 \\ 0 & -1 & 0 & 0 \\ -86.0000 & -0.8772 & -1 & -30.7692 \\ -2.9706 & -0.0256 & -0.0295 & -1 \end{bmatrix} \tag{2.32}$$

Therefore, using respectively equation (2.17a) and (2.17b), we get for the vector of experimental relative yields $\boldsymbol{y}^{\text{ex}}$ and for the matrix $\hat{\mathbf{A}} = [a_{ij}]$:

$$\boldsymbol{y}^{\text{ex}} = \begin{bmatrix} 0.010 \\ 0.912 \\ 0.120 \\ 0.001 \end{bmatrix} \tag{2.33}$$

$$\hat{\mathbf{A}} = \begin{matrix} & \text{Ap} & \text{Ch} & \text{Mi} & \text{Se} \\ \text{Ap} & -1 & -0.978 & -0.597 & -1.778 \\ \text{Ch} & 0 & -1 & 0 & 0 \\ \text{Mi} & -0.865 & -1.000 & -1 & -0.911 \\ \text{Se} & -1.009 & -0.987 & -0.998 & -1 \end{matrix} \tag{2.34}$$

If we consider that the focal species is for example *Ch*, then we obtain the focal matrix $\hat{\mathbf{A}}^{(Ch)}$ (with boxed elements unchanged):

$$\hat{\mathbf{A}}^{(Ch)} = \begin{array}{c|cccc} & Ap & Ch & Mi & Se \\ \hline Ap & -1 & \boxed{-1.133} & \mathbf{-1} & \mathbf{-1} \\ Ch & \boxed{0} & \boxed{-1} & \boxed{0} & \boxed{0} \\ Mi & \mathbf{-1} & \boxed{-1.003} & -1 & \mathbf{-1} \\ Se & \mathbf{-1} & \boxed{-1.155} & \mathbf{-1} & -1 \end{array} \tag{2.35}$$

The matrix (2.35) is not invertible (three columns are equal and thus its determinant is 0). Therefore, we cannot use equation (2.18) to get the equilibrium vector $\boldsymbol{y}$. Nevertheless, we can solve the differential equations numerically, until the equilibrium state is reached, and we obtain:

$$\boldsymbol{y} = \begin{bmatrix} 0.000 \\ 1.000 \\ 0.000 \\ 0.000 \end{bmatrix} \tag{2.36}$$

Thus $y_{Ap}^{(Ch)} = 1.000$, differs by 10% with respect to the experimental $y^{ex}{}_{Ch} = 0.912$ (second entry of equation (2.33)). So, not so bad. More interesting are the results when using the focal approximation for species *Ap* (see exercise 2.8).

Figure 2.6 compares the results of using this approximation (open big circles) with those when using the full set of LLVGE (filled small circles) for the same four experiments shown in figure 2.3. We can see that the accuracy with which the focal species approximation predicts the focal species is comparable to the one obtained when using the full set of LLVGE.

Remember that the fourth and last question we posed was about predicting the performance of a given particular species when our knowledge on the set of LLVGE parameters for the other species is incomplete.

Answer to question (4): It is possible to predict the yield of a given particular species when we just know the interaction coefficients involving this focal species, by means of the focal species approximation, with accuracy comparable to the one obtained when using the full set of LLVGE.

2.6 Conclusion

We have introduced the linear Lotka–Volterra generalized equation (LLVGE) as the simplest system of equations to model a community of species with all possible types of interspecific interactions: −/−, −/0, −/+, +/0 and +/+. Particular cases of such equations are the competition model (all species interactions −/−) and the mutualistic model (all species interactions +/+).

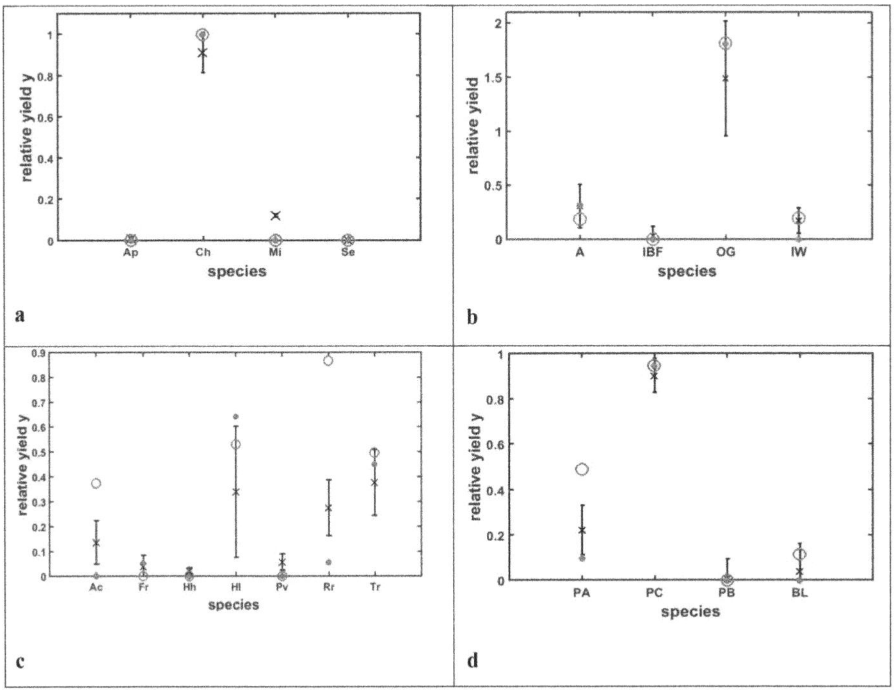

Figure 2.6. The focal species approximation compared with the full set of LLVGE. Experimental relative yields $R_i = Y_i/K_i$ with their error bars (crosses), predicted values *ex ante* by the LLVGE (red filled circles) and the corresponding values produced by the focal species approximation (open blue circles): (a) Algae (Huisman *et al* 1999). (b) Plants A–IBF–OG–IW mixture (Picasso *et al* 2008). (c) Plants (Roxburgh and Wilson 2000). (d) Protozoa (Vandermeer 1969). Reprinted from Fort 2018a.

Next, we have shown that the LLVGE can be used as a quantitative tool. If all the model parameters are known, the full set of LLVGE allows in general to predict the yields of all species making up the community with reasonable accuracy. More elaborate models may be necessary for describing and predicting species abundances for particularly complex communities (Brown *et al* 2001). However, according to our analysis, it seems that such cases would be more the exception than the rule. At any event, the LLVGE provide a benchmark against which the performance of more complex models can be evaluated.

Indeed, the LLVGE constitute a useful tool for assessing the effects of removing existing species or adding foreign species to an existing community. Also, using the LLVGE for quantitative analysis has shown to be helpful in predicting optimal combinations of crop mixtures (companion Application 2 chapter). Another application of the LLVGE in agriculture is for designing rational weed management practices. For example, by enhancing the competitive ability of crops and simultaneously avoiding environmental damage with the use of herbicides (Guglielmini *et al* 2017).

We have also seen that, for the majority of real communities we only know a fraction of the model parameters. In these circumstances with incomplete information, we still can make quantitative predictions. Firstly, we have the mean field approximation (MFA). The MFA allows making predictions on aggregate or waverage quantities, involving the whole community of coexisting species. Secondly, for such cases in which all the interaction parameters involving a particular species are available, we have the focal species approximation which can be used for predicting the yield of this focal species.

Exercises

Exercise 2.1. Redoing calculations for the experiment involving four species of winter annuals plants of Rees *et al* (1996)

(A) Compute the eigenvalues of the Jacobian matrix.
(B) Write a script for integrating the Lotka–Volterra dynamical equation (2.8) and check figure 2.1.

Exercise 2.2.

As we have mentioned, in many of the experimental treatments (21 out of 33) the interaction matrix obtained from biculture experiments lead to unfeasible equilibria (at least one yield is negative).

That is, the pair $\{\mathbf{A}, \mathbf{k}\}$ obtained—by equations (2.10)—is not fully consistent with the equilibrium in which all the species coexist i.e. this equilibrium is unfeasible and thus $n_l = 0$ for some species l.

Therefore, rather than equation (2.12), which is valid for an equilibrium in which all species coexist, we have to consider the full equilibrium equation (2.9). To find out to what equilibrium among this set of possible equilibria the system converges we can simulate the dynamical equation (2.9), starting from a random set of initial values n_i^0. Let us illustrate this latter situation in which the equilibrium state with all species coexisting is incompatible with the empirical interaction matrix by means of the Huisman *et al* (1999) experiment involving four species of algae: *Aphanizomenon floaquae* (Ap), *Chlorella vulgaris* (Ch), a *Microcystis* strain (Mi) and *Scenedesmus protuberans* (Se). The carrying capacity vector $\boldsymbol{k}^{\mathrm{Huis99}}$ and the interaction matrix $\boldsymbol{A}^{\mathrm{Huis99}}$ are given by:

$$\mathbf{k}^{\mathrm{Huis99}} = \begin{bmatrix} 0.180 \\ 20.40 \\ 17.90 \\ 0.531 \end{bmatrix} \tag{2.37a}$$

$$\mathbf{A}^{\mathrm{Huis99}} = \begin{bmatrix} -1 & -0.0086 & -0.0060 & -0.4000 \\ 0 & -1 & 0 & 0 \\ -86.0000 & -0.8772 & -1 & -30.7692 \\ -2.9706 & -0.0256 & -0.0295 & -1 \end{bmatrix} \tag{2.37b}$$

(A) Check that $\det(\mathbf{A}^{\mathrm{Huis99}}) \neq 0$, and thus the matrix is invertible.
(B) Check now that using equations (2.37a) and (2.37b) into equation (2.13) produces an unfeasible equilibrium.

(C) To solve these dynamical equations either work with a system of difference equations and write your own code from scratch or use an ordinary differential equations (ODEs) solver (e.g. the ode45 MATLAB ode solver).

(D) Verify that the system always converges to an equilibrium state with $n_3 = n_4 = 0$ (and additionally with $n_1 = 0$).

(E) Compare the equilibrium vector you obtained with the empirical one given by:

$$\mathbf{n}^{\text{ex Huis99}} = \begin{bmatrix} 0.0018 \pm 0.01 \\ 18.6000 \pm 2.25 \\ 2.15 \pm 0.25 \\ 0.0001 \pm 0.001 \end{bmatrix}.$$

What do you conclude?

(F) Verify that the equilibrium state $\mathbf{n}^{\text{Huis99}}$ is locally stable since the Jacobian matrix has only a non-null element on the diagonal, which therefore coincides with its single eigenvalue $\lambda_1 = -20.4$.

Exercise 2.3 (from Goh 1977)

Consider the set of three competing species governed by equations:

$$\begin{aligned} \dot{X}_1 &= X_1(2 - 0.8X_1 - 0.7X_2 - 0.5X_3) \\ \dot{X}_2 &= X_2(2.1 - 0.2X_1 - 0.9X_2 - X_3) \\ \dot{X}_3 &= X_3(1.5 - X_1 - 0.3X_2 - 0.2X_3). \end{aligned}$$

The non-trivial equilibrium is at 1,1,1 and it is locally stable; the eigenvalues are approximately −1.88 and −0.00985 ± 0.288i. Yet for many initial conditions the system does not return to 1,1,1 but to 0,0,7.5, which is an equilibrium with only X_3 present. Check for example that the solution from the initial state (0.5, 1, 2) tends rapidly to (0, 0, 7.5). Thus the equilibrium is locally stable but not globally stable.

Exercise 2.4. A Lyapunov function for the Lotka–Volterra competition equations with symmetric interaction coefficients

Given the Lotka–Volterra competition equations:

$$\frac{dy_i}{dt} = r_i y_i \left(1 - \sum_{j=1}^{S} a_{ij} y_j\right) \qquad i = 1, \dots, S \quad (a_{ij} \geqslant 0 \text{ for all } i \text{ and } j)$$

If $a_{ij} = a_{ij}$ then show that

$$Q(t) = \sum_{i,j=1}^{S} (y_i - y_i^*) a_{ij} (y_j - y_j^*) > 0 \text{ and } \frac{dQ}{dt} \leqslant 0.$$

(Here the y_i^* denote, as usual, equilibrium yields.) Therefore $Q(t)$ is a Lyapunov function and the full nonlinear global stability analysis of the system is described by the linearized neighborhood analysis.

Exercise 2.5. The reference point for the modified index of agreement d_1

Suppose two different sets of S values Y_i and Y_i' are completely uncorrelated, for example each randomly drawn from a uniform distribution in $[0,L]$. Convince yourself that in the limit of very large S we can replace these discrete variables by continuous variables and thus express this threshold value for the modified index of agreement d_1^* as:

$$
\begin{aligned}
d_1^* &= 1 - \frac{\int_0^L \int_0^L |Y - Y'| dY dY'/L^2}{\int_0^L |Y - \bar{Y}| dY/L + \int_0^L |Y' - \bar{Y}| dY'/L} \\
&= 1 - \frac{\int_0^L \int_0^L |Y - Y'| dY dY'/L^2}{2\int_0^L |Y - L/2| dY/L}.
\end{aligned}
\tag{2.38}
$$

(A) By a change of variables $X = |Y - Y'|$ show the numerator can calculated as:

$$
\int_0^L dY \int_0^Y X dX/L^2 = L/6. \tag{2.39a}
$$

(B) In a similar way, taking now $X = |Y - L/2|$, show that the denominator of (2.38) is given by:

$$
2\int_0^{L/2} X dX/L = L/4. \tag{2.39b}
$$

Thus substituting (2.39a) and (2.39b) into (2.38) we get $d_1^* = 1 - 2/3 = 1/3$.

Exercise 2.6. Working with relative yields rather than yields

Show that:

(A) The interaction matrix in terms of these relative yields, $y_i = n_i/\mathcal{K}_i$ can be rewritten as $[a_{ij}] = [\alpha_{ij}\mathcal{K}_j/\mathcal{K}_i]$.

(B) The equation (2.10b) can be rewritten as: $a_{ij} = (y_{i(j)}^{\text{ex*}} - 1)/y_{j(i)}^{\text{ex*}}$, $a_{ji} = (y_{j(i)}^{\text{ex*}} - 1)/y_{i(j)}^{\text{ex*}}$.

(C) An equation similar to equation (2.12) for the equilibrium in which all species coexist can be written as

$$\sum_{j=1}^{S} a_{ij} y_j = -1, \qquad i = 1, \dots, S, \text{ or in matrix form: } \hat{\boldsymbol{A}}\boldsymbol{y} = -\boldsymbol{1}.$$

Exercise 2.7. Using the mean field approximation for predicting the *RYT* of a BIODEPTH 32 plant species experiment

From the dataset of BIODEPTH experiments for grassland plant communities (Hector *et al* 2010) use the treatments with 32 species conducted in Switzerland to check the *RYT* calculations of box 2.3.

Exercise 2.8. An example of application of the focal approximation

We will consider the experiment of four algal species of Huisman *et al* (1999). The carrying capacity vector **k** and the interaction matrix $\boldsymbol{A}$ are given, respectively, by equations (2.31) and (2.32).

(A) Obtain the focal matrix for species *Ap*, $[a^{Ap}{}_{ij}]$.

(B) Show that if we use equation (2.18) and invert this matrix we get 0 for the relative yield of species *Ap*.

(C) $y_{Ap}^{(Ap)} = 0.00$ differs in 1% with respect to its experimental value of 0.01. Check that by numeric integration of the Lotka–Volterra equations you get for equilibrium a greater than 0 value that is also quite similar to that empirically observed.

References

Aarssen L W 1988 Pecking order of species from pastures of different ages *Oikos* **51** 3–12

Abrams P A 1983 Arguments in favor of higher order interactions *Am. Nat.* **121** 887–91

Arbiza J, Mirazo S and Fort H 2010 Viral quasispecies profiles as the result of the interplay of competition and cooperation *BMC Evol. Biol.* **10** 137

Brown J H, Whitham T G, Ernest S K M and Gehring C A 2001 Complex species interactions and the dynamics of ecological systems: long-term experiments *Science* **293** 643–50

Cardinale B J *et al* 2012 Biodiversity loss and its impact on humanity *Nature* **486** 59–67

Cardinale B J *et al* 2011 The functional role of producer diversity in ecosystems *Am. J. Bot.* **98** 572–92

Chase J M and Leibold M A 2002 Spatial scale dictates the productivity–biodiversity relationship *Nature* **416** 427–30

de Wit C T 1970 *Proc. Adv. Study Inst Dyn. Numbers Popul. (Oosterbeek)* 269

Downing A S *et al* 2012 The resilience and resistance of an ecosystem to a collapse of diversity *PLoS One* **7** e46135

Fort H 2018a On predicting species yields in multispecies communities: Quantifying the accuracy of the linear Lotka–Volterra generalized model *Ecol. Model.* **387** 154–62

Fort H 2018b Quantitative predictions from competition theory with an incomplete knowledge of model parameters tested against experiments across diverse taxa *Ecol. Model.* **368** 104–10

Fort H and Segura A 2018 Competition across diverse taxa: quantitative integration of theory and empirical research using global indices of competition *Oikos* **127** 392–402

Goh B S 1977 Global stability in many species systems *Am. Nat.* **111** 135–43

Guglielmini A C, Verdú A M C and Satorre E H 2017 Competitive ability of five common weed species in competition with soybean *Int. J. Pest Manag.* **63** 30–6

Halty V *et al* 2017 Modelling plant interspecific interactions from experiments of perennial crop mixtures to predict optimal combinations *Ecol. Appl.* **27** 2277–89

Hector A *et al* 2010 *Ecological Archives E091-155-S1* http://www.esapubs.org/archive/ecol/E091/155/

Hofbauer M and Sigmund K 1998 *Evolutionary Games and Population Dynamics* (Cambridge: Cambridge University Press)

Holmgren M, Scheffer M and Huston M A 1997 The interplay of facilitation and competition in plant communities *Ecology* **78** 1966–75

Hooper D U and Dukes J S 2004 Overyielding among plant functional groups in a long-term experiment *Ecol. Lett.* **7** 95–105

Hubbel S P 2002 *The Unified Neutral Theory of Biodiversity and Biogeography* (Princeton, NJ: Princeton University Press)

Huisman J, Jonker R R, Zonneveld C and Weissing F J 1999 Competition for light between phytoplankton species: experimental tests of mechanistic theory *Ecology* **80** 211–22

Kastendiek J 1982 Competitor-mediated coexistence: interactions among three species of macroalgae *J. Exp. Mar. Biol. Ecol.* **62** 201–10

Keddy P A 2001 *Competition* 2nd edn (Dordrecht: Kluwer)

Legates D R and McCabe G J 1999 Evaluating the use of 'goodness-of-fit' measures in hydrologic and hydroclimatic model validation *Water Resour. Res.* **35** 233–41

Levins R 1968 *Evolution in Changing Environments: Some Theoretical Explorations* (Princeton, NJ: Princeton University Press)

Li J 2017 Assessing the accuracy of predictive models for numerical data: Not r nor r^2, why not? Then what? *PLoS One* **12** e0183250

Li J and Heap A 2008 A review of spatial interpolation methods for environmental scientists *Geoscience Australia, Record* 2008/23

Loreau M 1998 Biodiversity and ecosystem functioning: A mechanistic model *Proc. Natl. Acad. Sci.* **95** 5632–6

Loreau M and Hector A 2001 Partitioning selection and complementarity in biodiversity experiments *Nature* **412** 72–6

Lotka A J 1925 *Elements of Physical Biology* (Baltimore, MD: Williams and Wilkins)

Morin P J 2011 *Community Ecology* (Chichester: Wiley-Blackwell)

Nash J E and Sutcliffe J V 1970 River flow forecasting through conceptual models, I, A discussion of principles *J. Hydrol.* **10** 282–90

Neill W E 1974 The Community Matrix and Interdependence of the competition coefficients *Am. Nat.* **108** 399–408

Novak M *et al* 2016 Characterizing species interactions to understand press perturbations: What is the community matrix? *Annu. Rev. Ecol. Evol. Syst.* **47** 409–32

Pastor J 2008 *Mathematical Ecology of Populations and Ecosystems* (Chichester: Wiley-Blackwell)

Picasso V, Brummer E C, Liebman M, Dixon P M and Wilsey B J 2008 Crop species diversity affects productivity and weed suppression in perennial polycultures under two management strategies *Crop. Sci.* **48** 331

Rees M P *et al* 1996 Quantifying the impact of competition and spatial heterogeneity on the structure and dynamics of a four-species guild of winter annuals *Am. Nat.* **147** 1–32

Roxburgh S H and Wilson J B 2000 Stability and coexistence in a lawn community: mathematical prediction of stability using a community matrix with parameters derived from competition experiments *Oikos* **88** 395–408

Schnitzer S A *et al* 2011 Soil microbes drive the classic plant diversity-productivity pattern *Ecology* **92** 296–303

Simberloff D 1982 The status of competition theory in ecology *Ann. Zool. Fenn.* **19** 241–53

Turner P E and Chao L 1999 Prisoner's dilemma in an RNA virus *Nature* **398** 441–3

Vandermeer J H 1969 The competitive structure of communities: an experimental approach with protozoa *Ecology* **50** 362–71

Vandermeer J H 1989 *The Ecology of Intercropping* (Cambridge: Cambridge University Press)

Vandermeer J H and Goldberg D E 2013 *Population Ecology: First Principles* 2nd edn (Princeton, NJ: Princeton University Press)

Volterra V 1931 *Leçons sur la Théorie Mathématique de la Lutte pour la Vie* (Paris: Gauthier-Villars)

Volterra V 1926 Fluctuations in the abundance of a species considered mathematically *Nature* **118** 558–60

Wiens J A 1984 On understanding a non-equilibrium world: myth and reality in community patterns and processes *Ecological Communities: Conceptual Issues and the Evidence* (Princeton, NJ: Princeton University Press) pp 439–57

Wilcox B P, Rawls W J, Brakensiek D L and Wight J R 1990 Predicting runoff from rangeland catchments A: comparison of two models *Water Resour. Res.* **26** 2401–10

Willmott C J 1984 On the evaluation of model performance in physical geography *Spatial Statistics and Models* ed G L Gaile and C J Willmott (Norwell, MA: D. Reidel) pp 443–60

Willmott C J 1981 On the validation of models *Phys. Geogr.* **2** 184–94

Wilson S D and Keddy P A 1986 Species competitive ability and position along a natural stress/disturbance gradient *Ecology* **67** 1236–42

Zarnetske P L, Gouhier T C, Hacker S D, Seabloom E W and Bokil V A 2013 Indirect effects and facilitation among native and non-native species promote invasion success along an environmental stress gradient *J. Ecol.* **101** 905–15

Chapter A2

Predicting optimal mixtures of perennial crops by combining modelling and experiments

If you can look into the seeds of time, and say which grain will grow and which will not, speak then unto me.

William Shakespeare, *Macbeth.*

1. A long standing issue in agriculture is the contribution of plant species richness to crop productivity and ecosystem functioning.
2. Experiments as well as quantitative modelling are fundamental to design sustainable agroecosystems and for the optimization of crop production. Here we will show how to integrate both in a single framework capable of making useful predictions.
3. We use the generalized Lotka–Volterra model we introduced in chapter 2 to model communities of perennial crop mixtures. That is, the model incorporates, in addition to mutual competition, interspecific interactions that can be facilitative. We estimate model parameters—carrying capacities and interaction coefficients—from, respectively, the observed biomass of monocultures and bicultures measured in a large diversity experiment of seven perennial forage species in Iowa, United States.
4. To test the model we simulate combinations of more than two species that were sown and compare the corresponding theoretical yields with the experimental ones. Overall, theoretical predictions are in good agreement with the experiments.
5. Once we have confidence in the model performance, we simulate species combinations that were not sown. From all possible mixtures (sown and not sown) we identify which are the most productive species combinations.
6. We conclude with some caveats and possible extensions of this framework.

A2.1 Background information

Population growth and the resulting greater demands for food are putting pressure upon agriculture to provide for mankind. Nowadays primary production is based mainly on annual plantings and harvestings of a single species over an extended area (Dewar 2007). Owing mainly to the size of the seed, or/and the ease of establishment combined with relatively quick production, annuals have been preferred over herbaceous perennial species and therefore have garnered almost all of the effort and resources for improvement (Cattani 2014).

However, current production agriculture with its ecosystems simplification, pesticide and fertilizer use often appears to be at odds with conservation biology (Banks 2004). Reduction in anthropogenic impacts of agriculture may be accomplished through perennials crops which should greatly reduce input demands, nutrient losses and the associated environmental impacts (Crews 2005). Indeed, current interest and research into the development of herbaceous perennial species for food is providing new options for food production systems.

Multi-species perennial polycultures are increasingly used for pastures, forage and biomass production, as well as for grain production. Benefits of such multi-species mixtures include *overyielding* (i.e. production in mixtures that exceeds expectations based on monoculture, Trenbath 1974), stability (lower variability in production), resilience (the ability to recover after a perturbation such as drought) and a reduction of soil erosion (Cox *et al* 2006, Jackson 2002, Schulte *et al* 2006). Furthermore, perennial polyculture farming could be instrumental in solving a wide variety of global problems, ranging from malnutrition and hunger, worries about fossil fuel consumption, environmental degradation and loss of biodiversity (Lomborg 2004, Dewar 2007).

Identifying the optimal combinations of species in perennial crop mixtures requires field experiments. We have seen in chapter 2 that the number of experimental treatments (species combinations) grows at least with the square number of the species involved. For this reason they quickly become unfeasible because of the time, labor, and costs they require (Hector *et al* 1999, Tilman *et al* 1997, 2001). Thus, in addition to field experiments, it was also identified as a key task the development of quantitative methods using mathematical modelling (Agrawal *et al* 2007, Morin 2011). Mathematical models can help in many ways: as tools to predict the outcome of different seeded polycultures; to reveal ecological relevant patterns; to summarize vast sources of information and facilitate comparisons; to provide a better evaluation of the effectiveness of management practices that would serve as a basis for management optimization (Vandermeer 2011). Nevertheless, applications of phenomenological models, in terms of interspecific interactions, to practical agricultural situations are few. Moreover, these models have rarely been capable of identifying the contribution of each particular interaction as opposed to measuring the net effect of all interactions (Kirwan *et al* 2007, Hector *et al* 2009).

A2.2 Overview

To develop the present application let us follow the list of questions we provided in box 0.1 as a guide for the process of mathematical modelling.

What is our goal? The general goal is to contribute to the development of quantitative tools for optimizing crop production.

What do we want to know? We want to identify, *a priori*, which combinations of species are the most productive.

What do we know? We have data from field experiments carried out from 2003 to 2005 in Iowa involving seven perennial crop species from three functional groups (Picasso *et al* 2008, 2011). In those experiments, all the monocultures, all except two bicultures, and several mixtures of three or more species were sown and the yield of each species was measured.

What can we assume? We will assume that (a) pairwise interactions are enough to describe the interspecific interactions in polycultures of $S > 2$ coexisting species, and (b) equilibrium was reached. Therefore, this allows us to estimate all the necessary model parameters.

How should we look at this model? We will use the Linear Lotka–Volterra Generalized Equations (LLVGE) of chapter 2 to model the dynamics of polycultures.

What will our model predict? It will predict the species yields of forage mixtures not included in the experiments; and, by computing overyielding indices and total biomass produced, to identify optimal mixtures.

Are the predictions valid? We will test our model quantitatively by comparing its predictions against empirical data for sown polycultures involving three or more species.

How can we improve the model? At the end of this chapter we will discuss some caveats and possible further developments.

A2.3 Experimental design and data

The data we will use were collected in field experiments carried out from 2003 to 2005 in Iowa involving seven perennial crop species from three different functional groups (Picasso *et al* 2008, 2011):

- Legumes: *Medicago sativa* L. aka alfalfa (A), *Trifolium repens* L. aka white clover (WC), *Desmanthus illinoensis* aka Illinois bundleflower (IBF).
- Cool-season grasses: *Dactylis glomerata* L. aka orchardgrass (OG) and *Thinopyrum intermedium* aka intermediate wheatgrass (IW).
- Warm-season grasses: *Panicum virgatum* L. aka switchgrass (SW) and *Tripsacum dactyloides* aka eastern gamagrass (EG).

Hereafter we number these species, following the above order, in such a way that species $i = 1$ is A, species $i = 2$ is WC, and so on so forth.

Field experiments included different treatments. Firstly, monocultures of each of the seven species. Secondly, almost all possible bicultures (19 out of 21[1]). Thirdly, some of the possible combinations with three, four, five and six species. All these

[1] The number of possible bicultures for seven species is immediately obtained as 7 (number of choices for the first species) × 6 (number of choices for the second species)/2 (we have to divide by 2 in order not to count pairs twice). Hence we arrive at the 21 bicultures.

combinations were sown in two different locations and in three replications in each location. The biomass of these plots was harvested once a year then sorted by species and weighted. This procedure was repeated during three years (2003, 2004 and 2005). Thus for each treatment we have 18 observations = 2 (number of locations) × 3 (number of replicates) × 3 (number of years).

The interested reader can find further details on the experiments—like environmental information, soil nutrient status, irrigation, etc—in Picasso *et al* (2008; 2011).

A2.4 Modelling

A2.4.1 Model equations

We want to predict the species yields of different sets of coexisting species of perennial plants. Therefore, we will use the LLVGE for S species we introduced in chapter 2, given by:

$$\frac{dn_i}{dt} = r_i n_i \left(1 + \frac{\sum_{j=1}^{S} \alpha_{ij}\, n_j}{\mathcal{K}_i} \right) \qquad i = 1, \ldots, S \quad (\alpha_{ii} = -1), \tag{A2.1}$$

where i denotes the species number, $r_i > 0$ is the maximum growth rate (with dimension time^{-1}), n_i is the biomass density (dimensions of mass area^{-1}), $\mathcal{K}_i > 0$ is the carrying capacity (dimensions of mass area^{-1}), and α_{ij} is the pairwise interaction coefficient (non-dimensional), measuring how species i is affected by species j.

As we have seen, the sign of the parameters α_{ij} may be positive or negative (and, of course, 0 meaning neutral interaction). If $\alpha_{ij} < 0$ then the net or dominating effect of species j over species i is a competition for resources. On the other hand, if $\alpha_{ij} > 0$ then the presence of species j facilitates species i. The cases $i = j$ correspond to intraspecific competition and we take $\alpha_{ii} = -1$ for all species in such a way to recover the logistic equation for monocultures (see chapter 2).

A2.4.2 Data curation

A common problem with this type of experiments is the appearance of alien species that were not seeded. That is, in almost all the experiments besides the seeded species there were either weeds or species not sown in this particular treatment but which were in the soil seed bank. So a problem is how to handle this sort of contamination. One possible recipe is if in an S-culture plot, $S = 1,\ldots,7$, the sum of the biomasses of the S seeded species was below 70% of the total, this entry was considered 'contaminated' and discarded. This threshold of 70% was a trade-off between accuracy and sufficient statistics. The amount of discarded entries varied along the different species with no clear pattern. In total, 213 entries (out of 993) were discarded.

A2.4.3 Initial parameter estimation from experimental data

As explained in section 2.2.3 of chapter 2, a straightforward procedure to obtain the model parameters is to perform: (a) the seven *monoculture* experiments, and from each of them to estimate the carrying capacities as the mean (over replicas) of the yield of the species i in monoculture m_i^{ex} and (b) all the pairwise experiments and for each of them, obtain the pair of mean (over replicas) of the *biculture* yields, $n_{i(j)}^{\text{ex}}$ and $n_{j(i)}^{\text{ex}}$. Using (a) we obtain $\mathcal{K}_i$ as:

$$\mathcal{K}_i = \overline{m_i^{\text{ex}}} = \frac{1}{M}\sum_{\mu=1}^{M} m_i^{\text{ex}(\mu)}, \tag{A2.2a}$$

where the bar over quantities denotes mean over the number of replicas or sample size M for the monoculture experiment for species i (table A2.1). Similarly, from (b) we obtain $n_{i(j)}^{\text{ex}}$ as:

$$\overline{n_{i(j)}^{\text{ex}}} = \frac{1}{B}\sum_{\mu=1}^{B} n_i^{\text{ex}(\mu)}, \tag{A2.3}$$

Table A2.1. Sample size, initial and final estimated values of the carrying capacities $\mathcal{K}$ and their errors.

Species	Sample size M	$\mathcal{K}$ (g m^{-2})		$\varepsilon_{\mathcal{K}}$ = 1.96 SE (g m^{-2})
		Initial	Final	
Alfalfa (A)	11	586	553	165
White clover (WC)	12	116	87	35
Illinois bundleflower (IBF)	2	418	385	102
Orchardgrass (O)	16	319	275	71
Intermediate wheatgrass (IW)	16	791	645	148
Switchgrass (SW)	9	522	491	65
E. gammagrass (EG)	2	277	243	78

Table A2.2. Mean yields of species in biculture treatments. The yield (g m^{-2}) corresponds to row species in pairwise experiment with column species.

Species	A	WC	IBF	OG	IW	SW	EG
A	—	639	625	173	243	538	633
WC	8	—	149	56	89	122	146
IBF	74	37	—	96	92	127	363
OG	497	408	321	—	321	301	NA
IW	554	699	568	183	—	NA	615
SW	180	122	438	41	NA	—	434
EG	9	6	180	NA	20	46	—

where B is the number of replicas for the biculture experiment for species i and j (table A2.2). In turn, we obtain α_{ij} and α_{ji} by solving equation (2.9) for $S = 2$, as:

$$\alpha_{ij} = \left(\overline{n_{i(j)}^{\text{ex}}} - \overline{m_i^{\text{ex}}}\right)/\overline{n_{j(i)}^{\text{ex}}}, \quad \alpha_{ji} = \left(\overline{n_{j(i)}^{\text{ex}}} - \overline{m_j^{\text{ex}}}\right)/\overline{n_{i(j)}^{\text{ex}}}. \tag{A2.2b}$$

A2.4.4 Adjustment of the initial estimated parameters to meet stability conditions

As we have seen in chapter 1, for $S = 2$ mutually competing species the required condition to ensure species' coexistence is:

$$|\alpha_{ij}| < \mathcal{K}_i/\mathcal{K}_j, \ |\alpha_{ji}| < \mathcal{K}_j/\mathcal{K}_i. \tag{A2.4}$$

Otherwise one species will exclude the other. However, it turns out that the estimated values of the parameters $\mathcal{K}_i$ and α_{ij} for the biculture combinations SW–EG, IBF–IW and IW–EG are such that stability conditions for the coexistence equilibrium (A2.4) break down and thus would imply that their coexistence is unstable. For example, for SW-EG (species number 6 and 7, respectively) we have $\mathcal{K}_6 = 522 \text{ g m}^{-2}$ and $\mathcal{K}_7 = 277 \text{ g m}^{-2}$ (table A2.1), whilst $\overline{n_{6(7)}^{\text{ex}}} = 434 \text{ g m}^{-2}$ and $\overline{n_{7(6)}^{\text{ex}}} = 46 \text{ g m}^{-2}$ (table A2.2); therefore, according to equation (A2.2b) we obtain $\alpha_{67} = (434 - 522)/46 = -1.913$, $\alpha_{76} = (46 - 277)/434 = -0.532$. And, since $\mathcal{K}_6/\mathcal{K}_7 = 522/277 = 1.885$ and $\mathcal{K}_7/\mathcal{K}_6 = 1/1.885 = 0.531$, conditions (A2.4) are not satisfied. This means that with these values of the competition coefficients between species SW and EG, the coexistence of these two species is not stable. However we still have the freedom to modify the LLVGE parameters around the computed mean values, provided we keep them within their confidence intervals, in such a way as to transform the unstable equilibrium into a stable one fulfilling conditions (A2.4). Then, varying the carrying capacities within their confidence intervals (see box 2.2 of chapter 2) we can determine a region in the space of these parameters for which the coexistence equilibria were mathematically stable for all the bicultures. Within this

Box A2.1. Summary: the recipe to build the model.

The recipe we used can be summarized as follows:

i. ***Data curation***. To avoid data contamination by species that were not seeded we discarded those experiments in which the sum of the biomasses of the seeded species was below a threshold (70% of the total).
ii. ***Parameter estimation***. For each species i we estimated $\mathcal{K}_i$ as the mean yield across all the corresponding monoculture samples. Once we had the $\mathcal{K}_i$ for all species we estimated the competition coefficients α_{ij} between a pair of species i–j using the yields of these species in the corresponding pairwise experiment through equation (A2.2b).
iii. ***Adjustment and optimization of model parameters***. To further improve model accuracy we varied the carrying capacities within their confidence intervals such that the errors of theoretical biomasses with respect to the experimental ones were minimized.

Table A2.3. Final estimated values of the interaction coefficients α_{ij} of column j species over row i species.

Species	A	WC	IBF	OG	IW	SW	EG
ALFALFA (A)	–1	10.42	0.98	–0.77	–0.56	–0.08	8.78
WHITE CLOVER (WC)	–0.12	–1	1.67	–0.08	0.003	0.28	9.07
I. BUNDLEFLOWER (IBF)	–0.50	–2.34	–1	–0.90	–0.51	–0.59	–0.12
ORCHARDGRASS (OG)	1.29	2.40	0.48	–1	0.25	0.63	NA
INT. WHEATGRASS (IW)	–0.38	0.60	–0.83	–1.44	–1	NA	–1.51
SWITCHGRASS (SW)	–0.58	–3.03	–0.42	–1.49	NA	–1	–1.24
E. GRAMMAGRASS (EG)	–0.37	–1.62	–0.17	NA	–0.36	–0.45	–1

region we can chose the set of seven carrying capacities $\{\mathcal{K}_1, \ldots, \mathcal{K}_7\}$ that minimizes the sum of square deviations between theoretical and experimental biomasses. Following this recipe we provide in box A2.1 both the initial and these final values together with their experimental errors $\varepsilon_{\mathcal{K}}$ calculated as the radius of the 95% confidence intervals.

Using the final estimates of $\mathcal{K}_i$ in table A2.1 and the biculture yields listed in table A2.2 we obtain, through equation (A2.2b), the interaction coefficients α_{ij} of column j species over row i species appearing in table A2.3. For example, the 10.42 in entry A–WC means that species WC exerts a facilitative interaction over species A and so on.

Since there were no plots with the combinations SW–IW and EG–OG, we cannot estimate their interaction parameters, respectively, $\{\alpha_{56}; \alpha_{65}\}$ and $\{\alpha_{47}; \alpha_{74}\}$ (denoted as non available in table A2.3). However, it turns out that for all the seeded mixtures of three or more species in which the pair (SW,IW) or/and (EG,OG) was included, either species SW or/and EG did not grow (Picasso *et al* 2008, 2011). This is a fortunate coincidence because in the simulations we will not need to make use of the corresponding non available interaction coefficients.

A2.4.5 On the types of interspecific interactions

What can we say about the types of interspecific interactions? From table A2.3 we observe that the most frequent interspecific interactions between the 21 pairs are either kind of amensalism, 0/– or –/0, and parasitism, +/– or –/+. There are no mutualistic +/+ interactions.

Something that is worth remarking is that a **large absolute value for a coefficient α_{ij} does not necessarily imply a dramatic effect of species *j* on species *i* in a mixture** involving this pair. Remember that α_{ij} measures the **per capita** effect of species j on species i, thus when the yield of species j, n_j, is very small it will have a mild effect on species i in absolute terms (or even null if $n_j = 0$).

Notice that:

a. The pair of legumes alfalfa and white clover are usually not strongly affected by other species, with the exception of cool season grasses (orchardgrass and intermediate wheatgrass).

b. In turn, orchardgrass was usually facilitated by legumes (alfalfa and white clover). It is known that legumes, by fixing nitrogen, tend to enhance soil fertility. In fact, it was argued legumes are a primary cause of overyielding (Huston *et al* 2000, Storkey *et al* 2015). The nitrogen transfer from the legumes to this grass species would explain why orchardgrass achieved a greater biomass in mixture than in monoculture and why it always ranked first in mixtures.

c. Illinois bundleflower, intermediate wheatgrass, switchgrass and eastern gamagrass are generally negatively affected by the other species. This effective competition they experience explains why they achieved lower biomass in mixture than in monoculture.

A2.5 Metrics for overyielding and equitability

To quantify the overyielding of polycultures there are several metrics. Here we will consider two of the most popular indices in agriculture science: the *land equivalent ratio* (*LER*), also known as the *relative yield total* (*RYT*) and the *transgressive overyielding factor* (*TOF*). Since the relationship between biodiversity and ecosystem functioning (BEF), either positive or negative, is currently a matter of discussion we will also compute the Shannon equitability (*SE*) index, measuring the diversity of mixtures, to find out if there is a relationship.

As we have seen in chapter 2, the *LER* or *RYT* is defined by the sum of relative yields $y_i = n_i/\mathcal{K}_i$ (i.e. the species yield in mixture normalized by its yield in monoculture):

$$LER = \sum_{i=1}^{S} y_i. \tag{A2.5}$$

If the $LER > 1$ then the polyculture is better than the associated monocultures and implies species complementarity (Vandermeer 1989).

A more stringent criteria for overyielding implies ***transgressive overyielding***, which occurs when a polyculture yields more than the highest yielding component species in monoculture (Trenbath 1974). This can be measured by the transgressive overyielding factor (TOF) (Trenbath 1974), defined as percentages by

$$TOF = 100\frac{\sum_{i=1}^{S} n_i - \max(m_i)}{\max(m_i)} = 100\frac{\sum_{i=1}^{S} n_i - \max(\mathcal{K}_i)}{\max(\mathcal{K}_i)}. \tag{A2.6}$$

From equation (A2.6) we see that there is transgressive overyielding whenever $TOF > 0$.

The most straightforward and roughest measure of biodiversity of a mixture is its species richness *S*. The more sophisticated Shannon equitability (*SE*) is defined as:

$$SE = \frac{-\sum_{i=1}^{S} p_i \ln(p_i)}{\ln(S)}, \tag{A2.7}$$

with $p_i = n_i / \sum_{j=1}^{S} n_j$. Therefore, SE varies between 0 and 1 corresponding, respectively, to total dominance of a single species (monoculture) and complete equitability between biomasses of the species in a mixture of S species.

A2.6 Model validation: theoretical versus experimental quantities

A2.6.1 Qualitative check: species ranking

First of all, the model does a good job at reproducing the empirical species yield ranking for three and four species mixtures (see tables A2.4 and A2.5).

Note that

Table A2.4. Species ranking for sown and not sown mixtures of three-species combinations and the percentage each species represents in the mixture, theory versus experiment (between parenthesis).

	First	Second	Third
A-WC-IBF	A 97% (A 97%)	WC 2% (IBF 3%)	IBF 1% (WC 0%)
A-OG-IW	OG 74% (OG 56%)	A 26% (A 25%)	IW 0% (IW 19%)
A-SW-EG	A 75% (A 68%)	SW 25% (SW 31%)	EG 0% (EG 1%)
WC-OG-IW	OG 77% (OG 63%)	IW 13% (IW 31%)	WC 10% (WC 6%)
WC-SW-EG	WC 51% (SW 57%)	SW 49% (WC 40%)	EG 0% (EG 3%)
IBF-OG-IW	OG 64% (OG 68%)	IW 35% (IW 24%)	IBF 1% (IBF 8%)
A-OG-SW	OG 74% (OG 64%)	A 26% (A 27%)	SW 0% (SW 9%)
A-IW-EG	IW 69% (IW 62%)	A 31% (A 38%)	EG 0% (EG 0%)
WC-OG-SW	OG 88% (OG 89%)	WC 12% (SW 6%)	SW 0% (WC 5%)
WC-IW-EG	IW 89% (IW 87%)	WC 11% (WC 11%)	EG 0% EG 2%
IBF-OG-SW	OG 77% (OG 72%)	IBF 23% (IBF 17%)	SW 0% (SW 11%)
IBF-IW-EG	IW 63% (IW 89%)	IBF 26% (IBF 10%)	EG 11% (EG 1%)
IBF-SW-EG	SW 62% (SW 84%)	IBF 29% (IBF 14%)	EG 9% (EG 2%)
A-WC-OG	OG 71% (NA)	A 27% (NA)	WC 2% (NA)
A-WC-IW	A 53% (NA)	IW 45% (NA)	WC 2% (NA)
A-WC-SW	A 92% (NA)	SW 6% (NA)	WC 2% (NA)
A-WC-EG	A 99% (NA)	WC 1% (NA)	EG 0% (NA)
A-IBF-OG	OG 74% (NA)	A 26% (NA)	IBF 0% (NA)
A-IBF-IW	IW 69% (NA)	A 31%(NA)	IBF 0% (NA)
A-IBF-SW	A 75% (NA)	SW 23% (NA)	IBF 2% (NA)
A-IBF-EG	A 90% (NA)	IBF 10% (NA)	EG 0% (NA)
WC-IBF-OG	OG 88% (NA)	WC 12% (NA)	IBF 0% (NA)
WC-IBF-IW	IW 89% (NA)	WC 11% (NA)	IBF 0% (NA)
WC-IBF-SW	WC 64% (NA)	SW 26% (NA)	IBF 10% (NA)
WC-IBF-EG	WC 81% (NA)	IBF 19% (NA)	EG 0% (NA)

Table A2.5. Species ranking for sown and not sown mixtures of four-species combinations and the percentage each species represents in the mixture, theory versus experiment (between parenthesis).

	First	Second	Third	Fourth
A-WC-OG-IW	OG 71% (OG 55%)	A 27% (A 25%)	WC 2% (IW 20%)	IW 0% (WC 0%)
A-WC-SW-EG	A 92% (A 86%)	SW 6% (SW 11%)	WC 2% (EG 2%)	EG 0% (WC 1%)
A-IBF-OG-IW	OG 74% (OG 58%)	A 26% (A 24%)	IW 0% (IW 16%)	IBF 0% (IBF 2%)
A-IBF-SW-EG	A 75% (A 83%)	SW 23% (SW 14%)	IBF 2% (IBF 3%)	EG 0% (EG 0%)
WC-IBF-OG-IW	OG 77% (OG 63%)	IW 13% (IW 31%)	WC 10% (IBF 4%)	IBF 0% (WC 2%)
WC-IBF-SW-EG	WC 65% (SW 61%)	SW 24% (WC 25%)	IBF 11% (EG 9%)	EG 0% (IBF 5%)
A-WC-IBF-OG	OG 71% (NA)	A 27% (NA)	WC 2% (NA)	IBF 0% (NA)
A-WC-IBF-IW	A 53% (NA)	IW 45% (NA)	WC 2% (NA)	IBF 0% (NA)
A-WC-IBF-SW	A 92% (NA)	SW 6% (NA)	WC 2% (NA)	IBF 0% (NA)
A-WC-IBF-EG	A 97% (NA)	WC 2% (NA)	IBF 1% (NA)	EG 0% (NA)
A-WC-OG-SW	OG 71% (NA)	A 27% (NA)	WC 2% (NA)	SW 0% (NA)
A-WC-IW-EG	A 53% (NA)	IW 45% (NA)	WC 2% (NA)	EG 0% (NA)
A-IBF-OG-SW	OG 74% (NA)	A 26% (NA)	IBF 0% (NA)	SW 0% (NA)
A-IBF-IW-EG	IW 69% (NA)	A 31% (NA)	IBF 0% (NA)	EG 0% (NA)
WC-IBF-OG-SW	OG 88% (NA)	WC 12% (NA)	IBF 0% (NA)	SW 0% (NA)
WC-IBF-IW-EG	IW 89% (NA)	WC 11% (NA)	IBF 0% (NA)	EG 0% (NA)

(a) Orchardgrass ranked first in all eight plots (and in seven of the simulations) involving this species.
(b) Intermediate wheatgrass ranked first in all three plots (and simulations) involving this species and excluding orchardgrass.
(c) Alfalfa ranked first in all four plots (and simulations) involving this species and excluding orchardgrass and intermediate wheatgrass.

The fact that orchardgrass and alfalfa dominate the majority of the polycultures in which they are present is an outcome well known by farmers.

A2.6.2 Quantitative check I: individual species yields

To assess the accuracy of our model (how close predicted values are to the experimental ones) we performed a linear regression analysis between experimental and theoretical values of species yields in seeded mixtures involving three or more species. Figure A2.1 shows the theoretical versus observed species yield for all these mixtures. The resulting linear fit is a line with an intercept very close to the origin, slope $a = 1.03$, i.e. very close to one, and a coefficient of determination $R^2 = 0.935$. This was the first check confirming that the model was working properly.

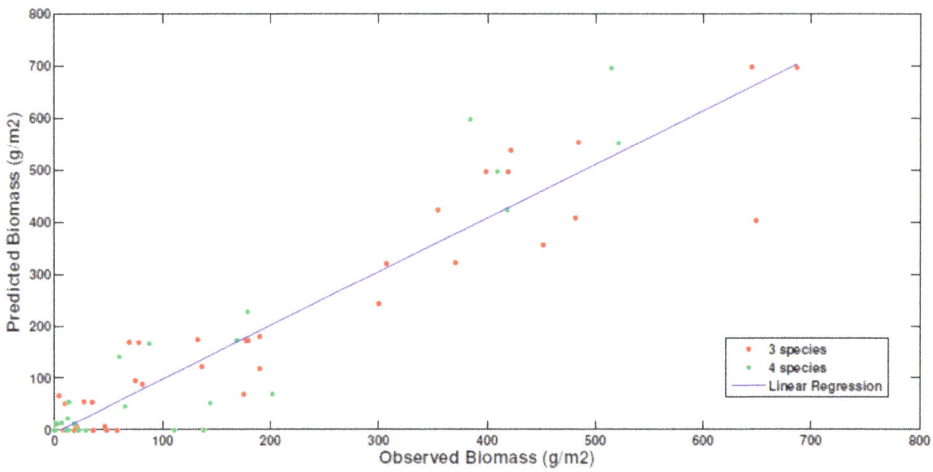

Figure A2.1. Predicted versus observed biomass for mixtures of more than two species. From Halty *et al* 2017.

Nevertheless, we have seen in chapter 2 important caveats of using the coefficient of determination as accuracy metric. In chapter 2 we also provided a set of four accuracy metrics that are more appropriate. These four error/accuracy metrics are:

- The relative mean absolute error (*RMAE*),
- *P*95 (the percentage of predictions which fall within the confidence intervals of 1.96σ,
- the modified version of the coefficient of efficiency E_1,
- the modified index of agreement d_1.

The reader is referred to chapter 2 for definitions, details and references about this metrics. In chapter 2 we also discussed the threshold values for each of these four metrics separating good agreement from bad agreement between experimental data and model predictions. The convention for these thresholds we adopted was: *RMAE* < 50% (comparable with the typical standard error of the experimental yields), *P*95 $\geqslant$ 66.7%, $E_1 > 0.0$, $d_1 > 1/3$. This is summarized in table A2.6, which indicates the model does a good job for 12 out of 18 experimental treatments. In fact, we can adopt more stringent thresholds for E_1 and d_1: $E_1 > 0.5$, $d_1 > 0.5$ and in this case the model is still reasonably accurate for 11 out of 18 experimental treatments (now it fails for the three-species polyculture WC–OG–IW for which $E_1 = 0.418$).

Since a picture's worth a thousand words we show in figures A2.2 and A2.3 for the above 18 seeded mixtures of three and four species, respectively, the experimental yields of each species versus the corresponding theoretical value produced by the model. As you can see, in general there is good agreement between the theoretical and empirical yields. The number of theoretical yields that fall within confidence intervals around the corresponding experimental values are: for experiments involving three species, 26 out of 36 (3 species × 12 experiments); for experiments involving four species, 15 out of 24 (4 species × 6 experiments). We also computed for each mixture the percentage difference (PD) between the predicted and

Table A2.6. Model accuracy for three- and four-species treatments measured by using four metrics *RMAE*, *P*95, E_1 and d_1. Marked in boldface are those treatments such that $RMAE > 50\%$ or $P95 < 66.7\%$ or $E_1 < 0.0$ or $d_1 \leqslant 1/3$ (poor agreement between model and experimental data).

Treatment	*RMAE* (%)	*P*95 (%)	E_1	d_1
A–WC–IBF	4.8	66.67	0.962	0.981
A–OG–IW	33.9	**33.33**	0.244	0.726
A–OG–SW	21.6	66.67	0.648	0.850
A–SW–EG	21.9	100.00	0.684	0.864
WC–OG–IW	34.5	66.67	0.418	0.763
IBF–IW–EG	**56.3**	**0.00**	0.502	0.668
IBF–SW–EG	40.4	**33.33**	0.590	0.742
IBF–OG–IW	25.9	66.67	0.622	0.796
IBF–OG–SW	19.3	66.67	0.747	0.881
A–IW–EG	16.1	100.00	0.758	0.884
WC–SW–EG	27.3	100.00	0.550	0.727
WC–OG–SW	25.2	100.00	0.772	0.880
A–WC–OG–IW	**58.6**	**25.00**	0.029	0.646
A–WC–SW–EG	36.9	75.00	0.699	0.842
A–IBF–SW–EG	18.5	75.00	0.841	0.897
A–IBF–OG–IW	29.6	75.00	0.540	0.788
WC–IBF–OG–IW	31.4	**25.00**	0.641	0.875
WC–IBF–SW–EG	**84.2**	**50.00**	**−0.17**	0.593

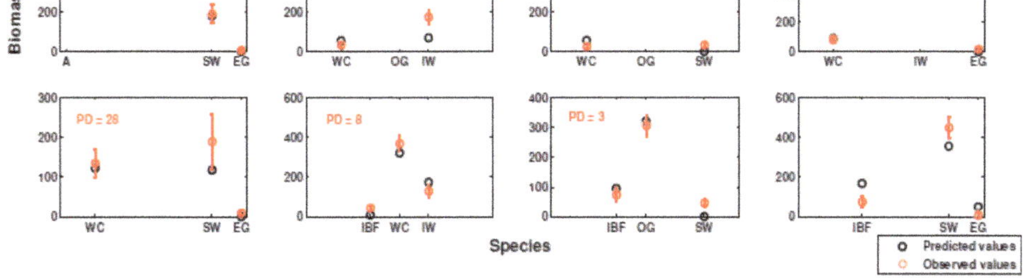

Figure A2.2. Yields for the different species in mixtures of three seeded species. Red circles correspond to experimental values. Error bars correspond to confidence intervals from the mean observed value. PD means percentage difference between model and experimental total yield (see text). In some cases the predicted yields (black circles) are completely covered by the experimental ones. From Halty *et al* 2017.

experimental *total yield* (shown as an inset in the corresponding panel). It turns out that the matching between total yield for these species mixtures is even better (the PD are generally quite small).

A2.6.3 Quantitative check II: overyielding, total biomasses and equitability

Let us now check how well the model is able to capture global metrics of interest not only to agriculture scientists, but also to farmers, like the *LER* and *TOF* and total yield. Experimental and model predictions of such quantities are shown in table A2.7. A first important observation is that all experiments verify that *LER* > 1. Furthermore, 74% (14 out of 19) of polyculture experiments exhibit transgressive overyielding, i.e. *TOF* > 0.

Interestingly, no mixture maximized *LER, TOF* and total yield (biomass density) simultaneously nor even a pair of these three quantities simultaneously (the three different polycultures appear in boldface in table A2.7). The experimental data showed that the maximum *LER* = 2.13 is attained for white clover–orchardgrass–switchgrass, and this polyculture exhibits a positive *TOF* = 11 (above the experimental *TOF* = 4) and a modest total yield of 543 g m^{-2} (below the experimental mean of 602 g m^{-2}). The maximum experimental *TOF* = 38 is attained by alfalfa–white clover–Illinois bundleflower, this polyculture exhibits a *LER* = 1.42 (below the experimental *LER* = 1.62) and a total yield of 709 g m^{-2} (well above the

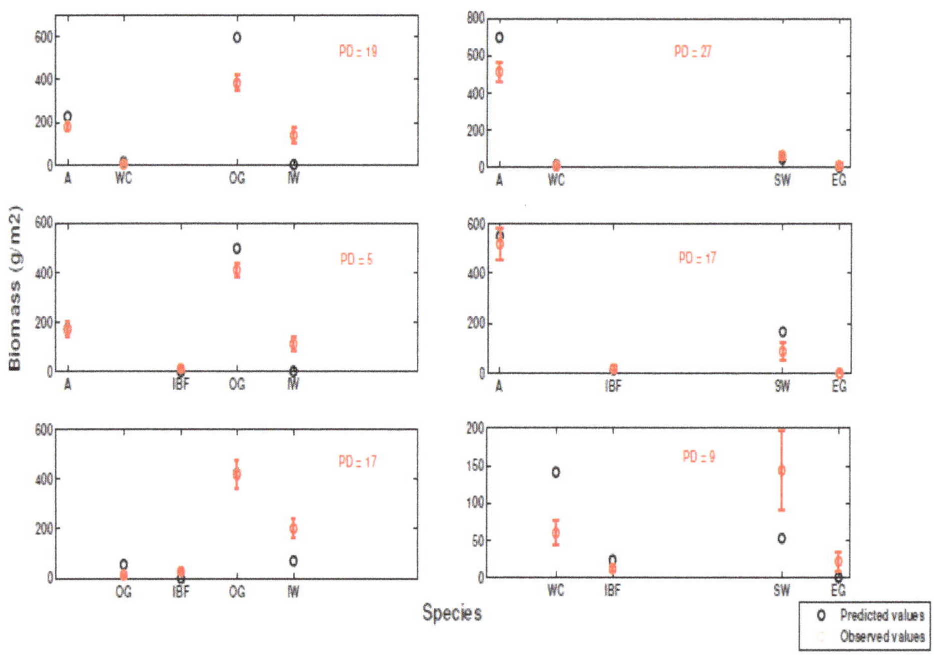

Figure A2.3. Yield for the different species in mixtures of four seeded species. Red circles correspond to experimental values. Error bars correspond to confidence intervals from the mean observed value predicted yields. Black circles are the values produced by the model. PD means percentage difference between model and experimental total biomass (see text). From Halty *et al* 2017.

Table A2.7. (From Halty *et al* 2017). Land equivalent ratio (*LER*), transgressive overyielding factor (*TOF*), total yield (g m^{-2}) and Shannon equitability (*SE*) for model calculations and experimental data (in this latter case values are means over replicas). Polycultures that reached maximum *LER*, *TOF* and total biomass densities in the experiments appear in boldface.

Polyculture	*LER* (theo)	*LER* (exp)	*TOF* (theo)	*TOF* (exp)	Total yield (theo)	Total yield (exp)	*SE* (theo)	*SE* (exp)
A–WC–IBF	1.43	1.42	30	38	717	709	0.13	0.07
A–WC–OG	2.73	NA	52	NA	838	NA	0.6	NA
A–WC–IW	1.93	NA	57	NA	1014	NA	0.71	NA
A–WC–SW	1.52	NA	37	NA	757	NA	0.29	NA
A–WC–EG	1.25	NA	17	NA	648	NA	0.06	NA
A–IBF–OG	2.12	NA	21	NA	670	NA	0.52	NA
A–IBF–IW	1.3	NA	24	NA	797	NA	0.56	NA
A–IBF–SW	1.37	NA	32	NA	731	NA	0.56	NA
A–IBF–EG	1.33	NA	27	NA	702	NA	0.31	NA
A–OG–IW	2.12	1.99	4	11	670	716	0.52	0.9
A–OG–SW	2.12	1.96	21	18	669	654	0.52	0.78
A–IW–EG	1.3	1.29	24	22	797	784	0.56	0.61
A–SW–EG	1.34	1.18	30	12	718	620	0.51	0.62
WC–IBF–OG	2.12	NA	21	NA	464	NA	0.33	NA
WC–IBF–IW	2.11	NA	22	NA	787	NA	0.32	NA
WC–IBF–SW	1.78	NA	−55	NA	219	NA	0.79	NA
WC–IBF–EG	1.8	NA	−52	NA	184	NA	0.45	NA
WC–OG–IW	2.27	1.95	−15	−12	548	564	0.63	0.75
WC–OG–SW	2.12	2.13	−6	11	464	543	0.33	0.39
WC–IW–EG	2.11	2.01	22	15	787	744	0.32	0.42
WC–SW–EG	1.65	1.99	−51	−31	241	336	0.64	0.72
IBF–OG–IW	1.46	1.67	−22	−15	505	549	0.65	0.74
IBF–OG–SW	1.42	1.41	−15	−12	417	430	0.49	0.72
IBF–IW–EG	1.34	1.2	−1	12	640	722	0.8	0.32
IBF–SW–EG	1.38	1.16	17	10	577	539	0.79	0.45
A–WC–IBF–OG	2.73	NA	52	NA	838	NA	0.47	NA
A–WC–IBF–IW	1.93	NA	57	NA	1014	NA	0.56	NA
A–WC–IBF–SW	1.51	NA	37	NA	756	NA	0.23	NA
A–WC–IBF–EG	1.43	NA	30	NA	717	NA	0.11	NA
A–WC–OG–IW	2.73	1.95	30	9	838	702	0.47	0.73
A–WC–OG–SW	2.73	NA	51	NA	838	NA	0.47	NA
A–WC–IW–EG	1.94	NA	57	NA	1014	NA	0.56	NA
A–WC–SW–EG	1.51	1.18	37	8	756	596	0.23	0.35
A–IBF–OG–IW	2.12	2	4	9	670	701	0.41	0.74
A–IBF–OG–SW	2.12	NA	21	NA	669	NA	0.41	NA
A–IBF–IW–EG	1.3	NA	24	NA	797	NA	0.44	NA
A–IBF–SW–EG	1.37	1.17	32	13	731.59	626	0.45	0.48
WC–IBF–OG–IW	2.27	2.05	−15	3	547.37	661	0.5	0.62
WC–IBF–OG–SW	2.12	NA	−6	NA	464	NA	0.26	NA

WC–IBF–IW–EG	2.11	NA	22	NA	788	NA	0.26	NA
WC–IBF–SW–EG	1.78	1.1	−56	−52	217.03	238	0.62	0.73
Mean	**1.83**	**1.62**	**15**	**4**	**664**	**602**	**0.46**	**0.59**
Max	**2.73**	**2.13**	**57**	**38**	**1014**	**784**	**0.8**	**0.9**

experimental total yield). The maximum total yield = 784 g m^{-2} is obtained for alfalfa–intermediate wheatgrass–eastern gamagrass, this polyculture exhibits a relatively low $LER = 1.29$ and a high $TOF = 22$.

The reader can check that in general there are no correlations between *LER* and *TOF* and between these two indices and the total yield for the experimental data, except between *TOF* and total yield which exhibit a significant and strong (Pearson) correlation ($r = 0.64$, $p = 0.003$).

Table A2.7 also includes, as a metric of biodiversity, the Shannon equitability *SE*. The *SE* serves as an indicator of the degree of domination in each polyculture. For example, we observe that $SE > 0.5$ in 12 out of 19 experiments, indicating that polycultures are not generally dominated by one species (in which case *SE* would be smaller than 0.5).

A2.6.4 Verdict: model validated

The model we are using rests on two main assumptions: **equilibrium**, i.e. the assemblage of species reached the steady state, and **additivity**, i.e. the interaction effects can be modelled as if it were pairwise and linear in each species. In the strict mathematical sense a community is at equilibrium only when the rate of change for all species is zero. This is a theoretical ideal that is never achieved in natural communities, due to seasonal and stochastic environmental variation (Wiens 1984). Regarding the additivity assumption, we saw in chapter 2 that the Lotka–Volterra equations can be thought as the first approximation in a Taylor series expansion about the equilibrium points of a more complex and general theory (Lotka 1925, Volterra 1931) involving higher order (non-linear) interaction terms. In applying this model to real communities the question becomes, what effect departures from these two assumptions has on the validity of predictions.

We have presented several evidences, both qualitative and quantitative, that the model reproduces reasonably well the experimental observations, confirming then *a posteriori* the validity of both assumptions. Indeed, observed values of yields for $S > 2$ species can be reproduced with quite good accuracy. Although table A2.7 shows that the model tends to overestimate the *LER, TOF* and total yield, the agreement between predicted and observed results for these aggregate or global properties is even better. This is particularly true for the *LER*: in three cases the difference was less than one percent and in four additional cases this difference was less than ten percent.

A final comment regarding model validation is about the stability of its solutions. For $S > 2$ species, theoretically, more complex equilibria e.g. *strange attractors* (associated with a *chaotic* behavior) are possible. Actually, the Lotka–Volterra generalized equations we are using, in spite of impressive progress, have not been completely classified yet in terms of their stability. One of the reasons that have made difficult such classification is the existence of *heteroclinic* cycles (Hofbauer and Sigmund 1998), related with cyclic dominance (e.g. when for $S = 3$ species, species 1 dominates species 2 which, in turn, dominates species 3 and species 3 dominates species 1). However, when cyclic dominance is ruled out the Lotka–Volterra equations are in general monotonic and hence admit no chaos (see for instance Hirsch 1988 and Sigmund 1998). In fact, as the reader can check, in all the simulations the system quickly converges to a stable equilibrium with no traces of chaos (Halty *et al* 2017).

A2.7 Predictions: results from simulation of not sown treatments

The validation of our model encourages us to use it to carry out 'virtual experiments' and test the performance of polycultures different from the ones that were sown, harvested, and measured. Therefore, let us see the results of simulating those treatments of three, four and five species that were not part of the experiment (not sown)[2].

A2.7.1 Similarities and differences between theoretical results for sown and not sown polycultures

The species ranking for mixtures of three and four species that were not sown appear in tables A2.4 and A2.5, respectively. The results seem consistent with the ones found for sown mixtures since the patterns for dominant species are very similar. Another similarity we observe, from table A2.7, is that all not sown polycultures of three and four species also verify that $LER > 1$, and most of them exhibit $TOF > 0$ (19 out of 22).

On the other hand, table A2.7 also shows that for most not sown polycultures $SE < 0.5$ (14 out of 22), indicating that most of these polycultures would be dominated by one (or two) species.

A2.7.2 Using the model for predicting optimal mixtures

As we have seen, simulations of treatments for which there were no experimental data to compare with, yield rankings of species which are consistent with the ones observed for treatments for which we do have empirical data. This suggests that the model is a useful tool for searching optimum crop mixtures for overyielding. Another thing we have also observed is that three different mixtures maximize each of the three metrics, *LER*, *TOF* and total yield. Hence which is the optimal mixture depends on which metric you want to maximize. For instance, we observe from table A2.7, that in theory any mixture including alfalfa, white clover and

[2] Treatments of six and seven species cannot be simulated as they all contained the pair switchgrass–intermediate wheatgrass and/or eastern gamagrass–orchardgrass, for which no experiments were carried out and thus we could not estimate the corresponding interaction parameters.

intermediate wheatgrass produces the maximum total yield and *TOF*, 1014 g m^{-2} and 57, respectively, and a $LER = 1.93$ (which is greater than the mean experimental $LER = 1.62$). This is consistent with the finding that these three species were weed suppressive (Picasso 2008) thus enhancing productivity.

Alternatively, if our goal consists in maximizing *LER*, we observe that this condition is achieved by any combination involving alfalfa, white clover and orchardgrass. Besides, such mixtures lead to the second highest total yield and *TOF* (table A2.7). In summary, these findings suggest that any combination including alfalfa, and white clover plus another species chosen from intermediate wheatgrass and orchardgrass is a promising candidate in terms of productivity. Actually this is further confirmed by the alfalfa–white clover–orchardgrass–intermediate wheatgrass experiment which produced *LER*, *TOF* and total yield all greater than the corresponding experimental means.

A2.8 Using the model attempting to elucidate the relationship between yield and diversity

This section is a sort of 'bonus track' addressing an interesting debate in plant ecology regarding whether or not community biodiversity enhances productivity (Hector 1999, Hector *et al* 2009, Huston 2000). Trying to elucidate this point, first we will use the simplest and more straightforward measure of biodiversity, the species richness. Second, we will look at the more sophisticated measure provided by the Shannon Equitability index *SE*.

A2.8.1 Positive correlation between productivity and species richness.

It turns out that both the experimental and theoretical total biomass density or yield increase log-linearly with species richness S (panel (a) of figure A2.4) with, respectively, $R^2 = 0.995$ and $R^2 = 0.998$. The same happens for the *LER* (panel (b) of figure A2.4) with, respectively, $R^2 = 0.957$ and $R^2 = 0.989$. On the other hand, the *TOF* does not exhibit any significant relationship with the species richness.

In addition, the positive correlation between theoretical yield and species richness holds when including the non-seeded treatments; if we consider the simulations for all possible treatments (seeded and non-seeded) the total theoretical yield and *LER* also increase log-linearly with the species richness ($R^2 = 0.996$ and $R^2 = 0.994$, respectively) as shown in panels (c) and (d) of figure A2.4.

How to interpret the above results? We mentioned that a $LER > 1$ can be interpreted as synonymous of species complementarity, which may be the result of ***resource partitioning***—i.e. when species divide a niche to prevent competition for resources—or **facilitation** among species (Loreau and Hector 2001). Distinguishing the effects of niche differentiation and facilitation is often a difficult task in practice. Indeed, the existence of interaction coefficients of both signs suggests that the two effects occur in the polycultures. This is something expected to happen between legumes, which have the ability to fix atmospheric nitrogen, and other plants, which only have access to soil nitrogen. For instance, a rather crude way to disentangle the effects of resource partitioning from those of facilitation is by simply counting the

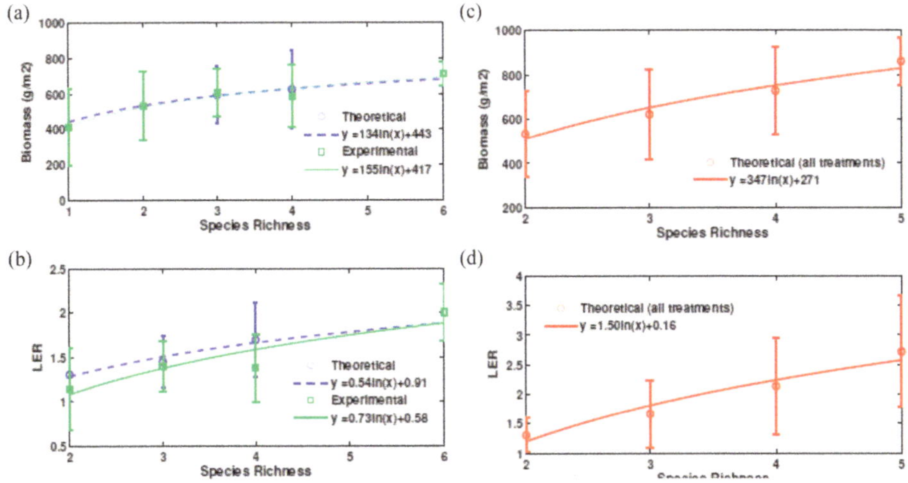

Figure A2.4. (a) Average of total yield for seeded species versus species richness. (b) Average LER for seeded species versus species richness. (Both for (a) and (b) green squares and filled fit lines are for experimental values; blue circles and dashed fit lines are for theoretical values.) (c) Average, over all treatments (seeded and non-seeded), theoretical total biomass versus species richness. (d) Average, over all treatments (seeded and non-seeded), theoretical LER versus species richness. From Halty *et al* 2017.

different types of interactions among the pool of seven species considered in this study. The 38 measured interspecific interactions comprise 25 negative, one null and 12 positive interactions (table A2.7). The recipients of facilitation are basically three species: alfalfa, white clover and orchardgrass. Interestingly, this is consistent with the finding that any combination including these species achieves an outstanding performance as measured by *LER*, *TOF* and total yield.

A2.8.2 No significant correlation between productivity and *SE*

Although total yield and *LER* exhibit a positive correlation with species richness, the experimental data show no significant correlation between any of the productivity metrics considered and the *SE*, i.e. $p > 0.05$. Therefore, the available data is not enough to support the claim that biodiversity generally enhances polyculture productivity.

A2.9 Possible extensions and some caveats

The model we built and analyzed is a minimally sufficient description for predicting yields of crop mixtures. All its parameters are directly estimated from experimental data from monocultures and bicultures. Therefore, in order to make it more realistic or accurate, this model can be extended or refined in several directions. One possibility would be through the introduction of parameters to the Lotka–Volterra equations to account for their deviation from additivity. However, as we mentioned, this would imply additional parameters to estimate from data, requiring additional experimental treatments.

Parsimonious modelling, a common theme of this book, is particularly compelling for developing quantitatively predictive tools for polycultures of plants. An important reason is that the outcome of experiments exhibits an important dependence on soil properties as well as on external conditions. Hence, the experimental errors are in general quite large precluding falsification of more elaborate models taking into account these non-linear and non-additive effects. It is important to bear in mind that, with such more realistic modelling, an adequate experimental design will become considerably more difficult.

This application assumes the studied polycultures have reached an equilibrium state. Nevertheless, in the real world, farmers harvest generally before this equilibrium state is reached, at a stage of maturity for desired quality and quantity. Or, they harvest at frequency and height to maintain a healthy plant community. To include these harvest management practices in the model it is necessary to work with the full dynamical LLVGE, rather than with their equilibrium states. In other words, the species intrinsic growth rate coefficients r_i need to be estimated. This in turn requires measuring temporal series for the yield monoculture. This issue will be addressed in application 3 when modelling the dynamics of tree abundances in tropical forests.

Let us conclude this section with some caveats regarding the application of this modelling methodology.

An important challenge for applying similar modelling efforts to other mixtures of crops and/or different soils is that the model parameters (e.g. the carrying capacities) are certainly going to be modified by climate factors as well as soil nutrient availability. Similarly, other edaphic conditions—like water content, texture or drainage—might also influence species interactions in mixtures, and mixture composition might vary with different environments and soil types. At any event, the results we used, which came from experiments in two separate locations with different soils, exhibited remarkable qualitative similarities. For example, the above mentioned positive complementarity effects were observed in both locations (Picasso 2011). Moreover, results of these experiments are consistent with ecological experiments at multiple sites throughout Europe (Hector *et al* 1999) where polycultures overyielded and complementarity effects were positive. Such robustness of patterns is certainly welcome.

A2.10 Bottom line

In summary, by making reasonable biological and ecological assumptions, well supported by the data of a thorough field experiment (Picasso *et al* 2008, 2011), we built a LLVGE model. We have shown that this model is able to reproduce with reasonable accuracy the yields of herbaceous polycultures involving three or more crop species.

Indeed, this methodology (Halty *et al* 2017) can be used to describe, explain and predict the yields of different crop mixtures with the goal of optimizing their productivity. In particular, it is applicable to several perennial crops common in modern industrial farming as well as in subsistence agriculture around the world. It is worth remarking that perennial polyculture farming seems to be the answer for a wide variety of global important problems of most of today's agronomic practices based on annual crops. Annuals require excessive water consumption, significant amounts of synthetic fertilizers and pesticides, labor, and disrupt natural biological processes. On the other hand, perennial crops are more rustic, improve soil structure, are more efficient for capturing and using water, and they also promote biodiversity and ecosystem functions (Batello *et al* 2013). Perennial crop mixtures could thus play a central role in ensuring food security and protection of biodiversity over the long term.

MATLAB code

```
% LV_FD_Ap2: LV is for Lotka-Volterra and FD for FINITE DIFFERENCES
% It uses Picasso's (2008) data for 7 species of grasses, thus N fixed = 7
clear all
prompt = {'simulation time = ','startv(=[1,0,..0,0]: only Sp 1,…,
         =[1,1,…1,1]: ALL the N species) = '}
titl = 'INPUT PARAMETERS';
lines= 1;
def = {'10000','1 1 0 1 1 0 0'};
answer = inputdlg(prompt,titl,lines,def);
fid1 = fopen('App2.dat','w+');
control=fprintf(fid1,'%s \n',answer{1});
control=fprintf(fid1,'%s \n',answer{2});
simtime=str2double(answer{1})
startv=str2num(answer{2})
r=1; % the net growth coefficient%is taken fixed and = 1 (we are not interested
% in the transient)
S=7;
K0_v=[586 116 418 319 791 522 277]; % unrefined or initial carrying capacities
eK_v=[83 18 28 37 76 32 22];

EmpiricPop_x2sp_M=[
K0_v(1) 639 625 173 243 538 633
8 K0_v(2) 149 56 89 122 146
74 37 K0_v(3) 96 92 127 363
497 408 321 K0_v(4) 321 301 0 % 0 for NA values
554 699 568 183 K0_v(5) 0 615
180 122 438 41 0 K0_v(6) 434
9 6 180 0 20 46 K0_v(7)
];          % yields for biculture experiments
```

```
EmpiricPop_x2spERR_M=[
eK_v(1) 45 55 25 35 69 68
5 eK_v(2) 34 24 22 22 43
22 14 eK_v(3) 28 30 76 321 % 0 for NA values
27 46 21 eK_v(4) 30 28 0
50 68 62 57 eK_v(5) 0 61
62 30 40 10 0 eK_v(6) 47
9 5 135 0 13 14 eK_v(7)
];

alpha0_M=-eye(S); % Initial interaction matrix

for im=1:S
 for jn=1:S
  if im ~= jn
  alpha0_M(im,jn)=(EmpiricPop_x2sp_M(im,jn)-K0_v(im))/…
         Empiric Pop_x2sp_M(jn,im);
  end
  end
end
alpha0_M(4,7)=0; alpha0_M(5,6)=0.; alpha0_M(6,5)=0; alpha0_M(7,4)=0.;

% Refined final parameters

alpha_M=[
 -1      10.425 0.978 -0.765 -0.559 -0.08   8.783
 -0.124 -1      1.672 -0.078  0.003  0.281 9.065
 -0.498 -2.343 -1     -0.899 -0.514 -0.589 -0.12
  1.287  2.399  0.479 -1      0.252  0.631 0
 -0.375  0.602 -0.828 -1.437 -1      0     -1.509
 -0.578 -3.027 -0.417 -1.493  0     -1     -1.242
 -0.369 -1.617 -0.174  0     -0.363 -0.454 -1  ]; % interacrion matrix

K_v=[552.89 87.33 384.67 274.97 644.81 490.96 242.85]

% Initialize values of populations
initP=startv.*round(0.1*K_v);
N0_M=zeros(simtime,S);
N_M=zeros(simtime,S);
N0_M(1,:)=initP;
N_M(1,:)=initP;

% Finite Difference Equations
```

```
dt=0.01
for is=2:simtime
 N0_M(is,:)=max(0, ( N0_M(is-1,:)'+dt*N0_M(is-1,:)'.*( 1+...
          alpha0_M*N0_M(is-1,:)'./K0_v' ) )' );
 N_M(is,:)=max(0, ( N_M(is-1,:)'+dt*N_M(is-1,:)'.*( 1+...
          alpha_M*N_M(is-1,:)'./K_v' ) )' );
end
figure,plot(N_M(1:1/dt:simtime,:)),xlabel('t'),...
          ylabel('Yields'),grid
% Local stability analysis

J0_M=zeros(S);   %A) Generation of Jacobian matrix J_M
 J_M=zeros(S);
for row=1:S
 J0_M(row,:)=alpha0_M(row,:)*N0_M(simtime,row);
 J_M(row,:)=alpha_M(row,:)*N_M(simtime,row);
end
[eigenVec0,eigenVal0]=eig(J0_M);  %B) Computation of eigenvalues of J_M
eigenVal0=sort(diag(eigenVal0))
[eigenVec,eigenVal]=eig(J_M);
eigenVal=sort(diag(eigenVal))
```

References

Agrawal A A *et al* 2007 Filling key gaps in population and community ecology *Front. Ecol. Environ.* **5** 145–52

Banks J E 2004 Divided culture: integrating agriculture and conservation biology *Front. Ecol. Environ.* **2** 537–45

Batello C, Wade L, Cox S, Pogna R, Bozzini A and Choptiany J 2013 Perennial crops for food security *Proc. of the FAO Expert Workshop* http://www.fao.org/agriculture/crops/thematic-sitemap/theme/spi/fao-expert-workshop-on-perennial-crops-for-food-security/en/

Cattani D J 2014 Perennial polycultures: how do we assemble a truly sustainable agricultural system? *FAO Expert Workshop on Perennial Crops for Food Security (Rome, Italy)* ch 24

Cox T S, Glover J D, Van Tassel D L, Cox C M and DeHaan L R 2006 Prospects for developing perennial-grain crops *Bioscience* **56** 649–59

Crews T E 2005 Perennial crops and endogenous nutrient supplies *Renew. Agric. Food Syst.* **20** 25–37

Dewar J P 2007 *Perennial Polyculture Farming Seeds of Another Agricultural Revolution?* Occasional paper, RAND Corp. www.rand.org

Halty V *et al* 2017 Modelling plant interspecific interactions from experiments of perennial crop mixtures to predict optimal combinations *Ecol. Appl.* **27** 2277–89

Hector A *et al* 1999 Plant diversity and productivity experiments in European grasslands *Science* **286** 1123–7

Hector A *et al* 2009 The analysis of biodiversity experiments: from pattern toward mechanism *Biodiversity, Ecosystem Functioning, and Human Wellbeing an Ecological and Economic*

Perspective ed S Naeem, D E Bunker, A Hector, M Loreau and C Perrings (Oxford: Oxford University Press) pp 3–13

Hirsch M W 1988 Systems of differential equations which are competitive or cooperative III: Competing species *Nonlinearity* **1** 51–71

Hofbauer M and Sigmund K 1998 *Evolutionary Games and Population Dynamics* (Cambridge: Cambridge University Press)

Huston M A 2000 No consistent effect of plant diversity on productivity *Science* **289** 1255a

Jackson W 2002 Natural systems agriculture: a truly radical alternative *Agric. Ecosyst. Environ.* **88** 111–7

Kirwan L *et al* 2007 Evenness drives consistent diversity effects in intensive grassland systems across 28 European sites *J. Ecol.* **95** 530–9

Lomborg B (ed) 2004 *Global Crises, Global Solutions* (Cambridge: Cambridge University Press)

Loreau M and Hector A 2001 Partitioning selection and complementarity in biodiversity experiments *Nature* **412** 72–6

Lotka A J 1925 *Elements of Physical Biology* (Baltimore, MD: Williams and Wilkins)

Morin P J 2011 *Community Ecology* (Chichester: Wiley)

Picasso V D, Brummer E C, Liebman M, Dixon P M and Wilsey B J 2008 Crop species diversity affects productivity and weed suppression in perennial polycultures under two management strategies *Crop. Sci.* **48** 331–42

Picasso V D, Brummer E C, Liebman M, Dixon P M and Wilsey B J 2011 Diverse perennial crop mixtures sustain higher productivity over time based on ecological complementarity *Renew. Agric. Food Syst.* **26** 317–27

Schulte L A, Liebman M, Asbjornsen H and Crow T R 2006 Agroecosystem restoration through strategic integration of perennials *J. Soil Water Conserv.* **61** 164A–9A

Sigmund K 1998 The population dynamics of conflict and cooperation *Doc. Math.* **ICM I** 487–506

Storkey J *et al* 2015 Engineering a plant community to deliver multiple ecosystem services *Ecol. Appl.* **25** 1034–43

Tilman D, Lehman C L and Thomson K T 1997 Plant diversity and ecosystem productivity: theoretical considerations *Proc. Natl. Acad. Sci. U.S.A.* **94** 1857–61

Tilman D, Reich P B, Knops J, Wedin D, Mielke T and Lehman C 2001 Diversity and productivity in a long-term grassland experiment *Science* **294** 843–5

Trenbath B R 1974 Biomass productivity of mixtures *Adv. Agron.* **26** 177–210

Vandermeer J H 1989 *The Ecology of Intercropping* (Cambridge: Cambridge University Press)

Vandermeer J H 2011 *The Ecology of Agroecosystems* (Sudbury, MA: Jones & Bartlett)

Volterra V 1931 *Leçons sur la Théorie Mathématique de la Lutte pour la Vie* (Paris: Gauthier-Villars)

Wiens J A 1984 On understanding a non-equilibrium world: myth and reality in community patterns and processes *Ecological Communities: Conceptual Issues and the Evidence* ed D R Strong *et al* (Princeton, NJ: Princeton University Press) pp 439–57

Part II

Ecophysics: methods from physics applied to ecology

Chapter 3

The maximum entropy method and the statistical mechanics of populations

'It would seem, then, that what is needed is an altogether new instrument; one that shall envisage the units of a biological population as the established statistical mechanics envisage molecules, atoms, and electrons; that shall deal with such average effects as population density, population pressure, and the like, after the manner in which thermodynamics deal with the average effects of gas concentration, gas pressures, etc; that shall accept its problems in terms of common biological data, as thermodynamics accepts problems stated in terms of physical data; and that shall give the answer to the problem in the terms in which it was presented.'

Alfred Lotka (1925)

'...although, as a matter of history, statistical mechanics owes its origin to investigations in thermodynamics, it seems eminently worthy of an independent development, both on account of the elegance and simplicity of its principles, and because it yields new results and places old truths in a new light in departments quite outside of thermodynamics.'

Josiah Willard Gibbs (1902)

Summary

Statistical physics originated in the mid-19th century in an effort to explain the empirical gas laws from the more fundamental Newton's laws. It has been used since then by physicists to analyze systems too complex to tackle by the canonical approach of solving the corresponding deterministic equations for their constituents.

doi:10.1088/978-0-7503-2432-8ch3

Later, statistical physics has shown to be a powerful method to analyze complex systems beyond the realm of physics, from stock markets, to societies or ecosystems. The connection between information entropy and thermodynamic entropy, provided by the maximum entropy (MaxEnt) method of inference from information theory, has been instrumental for extending statistical physics to these fields.

In the present chapter we start by reviewing Boltzmann's statistical thermodynamics. Next, we introduce the general MaxEnt formalism and show its power to make inferences with incomplete information. For example, MaxEnt allows building the so-called MaxEnt Theory of Ecology (METE). Thus, after introducing METE we review its main features and predictions.

3.1 Basics of statistical physics

3.1.1 The program of statistical physics

The essence of the statistical physics program is to explain a 'macroscopic' description of a system, involving aggregate variables and known phenomenological rules connecting them, in terms of the 'microscopic' components that constitute the system and the fundamental laws they obey. Statistical physics originated in the mid-19th century in an effort to explain the gas laws. The gas laws were crucial for understanding how to convert heat to work during the industrial revolution. As an illustrative example, imagine a mol of gas in a container of volume V, at pressure P and temperature T. Since the end of the 18th century we know these three macroscopic variables are connected by the phenomenological 'gas law', $PV = RT$ (R is the gas constant, a universal constant for all gases). The laws ruling gases were crucial for understanding how to convert heat to motive force during the industrial revolution. At a fundamental microscopic level a gas can be described by molecules of mass m moving with velocity v, obeying the known laws of particle mechanics (Newton's law, force = mass × acceleration). Given that there are $N_A = 6.02 \times 10^{23}$ (Avogadro's number) molecules in a mol of gas, it is clearly out of question to solve the corresponding equations of motion explicitly. Therefore, physicists resort to a statistical physics formulation of particle mechanics, *statistical mechanics*[1], to derive these phenomenological results of thermodynamics from a probabilistic examination of the underlying microscopic systems. The approach of statistical mechanics, as its name implies, is to express the thermodynamic macroscopic variables as statistical averages of the microscopic variables of particle mechanics. Assuming that: (1) the gas is composed of identical molecules moving in random directions, separated by distances that are large compared with their size and (2) the molecules undergo perfectly elastic collisions (no energy loss) with each other and with the walls of the container, but otherwise do not interact, James Clerk Maxwell (1867) was able to express $P = \frac{N_A}{3V} m\bar{v^2}$ and $T = \frac{N_A}{3R} m\bar{v^2}$, where $\bar{v^2}$ denotes the mean square velocity. It is immediate to check that these two expressions lead to

[1] Statistical mechanics is usually considered as a subfield of statistical physics, used to refer to only *statistical thermodynamics*. However both terms, statistical physics and statistical mechanics are generally used as synonyms and so we will do so hereafter.

the gas law. Statistical mechanics thus explains the gas law by deriving this equation from the more fundamental microscopic laws of particle mechanics. Furthermore, statistical mechanics allows a microscopic interpretation for the temperature as a quantity proportional to the mean kinetic energy of molecule $\varepsilon_k = 1/2m\overline{v^2} = 1/2\overline{mv^2}$. Furthermore, Maxwell provided the probability velocity distribution $f_{\text{Maxwell}}(v_x, v_y, v_z)dv_x dv_y dv_z$ which is proportional to $e^{\frac{mv^2}{2k_BT}}dv_x dv_y dv_z$, where $k_B \equiv R/N_A$.

The model used by Maxwell describes a *perfect* or *ideal* gas and is a reasonable approximation to a real gas, particularly in the limit of extreme dilution and high temperature. Such a simplified description, however, is not sufficiently precise to account for the behavior of gases at high densities. Ludwig Boltzmann later extended these ideas to real gases (with interacting molecules), liquids and solids providing a general statistical connection between the distribution of molecular energies and temperature.

As we will see next, a quantity which played an instrumental role in the development of statistical physics since Boltzmann is the **entropy**. In physics the entropy is associated with 'disorder'. For instance, let us consider the three phases of water, ice, liquid and vapor. Ice, with atoms assembled in a regular crystal lattice displays long-range order, is clearly the most ordered state. In the other extreme, gas, with its molecules moving randomly is the most disordered state. Finally, the molecules in liquids are not arranged in a repeating three-dimensional array, like the ions in solids but, unlike the molecules in gases, the arrangement of the molecules in a liquid is not completely random. So the entropy is low for ice, medium for liquid and high for vapor.

3.1.2 Boltzmann–Gibbs maximum entropy approach to statistical mechanics

A general procedure to obtain the probability distribution for the particle energies, devised by Boltzmann in 1877, is to take the distribution that maximizes the entropy of the gas. Boltzmann imagined gas molecules occupying s different energetic states $i = 1,2,3,\ldots,s$ of definite discrete energy[2] ε_i. Thus the total energy of the gas E would be equal to

$$E = \sum_{i=1}^{s} n_i \varepsilon_i, \tag{3.1}$$

where n_i denotes the number of molecules with energy ε_i. Actually equation (3.1) is an approximate formula that is valid only when the interaction energy between molecules is negligible (otherwise we would have to include in the sum these interaction energies). This is the case for a perfect or ideal gas defined as a gas for which the intermolecular potential energy of interaction is negligible. Clearly, if N is the number of gas molecules, we would also have:

[2] Boltzmann found it easier to think about discrete energies than continuous ones.

$$N = \sum_{i=1}^{s} n_i. \tag{3.2}$$

Now any set $\{n_i\}$ of occupation numbers for which E, N agree with the given information, represents a possible distribution, compatible with all that is specified. At this point it is convenient to distinguish between a **microscopic state** or ***microstate*** and **macroscopic state** or ***macrostate***. The microstate i is the actual set of velocities, and positions, of all the molecules. Thus the definition of a microstate requires the specification of $3N$ position coordinates (three per molecule) and $3N$ velocity components (also three per molecule). Geometrically, this set of $6N$ quantities may be regarded as a point in a space of $6N$ dimensions, called the ***phase space***. Thus Boltzmann imagined each gas molecule occupying small six-dimensional volumetric cell i ($i = 1, 2, 3,\ldots, s$) of phase space, with a well-defined energy ε_i. On the other hand, a **macroscopic state** or ***macrostate*** is characterized by a vector of occupation numbers $\vec{n} = [n_1, n_2, \ldots]$, i.e. n_1 molecules in the molecular energy level ε_1, n_2 molecules in the ε_2, and so on. Since we are assuming that molecules are indistinguishable, there are many microstates corresponding to a given macrostate of the gas: permutations of indistinguishable molecules between different single molecular energetic levels i of a microstate correspond to the same macrostate. Inasmuch as the number of microstates for a macroscopic system (i.e. N of the order of N_A) is huge, the number of macrostates which is smaller than the number of microstates is still very large (box 3.1 illustrates the growing of both microstates and macrostates with the number N of particles). Hence, out of the millions of such possible macrostates, which is the one most likely to be realized? Boltzmann's answer was that the 'most probable' macrostate is the one that can be realized in the greatest number of ways. The number of microstates corresponding to a given macrostate, or ***multiplicity*** W, is given by the multinomial formula:

$$W = \frac{N!}{n_1! n_2! \cdots n_s!} \tag{3.3}$$

Boltzmann recognized that the entropy $\mathcal{S}$ was proportional to the logarithm of the number of microstates, i.e. $\mathcal{S} = k_{\mathrm{B}} \ln W$, where k_{B} is the so-called Boltzmann's constant (connected with the the gas constant R by $k_{\mathrm{B}} = R/N_a$). Taking the logarithm of equation (3.3) we get (figure 3.1):

$$\mathcal{S} = k_{\mathrm{B}} \ln W = k_{\mathrm{B}}(\ln N! - \ln(n_1! n_2! \cdots n_s!)) = k_{\mathrm{B}}\left(\ln N! - \sum_{i=1}^{s} \ln n_i!\right). \tag{3.4}$$

Since the n_i are large we can approximate the factorial by Stirling's formula ($\ln n! \approx n \ln n - n$), obtaining the ***Gibbs entropy formula***:

$$\mathcal{S} \approx -k_{\mathrm{B}} N \sum_{i=1}^{s} (n_i/N) \ln(n_i/N) = -k_{\mathrm{B}} N \sum_{i=1}^{s} p_i \ln p_i, \tag{3.5}$$

where $_{pi} = n_i/N$ is the probability that a molecule has energy ε_i.

Box 3.1. The number of microstates and macrostates for particles with two possible states.

To simplify the counting of states imagine that each molecule or particle has only two possible accessible states, and thus can be represented as a coin with states either head (H) or tail (T).

Let us start with an example. If we toss up two coins, the possible outcomes are:

coin 1	coin 2
H	H
H	T
T	H
T	T

The four possible outcomes of this set of two coins with two states H and T are the ***microstates***: HH, HT, TH, TT.

If the two coins are exactly the same; i.e. ***indistinguishable***, then the two microstates HT and TH constitute one ***macrostate***, in such a way we have:

1 Head macrostate: $W_1 = 2$ microstates (in agreement with equation (3.3): $2!/1!1! = 2/1 = 2$)

2 Heads macrostate: $W_2 = 1$ microstate (in agreement with equation (3.3): $2!/2!0! = 2/2 = 1$)

0 Heads macrostate: $W_0 = 1$ microstate (in agreement with equation (3.3): $2!/0!2! = 2/2 = 1$)

If we denote the number of distinguishable microstates by Ω and the number of indistinguishable macrostates by M, we have:

$\Omega = W_0 + W_1 + W_2 = 4$ (alternatively, the number of microstates can be obtained as $\Omega = 2^2 = 4$), $M = 3$.

If we now consider $N = 100$ coins, the total number of microstates is $\mu = 2^{100} \approx 1.2677 \times 10^{30}$ and the number of macrostates (0 H, 1 H, 2 H,..., 100 H) is $M = 101$.

Therefore, we see that, for binary particles, $M \sim N$ and $\tilde{\Omega}\ 10^{N/3.3}$.

Boltzmann then argued that the occupation probabilities p_i of the most probable state at equilibrium are those that maximize the entropy, $\mathcal{S} = -k_B N \sum_{i=1}^{s} p_i \ln p_i$, and which also satisfy the constraint on the total particle number N (3.2) and the constraint on the total energy E (3.1). These two constraints can be written in terms of occupation probabilities p_i as:

$$\begin{cases} \sum_{i=1}^{s} p_i = 1, & (3.6a) \\ \sum_{i=1}^{s} p_i \varepsilon_i = \bar{\varepsilon}, & (3.6b) \end{cases}$$

where $\bar{\varepsilon} = E/N$ is the average energy (per particle). The state of equilibrium is computed by maximizing the entropy subject to the constraints. The mathematical

Figure 3.1. Boltzmann's tombstone in Vienna, Austria. The formula $\mathcal{S} = k \log W$ (equation (3.4)) is his equation for the entropy of a system of particles. It states that entropy depends only on a constant number k (now named Boltzmann's constant) and the number of possible 'microstates' W. Microstates essentially describe the configurations of the locations and velocities of particles in a system. Image from Martin Röll, Dresden, Germany.

procedure to do this is the ***Method of Lagrange Multipliers***, which is equivalent to maximizing the variational function $\mathbf{v}(\{p_j\})$ defined as:

$$\mathcal{V}(\{p_j\}) \equiv \frac{\mathcal{S}(\{p_j\})}{Nk_\mathrm{B}} - \lambda_0\left(\sum_{j=1}^{s} p_j - 1\right) - \lambda_1\left(\sum_{j=1}^{s} p_j\varepsilon_j - \bar{\varepsilon}\right), \tag{3.7}$$

where λ_0 and λ_1 are new variables called *Lagrange multipliers*. This variational function has $s + 2$ unknowns: s different p_j's and the two Lagrange multipliers. The maximum condition implies varying equation (3.7) with respect to each $_{pi}$ and equating these variations to zero. This produces a set of equations that uniquely determines the $s + 2$ unknowns. Let us do this; the variation of equation (3.7) with respect to each $_{pi}$ yields:

$$\frac{d\mathcal{V}}{dp_i} = -\ln p_i - 1 - \lambda_0 - \lambda_1\varepsilon_i, \tag{3.8}$$

and setting this to zero gives that the $_{pi}$ values that maximize the entropy subject to the constraints equation (3.6) are given by:

$$p_i{*}{=}e^{-1-\lambda_0-\lambda_1\varepsilon_i}. \tag{3.9}$$

To obtain λ_0 we use the normalization constraint (3.6a), and summing over i on both sides of equation (3.9) we get: $e^{1+\lambda_0} = z$, where $z \equiv \sum_{i=1}^{S} e^{-\lambda_1 \varepsilon_i}$ is called the molecular or single-particle *partition function* (Glazer and Wark 2006); the z stands for the German *Zustandsumme* ('sum of state'). Thus equation (3.9) can be re-written as:

$$p_i{}^* = \frac{e^{-\lambda_1 \varepsilon_i}}{z}. \tag{3.9$'$}$$

In turn, λ_1 can be identified by making contact with thermodynamics. Indeed, the first law of thermodynamics (principle of energy conservation) applied for a reversible process over a gas that keeps the volume constant allows connecting variations of entropy and energy at a given temperature T as $dE = TdS$. Using equations (3.1) and (3.4), this relation can be written as:

$$Nd\left(\sum_{i=1}^{S} p_i \varepsilon_i\right) = -Nk_{\mathrm{B}}Td\left(\sum_{i=1}^{S} p_i \ln p_i\right). \tag{3.10}$$

The differential variation of $\sum_{i=1}^{S} p_i \varepsilon_i$ is given by:

$$d\left(\sum_{i=1}^{S} p_i \varepsilon_i\right) = \sum_{i=1}^{S} (p_i d\varepsilon_i + \varepsilon_i dp_i).$$

Nevertheless, since the volume is fixed there is no work over the gas and therefore the energy levels ε_i are kept constant, i.e. $d\varepsilon_i = 0$ and so the above equation simplifies to:

$$d\left(\sum_{i=1}^{S} p_i \varepsilon_i\right) = \sum_{i=1}^{S} \varepsilon_i dp_i. \tag{3.11}$$

The differential variation of $\sum_{i=1}^{S} p_i \ln p_i$ in turn produces:

$$d\left(\sum_{i=1}^{S} p_i \ln p_i\right) = \left(\sum_{i=1}^{S} dp_i \ln p_i + p_i 1/p_i dp_i\right) = \sum_{i=1}^{S} (\ln p_i + 1) dp_i. \tag{3.12}$$

Substituting equation (3.9′) into equation (3.12) we get:

$$d\left(\sum_{i=1}^{S} p_i \ln p_i\right) = -\lambda_1 \sum_{i=1}^{S} \varepsilon_i dp_i + (1 - \ln z) \sum_{i=1}^{S} dp_i = -\lambda_1 \sum_{i=1}^{S} \varepsilon_i dp_i, \tag{3.13}$$

where the last equality in equation (3.13) holds because $\sum_{i=1}^{S} dp_i = d\sum_{i=1}^{S} p_i$ and, by equation (3.5a) $d\sum_{i=1}^{S} p_i = 0$.

Then, substituting equations (3.11) and (3.13) into equation (3.10) we get

$$\lambda_1 = 1/k_{\mathrm{B}}T, \tag{3.14}$$

and thus equation (3.9′) becomes the Boltzmann probability distribution:

$$p_i{}^* = \frac{e^{-\frac{\varepsilon_i}{k_B T}}}{z}. \tag{3.15}$$

Notice that equation (3.15) tells us that:

(i) The relevant scale for molecular energies is $k_B T = RT/N_a$, which, as we have seen, is (aside from a multiplicative factor) the mean molecular kinetic energy.

(ii) The probability of occupation of a state of energy ε_i increases when the **mean** molecular kinetic energy grows (i.e. when the temperature grows).

Although in the derivation of the Boltzmann probability distribution equation (3.15) we have assumed that the interaction energy between molecules is negligible (equation (3.1)), Josiah Willard Gibbs (1902) later showed that it holds in general when interactions between molecules are non negligible. Gibbs resorted in his derivation to the concept of ***statistical ensemble***, an idealization consisting of a large number of virtual copies (sometimes infinitely many) of a system, considered all at once, each of which represents a possible state that the real system might be in. In other words, a statistical ensemble is a probability distribution for the state of the system. Gibbs argued that, at equilibrium, the probability distribution of states $p_i{}^*$ in the ***phase space*** of all coordinates and momenta of molecules, must depend on these only through conserved quantities such as energy E_i, and governed by ***Liouville's equation*** (Landau and Lifshitz 1980).

Note that E_i denotes energy levels of a system of particles as a whole in contrast to ε_i, used earlier, that denote single-particle energy levels. This equilibrium probability distribution for a system in thermodynamic equilibrium at a temperature T can be written as:

$$P_i^* = \frac{e^{-\frac{E_i}{k_B T}}}{Z}, \tag{3.15'}$$

with Z, known as the ***canonical partition function*** given by $Z \equiv \sum_{i=1} e^{-\frac{E_i}{k_B T}}$.

Returning to Boltzmann's distribution equation (3.15), in the case of an ideal gas of free (non interacting) molecules, $\varepsilon_i = 1/2\, m v_i^2$ and equation (3.15) reduces to Maxwell's velocity distribution law. Alternatively, for a gas in a gravitational field, $\varepsilon_i = 1/2\, m v_i^2 + mgh$, where h stands for the height above sea level, equation (3.15) gives the usual *barometric formula* for decrease of the atmospheric density with height: $\rho(h) = \rho(0)\exp(-mgh)$.

Remarkably, Boltzmann's calculation did not take into account the millions of intricate dynamical details of the motion of molecules, the only dynamic information that was included is that the total energy is conserved. In addition Boltzmann assumed ***ergodicity***, i.e. that all accessible microstates are equiprobable over a long period of time. However, the fact that it is enough to predict the correct spatial and velocity distribution of the molecules shows that those microscopic details are actually irrelevant to the predictions (they cancel out when performing averages).

In the next section we will discuss the generalization of the Boltzmann's method to a more abstract and general Principle of Maximum Entropy, for inferring probability distributions from limited data.

3.2 MaxEnt in terms of Shannon's information theory as a general inference approach

MaxEnt as a general principle was first formulated by E T Jaynes (1957a, 1957b) who emphasized a natural correspondence between statistical mechanics and Shannon's *information theory*, the theory that studies the quantification, storage, and communication of information. In particular, Jaynes argued that **the entropy of statistical mechanics and the *information entropy* of information theory** (Shannon 1948) **are basically the same thing**. Consequently, statistical mechanics should be seen just as a particular application of a general tool of logical inference and information theory. Interestingly this idea was anticipated by Gibbs (1902): 'But although, as a matter of history, statistical mechanics owes its origin to investigations in thermodynamics, it seems eminently worthy of an independent development, both on account of the elegance and simplicity of its principles, and because it yields new results and places old truths in a new light in departments quite outside of thermodynamics.'

Jaynes' MaxEnt formulation can be regarded as a method of making predictions from limited data by assuming maximal ignorance about the unknown degrees of freedom. In other words, according to Jaynes, statistical physics is a way to draw inferences from incomplete information.

3.2.1 Shannon's information entropy

An alternative to Boltzmann's entropy is the information entropy introduced in 1948 by Claude Shannon. He was interested in a very different problem, namely the capacities of telecommunication lines to transmit information (Shannon 1948). He wanted to minimize the average number of bits needed to encode characters in messages that were sent through noisy channels. So Shannon discussed the problem of encoding a message, say English text, into binary digits in the most efficient way. The essential step is to assign probabilities to each of the conceivable messages in a way which incorporates the prior knowledge we have about the structure of English. Nevertheless, probably we shall never know the 'true' probabilities of English messages; and so Shannon suggested the principle by which we may construct the probability distribution actually used for applications. His recipe was to choose to use some of our statistical knowledge of English in constructing a code, but not all of it. In such a case we consider the source with the maximum information subject to the statistical conditions we wish to retain. The information of this source determines the channel capacity which is necessary and sufficient.

Suppose the message is a linear string of symbols that is L characters long, where each character is drawn independently from an n-letter alphabet, with probability $_{pi}$. Let l_i $(i = 1,2 \ldots,n)$ represent the number of times that the ith type of character is observed in the message. When L is large, the most likely L-letter message will have the composition $l_i = Lp_i$, and this occurs with probability

$$P = p_1^{l_1} \cdots p_n^{l_n} = p_1^{Lp_1} \cdots p_n^{Lp_n}. \tag{3.16}$$

In the limiting case of an alphabet having only a single letter, $P = 1$ and thus the probability of guessing the message is 100%. The larger the alphabet is, the smaller the value of P and the more unlikely we guess the message (and the greater the uncertainty of receiving a particular message). The first goal of Shannon was to construct a measure for the amount of information associated with a message. His idea was that the amount of information gained from the reception of a message depends on how *likely* it is; the less likely a message is, the more information is gained upon its reception. Thus the metric Shannon proposed for the amount of information of a message was the logarithm of $1/P$:

$$H \equiv \log_b(1/P), \tag{3.17}$$

which can be expressed, using equation (3.16), as

$$H = -L\sum_{i=1}^{n} p_i \log_b p_i, \tag{3.17$'$}$$

where b is the base of the logarithm used. Common values of b are 2, Euler's number e, and 10, and the corresponding units of entropy are the ***bits*** for $b = 2$, ***nats*** for $b = e$, and ***bans*** for $b = 10$. The mathematical form of H given above is the same as the entropy function $\mathcal{S}(\{p_i\})$ of Gibbs and Boltzmann, and so Shannon called it ***information entropy***, sometimes also called the ***uncertainty***.

To illustrate this, suppose messages which are linear strings of length $L = 8$ chosen from the $n = 4$ alphabet {1,2,3,4}, i.e. characters are chosen among these four numbers. For example let us consider these three messages:

(A) 1, 1, 1, 1, 1, 1, 1, 1
(B) 1, 1, 1, 1, 2, 2, 3, 4
(C) 1, 1, 2, 2, 3, 3, 4, 4

How do we order these three messages in terms of their information entropy H? After a short inspection we can say by intuition that message (A) has low information entropy, message (B) has medium information entropy and message (C) has high information entropy. But here is another question that is actually equivalent. Suppose a character is picked at random; how do you order these messages in terms of how easy it is for you to guess the number? So here is where knowledge comes into play. For sequence (A) you always are going to hit the number 1, so that's easy. For sequence (B) your best guess is to bet that it is a 1, let us say that it is of medium difficulty. Finally, in sequence (C) all the four numbers are equally likely and so it is hard to guess the number. More quantitatively, we can compute P, using equation (3.16), and then obtaining H, using equation (3.17). Thus, for these sequences we have:

(A) $p_1 = 1, p_2 = 0, p_3 = 0, p_4 = 0$, then $P = 1^8{\cdot}0^0{\cdot}0^0{\cdot}0^0 = 1{\cdot}1{\cdot}1{\cdot}1 = 1$ and thus $H = \log_2 1 = 0$;
(B) $p_1 = \frac{1}{2}, p_2 = \frac{1}{4}, p_3 = \frac{1}{8}, p_4 = \frac{1}{8}$, then $P = (\frac{1}{2})^4{\cdot}(\frac{1}{4})^2{\cdot}(\frac{1}{8}){\cdot}(\frac{1}{8}) = \frac{1}{2^{14}}$ and thus $H = \log_2 2^{14} = 14$;
(C) $p_1 = \frac{1}{4}, p_2 = \frac{1}{4}, p_3 = \frac{1}{4}, p_4 = \frac{1}{4}$, then $P = (\frac{1}{4})^2{\cdot}(\frac{1}{4})^2{\cdot}(\frac{1}{4})^2{\cdot}(\frac{1}{4})^2 = \frac{1}{2^{16}}$ and thus $H = \log_2 2^{16} = 16$.

Table 3.1 summarizes the properties of each of the above three messages.

We can think of the information content of a message per character as the number of yes/no questions we would have to ask to guess what character was picked. So imagine a guessing game, where someone picks one of these numbers and the questioner has to guess the number. Asking those questions is equivalent to pinpointing or decoding the message received. A naive approach for the above alphabet {1,2,3,4} would be to ask four questions: Question 1: Is it 1? If not, question 2: Is it 2? If not, question 3: Is it 3? If not, question 4: Is it 4? As the reader has noticed, the fourth question is unnecessary since if the number is different from 1, 2 and 3, the only chance it has is to be equal to 4. Therefore, it seems we would need three questions. Actually we can do even better: a general procedure to find the minimal average number of questions per character necessary to guess the character is to use a binary tree. The binary tree for the alphabet {1,2,3,4} looks like figure 3.2(a). We see that such a binary tree would reveal the label of the digit

Table 3.1. Difficulty, probability of guessing P, information entropy H and rate $r = H/L$ for messages (A)–(C).

Message	Difficulty	P	H (in bits)	$r = H/L$ (in bits/character)
1, 1, 1, 1, 1, 1, 1, 1	easy	1	0	0
1, 1, 1, 1, 2, 2, 3, 4	medium	$1/2^{14}$	14	14/8 = 1.75
1, 1, 2, 2, 3, 3, 4, 4	hard	$1/2^{16}$	16	16/8 = 2

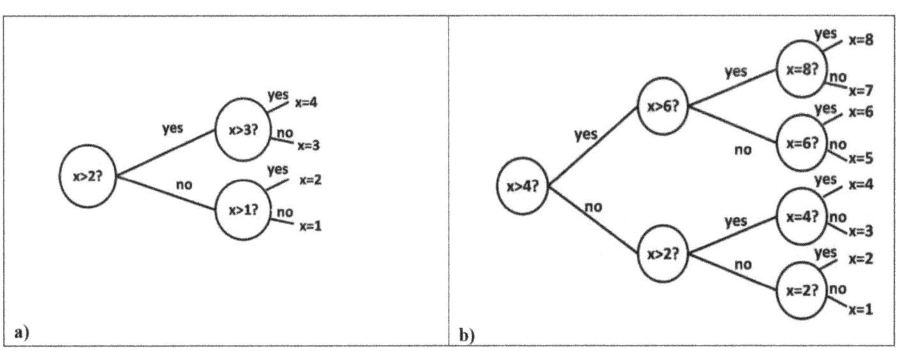

Figure 3.2. Guess binary trees. (a) For an alphabet of $n = 4$ characters. (b) For an alphabet of $n = 8$ characters.

using a minimum number of questions, that is, just two. If instead of an alphabet of four characters we have one of eight characters, i.e. a message can be any combination of the numbers {1, 2, 3..., 8} and all eight numbers are equally probable, the binary tree looks like figure 3.2(b). Likewise, this binary tree would reveal the label of the digit using three questions.

The ***entropy rate*** r of a data source means the average number of bits per symbol needed to encode it, i.e. $r = H/L$. For a given language, the maximum number of bits that can be coded in each character, assuming each character of the alphabet sequence is equally likely, is called its absolute rate R. As we have just seen, for $n = 4$ we get $R = 2 = \log_2 4$ bits/character while for $n = 8$ we get $R = 3 = \log_2 8$ bits/character. Indeed, from equation (3.17) we see that $R = \max(H)/L = -n \cdot 1/n \cdot \log_2 (1/n) = \log_2 n$. For English, with a 26 letter alphabet, the absolute rate is $\log_2 26$, or about 4.7 bits/letter. It turns out that English text, treated as a string of characters, has fairly low entropy, i.e. is fairly predictable. If we do not know exactly what is going to come next, we can be fairly certain that, for example, 'e' will be far more common than 'z'. The same happens for combinations of letters. E.g. the combination 'qu' will be much more common than any other combination with a 'q' in it, etc. Indeed, the typical entropy rate of English text is 1.3 bits per letter of the message (Schneier 1996), compared with an absolute rate of 4.7 bit/letter, meaning that the redundancy is 3.4 bits/letter.

3.2.2 MaxEnt as a method of making predictions from limited data by assuming maximal ignorance

Shannon's information entropy H reduces to Boltzmann's expression when all the probabilities are equal, but can, of course, be used when they are not. One of its virtues is that it yields immediate results without resorting to factorials or Stirling's approximation (as the reader can check going back to the previous subsection). Additionally, it shows that the expression of entropy as the mean of logs of probabilities (equation (3.17′)) has a deeper meaning quite independent from thermodynamics. With the work of E T Jaynes (1957a, 1957b), based on Shannon's information entropy, MaxEnt gained in **formalism** and precision as well as it became more widely understood, and used. More recently, MaxEnt has been regarded in a much broader light, as a fundamental requirement for ensuring that inferences drawn from data satisfy basic self-consistency requirements of probabilities (Shore and Johnson 1980). Indeed, although the information entropy of Shannon was intended for use in communication theory, it is applicable in many different areas; from problems in the fields of physics, biology, chemistry, neuroscience, to even some social sciences, such as sociology.

Before Jaynes statistical physics was completely based on the so-called ***frequentist*** or ***objective interpretation*** of probabilities. In the frequentist interpretation probabilities are associated with random physical systems such as flipping coins, rolling dice and radioactive atoms. Such probabilities are estimated as the fraction of times an outcome is observed in a large number of random trials. For instance, the frequency of appearance of each face of a die is readily determined by rolling a die many times, then dividing the number of appearances of each outcome by the total number of dice rolls. Large sequences of many repeated trials demonstrate that the empirical frequency converges to the limit 1/6 as the number of trials goes to infinity. Thus, a limitation of this frequentist view is that it is of course impossible to actually perform infinity repetitions of a random experiment to determine the probability of an event. For example, when we wrote the probability P of a message, equation (3.16), we assumed that we see a large number of messages, from which we can compute the frequencies $p_i = l_i/L$ of the different letters. But what if only one message is seen? Actually, Bernoulli realized long ago that the enumeration of options may be done in very few cases and almost nowhere else than in games of chance.

An alternative to the frequentist interpretation is the ***Bayesian*** or ***subjective interpretation*** of probabilities, in which the concept of probability is not limited to situations that are replicable. Rather, in the subjective interpretation, a probability characterizes an observer's inference based on prior knowledge (Cox 1961). That is, probability corresponds to a state of knowledge, and the rules of probability are simply ways to draw inferences from premises. For instance, the probability of rain tomorrow, say p_i, is a quantity that can be estimated, even though it is not describable by a repeatable experiment. In this instance it makes no sense to speak of a ratio such as $p_i = l_i/L$ because the number of times it rains tomorrow is not an enumerable quantity. The subjective view is evidently the broader one since it is always possible to interpret frequency ratios in this way.

Jaynes (1957a, 1957b) adopted this more broadly ***subjective*** viewpoint as a description of a state of knowledge. He regarded MaxEnt as a way of minimizing the extent to which we are incorporating into our inferences assumptions that we have no basis for making. That is, Jaynes discovered that the MaxEnt principle was powerful enough to be used as a statistical tool in its own respect, independent of thermodynamics by using information entropy. The idea is that in most real systems, the information entropy is not free to be at its maximum unconstrained value. Instead, it is checked by many constraints (e.g. the first law of conservation of energy in thermodynamics). The state of a system is thus determined by the balance of these two opposite forces: the entropic part (the natural tendency of the system towards the configuration with the highest possible entropy) and the constraints that the system has to obey. These two operating forces are taken into account by maximizing the entropy $H = -N\sum_{i=1} P_i \ln P_i$ subject to constraints. A common problem is that the data are limited, and thus many different probability distributions are compatible with these data. Of all such distributions, the choice is made by finding the set $\{p_i\}$ that both maximizes the entropy and satisfies the constraints. Such distribution is certainly the least biased distribution compatible with the data. For instance, to infer the canonical distribution equation (3.15′), two constraints are imposed. First, the normalization

$$\sum_{i=1} P_i = 1, \tag{3.18}$$

and second, the given value of the average energy $\bar{E}$ estimated as $\sum_{i=1} P_i E_i$, i.e.

$$\sum_{i=1} P_i E_i = \bar{E} \tag{3.19}$$

This is done straightforwardly by solving the equation

$$\delta\left[\frac{H}{N} - \lambda_0\left(\sum_{i=1} P_i - 1\right) - \lambda_1\left(\sum_{i=1} P_i E_i - \bar{E}\right)\right] = 0, \tag{3.20}$$

where the variation is with respect to each p_i. As before, the Lagrange multiplier λ_0 assures the normalization of the p_i's and λ_1 enforces the known value of the average energy (which is identical to fixing the temperature in the canonical ensemble). The solution is equation (3.15′).

Jaynes (1957a) summarized his contribution very clearly: '...we accept the Shannon expression for entropy, very literally, as a measure of the amount of uncertainty represented by a probability distribution; thus entropy becomes the primitive concept with which we work, more fundamental even than energy. If in addition we reinterpret the prediction problem of statistical mechanics in the subjective sense, we can derive the usual relations in a very elementary way without any consideration of ensembles or appeal to the usual arguments concerning ergodicity or equal *a priori* probabilities. The principles and mathematical methods of statistical mechanics are seen to be of much more general applicability than conventional arguments would lead one to suppose. In the problem of prediction, the maximization of entropy is not an application of a law of physics, but merely a method of reasoning which ensures that no unconscious arbitrary assumptions have been introduced.'

Therefore, the MaxEnt principle can be stated in a general and simple way as:

To describe some partly specified system one should prefer the probability distribution which is compatible with the known information (the system's constraints) and that maximizes the Shannon information entropy.

3.2.3 Inference of model parameters from the statistical moments via MaxEnt

We have seen that the maximum thermodynamic entropy distribution consistent with a known average energy $\bar{E}$ is the Boltzmann distribution, $P = \exp(-E/k_B T)/Z$, where k_B is Boltzmann's constant and T is the temperature. Using the Shannon information entropy this generalizes, so that if we know the average values of many variables $V_k(n)$ describing the system, then the maximum entropy distribution is

$$\boxed{P(n) = \exp\left(-\sum_{k=1}^{K} \lambda_k V_k(n)/Z\right),} \tag{3.21}$$

where there is a separate Lagrange multiplier λ_k for each constraint $k = 1,2,\ldots,K$; $k = 0$ is already considered by writing the denominator as the normalizing partition function constant Z which is given by:

$$Z(\lambda_1, \ldots, \lambda_K) = \sum_{n=1} e^{-\sum_{k=1}^{K} \lambda_k V_k(n)}. \tag{3.22}$$

The λ_k are formally given by the solutions to:

$$\bar{V}_k = -\frac{\partial \ln Z}{\partial \lambda_k}. \tag{3.23}$$

To illustrate MaxEnt at work we will consider a general situation (which will be used in the next Application chapter): let us suppose we have a set of S real variables, $x_1 \ldots x_S$, say population densities just to fix ideas, and that we have M different measurements of these variables. Hence we can compute the first two statistical moments, the mean densities and covariances, that we denote by the vector m_i and matrix C_{ij}, respectively. These constraints on the distribution (in addition to the normalization constraint equation (3.17)) are then expressed as:

$$m_i = \bar{x}_i = \sum_{\mu=1}^{M} x_i^{\mu}/M, \tag{3.24a}$$

$$C_{ij} = \overline{(x_i - \bar{x}_i)(x_j - \bar{x}_j)} = \overline{x_i x_j} - \bar{x}_i\, \bar{x}_j = \sum_{\mu=1}^{M} x_i^{\mu} x_i^{\mu}/M - \sum_{\mu=1}^{M} x_i^{\mu} \sum_{\nu=1}^{M} x_j^{\nu}/M^2, \tag{3.24b}$$

where the indices μ and ν denote different measurements. Thus, according to equation (3.21), the maximum entropy probability distribution $P(\{x_i\})$ can be written, using a vector of Lagrange multipliers h_i and a matrix of Lagrange multipliers J_{ij} for, respectively, constraints equations (3.24a) and (3.24b) as proportional to:

$$\mathrm{e}^{-\sum_i h_i x_i - \frac{1}{2}\sum_{ij} J_{ij}(x_i - \bar{x}_i)(x_j - \bar{x}_j)}.$$

Therefore, after redefining the multipliers h_i, $P(\{x_i\})$ can be written as (exercise 3.3):

$$P(\{x_i\}) = \frac{\mathrm{e}^{-\sum_i h_i x_i - \frac{1}{2}\sum_{ij} J_{ij} x_i x_j}}{Z}, \tag{3.25}$$

where the partition function Z in the denominator of equation (3.25) ensures that the distribution is normalized and is given by:

$$Z = \int \prod_i dx_i \mathrm{e}^{-\sum_i h_i x_i - \frac{1}{2}\sum_{ij} J_{ij} x_i x_j}, \tag{3.26}$$

or, in matrix form:

$$\mathrm{Z} = \int \prod_i dx_i \mathrm{e}^{-\mathbf{h}^{\mathrm{T}}\mathbf{x} - 1/2\mathbf{x}^{\mathrm{T}}\mathbf{J}\mathbf{x}}. \tag{3.26'}$$

It turns out that the exponent of equation (3.25) is reminiscent of the energy of a widely used model in statistical physics, the ***Ising model*** (see box 3.2). The variables x_i correspond to microscopic magnetic dipole moments in the Ising model, while h_i

Box 3.2. The Ising model.

The Ising model, was invented by the physicist Wilhelm Lenz (1920) as a mathematical model of ferromagnetism in statistical mechanics. It is named after Ernst Ising, a student of Lenz, who solved this model in one spatial dimension in his 1924 thesis. The model consists of discrete variables that represent magnetic dipole moments of atomic *spins* s_i that can be in one of two states (+1 or −1). The spins, represented by arrows (up if they are positive and down if negative), are arranged in a graph, usually a lattice. A two-dimensional schematic representation is shown in figure 3.3. Each spin interacts only with its four nearest neighbors; the ones above, below, to its left and to its right. For example, in figure 3.3 the red spin interacts with the four blue spins.

The total energy of the system of spins is the sum of two terms. First, we have the interaction energy ε_{ij} between two spins or particles i and j, which depends on their states, and can be written as:

$$\varepsilon_{ij} = J_{ij}s_is_j = \begin{cases} -J_{ij} & \text{for } s_i = s_j, \\ +J_{ij} & \text{for } s_i = s_j, \end{cases}$$

with the matrix J_{ij} defined as

$$J_{ij} = \begin{cases} J & \text{if nodes } i \text{ and } j \text{ are neighbors,} \\ 0 & \text{otherwise.} \end{cases}$$

Second, there is a second term for the energy corresponding to the interactions of the magnetic dipole moments with an external magnetic field h_i, which is given by:

$$\varepsilon_i = -h_is_i = \begin{cases} -h_i & \text{for } s_i = 1, \\ +h_i & \text{for } s_i = -1. \end{cases}$$

Therefore, the total energy of the Ising model can be written as: $E = -\frac{1}{2}\sum_{<ij>}J_{ij}s_is_j - \sum_i h_is_i$, where the brackets < > for the sum indices in the first sum denote that sites i and j are nearest neighbors. The Ising model is widely used in statistical mechanics to describe a large variety of different systems and phenomena. From its original motivation which was the phenomenon of ferromagnetism, to a statistical model for the motion of atoms (*lattice gas*) and applications in neuroscience like modelling the activity of neurons in the brain.

In the exponential of equation (3.25) (a) the variables x_i are real rather than binary ($= \pm 1$), (b) the matrix connects all pairs of variables and (c) it varies from pair to pair; this is why it is a generalization of the Ising model (Landau and Lifshitz 1980).

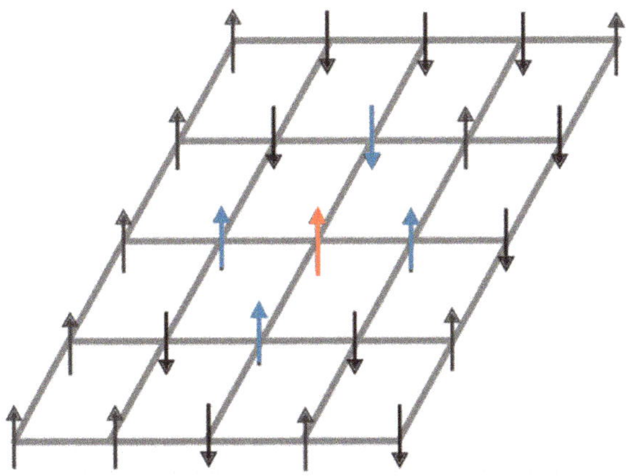

Figure 3.3. A two-dimensional square lattice Ising model. Arrows represent microscopic magnetic dipoles that can point either upward or downward.

and J_{ij} are parameters describing, respectively, a local magnetic field, that aligns the microscopic magnetic dipole moments, and the interaction coefficients between pairs of them x_i and x_j.

Equation (3.26) is a Gaussian integral, whose result is given by:

$$Z = \frac{1}{\sqrt{\det \mathbf{J}/2\pi}} \mathrm{e}^{-1/2\sum_{ij} J^{-1}_{ij} h_i h_j} = \frac{1}{\sqrt{\det \mathbf{J}/2\pi}} \mathrm{e}^{-1/2 \mathbf{h}^{\mathrm{T}} \mathbf{J}^{-1} \mathbf{h}}. \tag{3.27}$$

Now we will show how to relate the vector $\mathbf{h} = [h_i]$ and the matrix $\mathbf{J} = [J_{ij}]$ to the vector $\mathbf{m}$ and the matrix $\mathbf{C}$, which allows us to infer model parameters h_i and J_{ij} from empirical observations such as the means and covariances of the abundances. These relationships can be conveniently obtained from the derivatives of the partition function, which is the standard approach in statistical physics. Indeed, the mean abundances can be expressed as

$$\begin{aligned} \bar{x}_k &= \frac{\int \prod_i dx_i x_k \mathrm{e}^{-\sum_i h_i x_i - \frac{1}{2}\sum_{ij} J_{ij} x_i x_j}}{Z} \\ &= \frac{-\frac{\partial}{\partial h_k} \int \prod_i dx_i \mathrm{e}^{-\sum_i h_i x_i - \frac{1}{2}\sum_{ij} J_{ij} x_i x_j}}{Z} \\ &= -\frac{\partial \ln Z}{\partial h_k}. \end{aligned} \tag{3.28a}$$

And, a similar relationship holds for the covariance matrix:

$$\overline{x_i x_j} - \overline{x_i}\,\overline{x_j} = \frac{\partial^2 \ln Z}{\partial h_i \partial h_j}. \tag{3.28b}$$

Substituting equation (3.27) into equation (3.28a) and (3.28b) we get:

$$\mathbf{m} = \mathbf{J}^{-1}\,\mathbf{h}, \tag{3.29}$$

$$\mathbf{C} = -\mathbf{J}^{-1}, \tag{3.30}$$

which can be inverted to give:

$$\mathbf{h} = -\mathbf{C}^{-1}\,\mathbf{m}, \tag{3.31}$$

$$\mathbf{J} = -\mathbf{C}^{-1}. \tag{3.32}$$

Therefore, the matrix **J**, which has the natural interpretation of an interaction matrix, can be obtained as the negative of the covariance matrix. This is a very powerful result that we will use in the next Application chapter to model the dynamics of species of trees in tropical forests.

3.3 The statistical mechanics of populations

3.3.1 Rationale and first attempts

As we have seen, there are numerous situations in physics and other disciplines which can be described at different levels of detail in terms of probability distributions. Such descriptions arise because of the vast amount of details necessary for a complete description as, for example, in a gas. Statistical mechanics allows us to explain known laws connecting macroscopic variables such as the temperature and pressure in terms of the motion of microscopic molecules obeying the known Newton's laws of motion. In the absence of detailed information about the interaction between the species, a statistical mechanical treatment of a population seems to be a useful approach.

In ecology the microscopic variables, corresponding to the mass and velocity of a molecule, can be taken as the population densities of individual species. It is less clear what should be the macroscopic variables of ecology, analogous to pressure and temperature. Possible macroscopic variables are the primary productivity, the total biomass, the number of species, etc (Maynard Smith 1978). An example of a law connecting macroscopic variables, analogous to the gas law is the ***species–area relationship*** (SAR) connecting the number of species S in a given taxon on an island with the area A of the island. However, in ecology there is not a set of macroscopic variables generally accepted as appropriate for describing ecosystems, as pressure and temperature were accepted as appropriate to the description of gases before they were given a microscopic interpretation.

Another difficulty is that in ecology there is not anything comparable to universal microscopic laws of the kind of Newton's laws. The most similar thing to universal microscopic laws in ecology would be the Lotka–Volterra equations (LVE). Indeed, as we have seen in chapters 1 and 2, the LVE have been criticized from multiple angles.

The idea of adapting the methods of statistical physics to ecology is far from new and can be traced almost a century ago to Lotka's 1925 *Elements of Physical Biology*

(Lotka 1925) where he proposed that population ecology ought to be capable of some description in statistical mechanics-like terms. According to Lotka (1925) we should envisage the units of a biological population as the established statistical mechanics envisage molecules, atoms and electrons; that is, dealing with such average effects as population density, population pressure, and the like.

The first statistical mechanical formulation of the Lotka–Volterra model was made by Kerner (1957, 1959, 1972). His theory, though elegant, has been criticized because he has given no justification for the validity of the application of the statistical mechanics to the problem of population of interacting species. But a most important limitation of Kerner's formulation is that the Lotka–Volterra interaction matrix $[a_{ij}]$ must be antisymmetric and therefore it cannot accommodate mutually either intraspecific competition or competitive or mutualistic interspecific interactions. This is because Kerner used the formulation of statistical mechanics that was customary in physics at that time, elaborated by Gibbs (1902), which requires a conserved quantity that remains constant throughout the motion, namely the total energy. Kerner showed that if $a_{ij} = -a_{ij}$ then the quantity $\Phi \equiv \sum_{i=1}^{S}[N_i(t) - N_i^* \ln N_i(t)]$ is conserved (exercise 3.4).

Therefore, here we will consider the MaxEnt approach which doesn't explicitly relies on the existence of conserved quantities

3.3.2 Harte's MaxEnt theory of ecology (METE)[3]

As we have seen MaxEnt is a powerful method to link the fundamental or 'microscopic' variables with 'macroscopic' or state variables. The fundamental entities (for example, molecules in the thermodynamic example) are of two kinds: 'individuals' and 'species'. The macroscopic or ***state*** variables of ecology are those of ***macroecology***, the subfield of ecology that deals with the study of relationships between organisms and their environment at large spatial scales to characterize and explain statistical patterns of abundance, distribution and diversity (Brown and Maurer 1989). These state variables would play a similar role to that of volume, temperature, and number of molecules in thermodynamics. An example of a state variable is the ecosystem area, A_0.

In this subsection we will consider four state variables to describe an ecosystem: the total area, A_0, of the ecosystem; the total number of species from any specified taxonomic group, such as plants, within A_0, S_0; the total number of individuals in all of those species within A_0, N_0; and the summed metabolic energy rate of all the trees within A_0, E_0. We will show that these four parameters (A_0, S_0, N_0, and E_0), coupled with the MaxEnt method, are sufficient to allow us to infer numerous macroecological relationships or 'metrics'. Examples of such macroecological metrics are the species–area relationship (SAR) $S(A)$ relating average species richness found in a census cell to the area of that cell within A_0, the species-abundance distribution (SAD) describing the fraction of species with specified abundance, and the abundance–energy relationship (AER) relating a species metabolic energy requirement to its abundance.

[3] This subsection devoted to METE is based on chapter 7 of Harte's *Maximum entropy and ecology* (2011).

To accomplish this, we have to apply MaxEnt to two probability distributions from which all the macroecological metrics listed above can be derived. The first is a joint probability density, $R(n, \varepsilon)$, over the S_0 species in A_0. R is a discrete distribution in abundance, n, and a continuous distribution in metabolic energy rate ε. In particular:

$R(n, \varepsilon)d\varepsilon$ is the probability that if a species is picked at random from the species list in A_0, then it has abundance n and if an individual is picked at random from that species, then its metabolic requirement is in the interval ε, $\varepsilon + d\varepsilon$.

Note that ε is a rate of energy use, and so it is a measure of power, not energy.

$R(n, \varepsilon)$ satisfies the following constraints:

$$\sum_{n=1}^{N_0} \int_1^{E_0} R(n, \varepsilon)d\varepsilon = 1, \tag{3.33a}$$

$$\sum_{n=1}^{N_0} \int_1^{E_0} nR(n, \varepsilon)d\varepsilon = \frac{N_0}{S_0}, \tag{3.33b}$$

$$\sum_{n=1}^{N_0} \int_1^{E_0} n\varepsilon R(n, \varepsilon)d\varepsilon = \frac{E_0}{S_0}, \tag{3.33c}$$

Equation (3.33a) is a normalization condition, equation (3.33b) expresses the constraint on the average number of individuals per species, and equation (3.33c) expresses the constraint on the average total energy requirement of all the individuals within a species. Notice that the lower limit of the integrals (3.33) is 1. This is because we assume that there is a minimum metabolic rate, $\varepsilon_{\min}$, below which an individual, no matter how small you choose it to be from the pool of N_0 individuals, cannot exist. We define $\varepsilon_{\min}$ to be 1 by choice of energy units.

The second core probability distribution is the spatial abundance distribution $P_A^{(j)}(n)$ defined as the probability of finding n individuals of species j in a randomly selected habitat cell of area A within A_0. Let us denote the total abundance of species j in A_0 by n_0, then we have the normalization condition

$$\sum_{n=0}^{n_0} P_A^{(j)}(n) = 1. \tag{3.34a}$$

In addition, because the mean value of n, across cells of area A, is given by $\bar{n} = n_0 A/A_0$ we also have the constraint equation:

$$\sum_{n=0}^{n_0} nP_A^{(j)}(n) = n_0 \frac{A}{A_0}. \tag{3.34b}$$

From knowledge of R, and by integrating over all possible values of the energy variable, we can derive the species-abundance distribution $\Phi(n)$ and the energy distribution $\Psi(\varepsilon)$:

$$\Phi(n) = \int_1^{E_0} R(n, \varepsilon)d\varepsilon, \tag{3.35a}$$

$$\Psi(\varepsilon) = \frac{S_0}{N_0}\sum_{n=1}^{N_0} nR(n, \varepsilon), \tag{3.35b}$$

where $\Phi(n)$ is the fraction of species with abundance n, while $\Psi(\varepsilon)d\varepsilon$ is the probability that an individual picked from the entire individuals pool (N_0) has an energy requirement between ε, $\varepsilon + d\varepsilon$.

The SAR in turn can be derived from knowledge of $P_A^{(n_0)}(n)$ and $\Phi(n)$. In particular, the expected number of species on a patch of area A is the total number of species in A_0 times the sum over abundances, n_0, of the product of (the probability a species, with abundance n_0 in A_0, is present on the patch) × (the fraction of species with abundance n_0). Noting that the probability of presence is 1 minus the probability of absence, we have

$$S(A) = S_0 \sum_{n_0=1}^{N_0} [1 - P_A^{(n_0)}]\Phi(n_0). \tag{3.36}$$

The testable metrics predicted by the theory thus include $\Phi(n)$, $\Psi(\varepsilon)$, $P_A^{(n_0)}(n)$, and $S(A)$.

Predictions of the theory

From the constraint equations (3.33a)–(3.33c), MaxEnt equation (3.21) for the first core probability distribution results in

$$R(n, \varepsilon) = \frac{e^{-\lambda_1 n} e^{-\lambda_2 n\varepsilon}}{Z}, \tag{3.37}$$

where the partition function Z is given by:

$$Z = \sum_{n=1}^{N_0} \int_1^{E_0} e^{-\lambda_1 n} e^{-\lambda_2 n\varepsilon} d\varepsilon. \tag{3.38}$$

The Lagrange multipliers λ_i are given by application of equation (3.23) and become completely specified by the values of N_0, E_0, and S_0. Let us show this. We have $V_1 = n$, and then $\bar{n} = N_0/S_0$ (corresponding to equation (3.33b)), and $V_2 = n\varepsilon$, and then $\overline{n\varepsilon} = E_0/S_0$ (corresponding to equation (3.33c)). Thus, substituting equation (3.37) into equations (3.33b) and (3.33c) we get:

$$\sum_{n=1}^{N_0} \int_1^{E_0} n\frac{e^{-\lambda_1 n} e^{-\lambda_2 n\varepsilon}}{Z} d\varepsilon = \frac{N_0}{S_0}, \tag{3.39a}$$

$$\sum_{n=1}^{N_0} \int_1^{E_0} n\varepsilon\frac{e^{-\lambda_1 n} e^{-\lambda_2 n\varepsilon}}{Z} d\varepsilon = \frac{E_0}{S_0}, \tag{3.39b}$$

The integrals in equations (3.39a) and (3.39b) can be done exactly (exercise 3.8), yielding:

$$\frac{N_0}{S_0} = \frac{\sum_{n=1}^{N_0}(e^{-\beta n} - e^{-\sigma n})}{Z\lambda_2}, \tag{3.40}$$

$$\frac{E_0}{S_0} = \frac{\sum_{n=1}^{N_0}(e^{-\beta n} - E_0 e^{-\sigma n})}{Z\lambda_2} + \frac{\sum_{n=1}^{N_0}(e^{-\beta n} - E_0 e^{-\sigma n})}{nZ\lambda_2^2}, \tag{3.41}$$

where

$$\beta = \lambda_1 + \lambda_2, \tag{3.42a}$$

$$\sigma = \lambda_1 + E_0\lambda_2 \tag{3.42b}$$

Likewise, the integral of the partition function equation (3.38) yields:

$$Z = \sum_{n=1}^{N_0}\frac{(e^{-\beta n} - e^{-\sigma n})}{n\lambda_2}. \tag{3.43}$$

Substituting equation (3.43) into equations (3.30) and (3.41) we obtain (exercise 3.9):

$$\frac{N_0}{S_0} = \frac{\sum_{n=1}^{N_0}(e^{-\beta n} - e^{-\sigma n})}{\sum_{n=1}^{N_0}\frac{e^{-\beta n} - e^{-\sigma n}}{n}}, \tag{3.44}$$

$$\frac{E_0}{S_0} = \frac{\sum_{n=1}^{N_0}(e^{-\beta n} - E_0 e^{-\sigma n})}{\sum_{n=1}^{N_0}\frac{e^{-\beta n} - e^{-\sigma n}}{n}} + \frac{1}{\lambda_2}. \tag{3.45}$$

Equations (3.44) and (3.45) are two equations with two unknowns, λ_1 and λ_2 (remember that by equations (3.42) β and σ are functions of λ_1 and λ_2), and therefore they determine the two Lagrange multipliers as function of the state variables S_0, N_0, and E_0.

These equations are complicated and for realistic values of the state variables have to be solved numerically. However, we can still derive some further exact results. Firstly, notice that from equation (3.44) we can re-write the partition function equation (3.43) as:

$$Z = \frac{S_0}{N_0\lambda_2}\sum_{n=1}^{N_0}(e^{-\beta n} - e^{-\sigma n}),$$

and, recalling the expression for a geometric sum:

$$\sum_{n=1}^{N_0} e^{-\beta n} = \sum_{n=0}^{N_0} e^{-\beta n} - 1 = \frac{1 - e^{-\beta(N_0+1)}}{1 - e^{-\beta}} - 1 = \frac{e^{-\beta} - e^{-\beta(N_0+1)}}{1 - e^{-\beta}}, \tag{3.46}$$

we get:

$$Z = \frac{S_0}{N_0\lambda_2}\left(\frac{e^{-\beta} - e^{-\beta(N_0+1)}}{1 - e^{-\beta}} - \frac{e^{-\sigma} - e^{-\sigma(N_0+1)}}{1 - e^{-\sigma}}\right). \tag{3.47}$$

Secondly, once we have the expression (3.47) for the partition function we can derive, from equations (3.35a) and (3.35b), respectively, the species-abundance distribution $\Phi(n)$ and the energy distribution $\Psi(\varepsilon)$ (3.10):

$$\Phi(n) = e^{-\lambda_1 n}\int_1^{E_0} \frac{e^{-\lambda_2 n\varepsilon}}{Z} d\varepsilon = \frac{e^{-\lambda_1 n}}{\lambda_2 Zn}(e^{-\lambda_2 n} - e^{-\lambda_2 nE_0}) = \frac{e^{-\beta n} - e^{-\sigma n}}{\lambda_2 Zn},$$

$$\begin{aligned}\Psi(\varepsilon) &= \frac{S_0}{N_0}\sum_{n=1}^{N_0} ne^{-\lambda_1 n}\frac{e^{-\lambda_2 n\varepsilon}}{Z} = \frac{S_0}{N_0}\sum_{n=1}^{N_0} n\frac{e^{-\gamma n\varepsilon}}{Z} \\ &= \frac{S_0}{ZN_0}\left(-\frac{(N_0 + 1)e^{-\gamma(N_0+1)}}{1 - e^{-\gamma}} + \frac{e^{-\gamma}(1 - e^{-\gamma(N_0+1)})}{(1 - e^{-\gamma})}\right),\end{aligned}$$

where

$$\gamma = \lambda_1 + \lambda_2\varepsilon \tag{3.48}$$

Finally substituting Z by its expression equation (3.47) we obtain:

$$\Phi(n) = \frac{e^{-\beta n} - e^{-\sigma n}}{\frac{S_0}{N_0}\left(\frac{e^{-\beta} - e^{-\beta(N_0+1)}}{1 - e^{-\beta}} - \frac{e^{-\sigma} - e^{-\sigma(N_0+1)}}{1 - e^{-\sigma}}\right)n}, \tag{3.49}$$

$$\Psi(\varepsilon) = \lambda_2\frac{-\frac{(N_0 + 1)e^{-\gamma(N_0+1)}}{1 - e^{-\gamma}} + \frac{e^{-\gamma}(1 - e^{-\gamma(N_0+1)})}{(1 - e^{-\gamma})^2}}{\frac{e^{-\beta} - e^{-\beta(N_0+1)}}{1 - e^{-\beta}} - \frac{e^{-\sigma} - e^{-\sigma(N_0+1)}}{1 - e^{-\sigma}}} \tag{3.50}$$

So far, all results are exact. The Lagrange multipliers, λ_1 and λ_2, can be solved for numerically, using equations (3.44) and (3.45) and the relations connecting them with β, σ and γ (respectively equations (3.42a), (3.42b), and (3.48)) and the distribution $R(n,\varepsilon)$ is uniquely determined from the values of the state variables and the MaxEnt criterion.

Let us now apply MaxEnt equation (3.21) to the second core probability distribution. From the constraint equations (3.34a) and (3.34b), we obtain:

$$P_A^{(n_0)}(n) = \frac{e^{-\lambda'_1 n}}{Z'}. \tag{3.51}$$

The corresponding normalization summation or partition function can be carried out exactly, yielding:

$$Z' = \sum_{n=0}^{n_0} e^{-\lambda'_1 n} = \frac{1 - e^{-\lambda'_1(n_0+1)}}{1 - e^{-\lambda'_1}} \tag{3.52}$$

Substituting equation (3.52) into equation (3.51) we get:

$$P_A^{(n_0)}(n) = \frac{(1 - e^{-\lambda'_1})e^{-\lambda'_1 n}}{1 - e^{-\lambda'_1(n_0+1)}}. \tag{3.53}$$

Solving equation (3.34b) then yields the Lagrange multiplier λ'_1. To simplify the notation we define $x = \exp(-\lambda'_1)$, leading to:

$$n_0 \frac{A}{A_0} = \frac{\sum_{n=0}^{n_0} n x^n}{\sum_{n=0}^{n_0} x^n} = \frac{x}{1 - x} - \frac{(n_0 + 1)x^{n_0+1}}{1 - x^{n_0+1}}. \tag{3.54}$$

Some approximate results

It turns out that for realistic values of the state variables (Harte 2011), $|\beta| \ll 1$ and $\sigma \gtrsim S_0 > 1$, and λ_2 is $\sim S_0/E_0$. As a consequence, terms of order $e^{-\sigma}$ are very small compared to terms of order $e^{-\beta}$ and terms of order $E_0 e^{-\sigma}$ are very small compared with terms of order $1/\lambda_2$. Then, dropping the terms in equations (3.44) and (3.45) with $e^{-\sigma n}$ we can combine the resulting equations and obtain (exercise 3.11):

$$\lambda_2 \approx \frac{S_0}{E_0 - N_0}. \tag{3.55}$$

It is also possible to derive another approximation for S_0/N_0 (exercise 3.12):

$$\frac{S_0}{N_0} \approx \frac{1 - e^{-\beta}}{e^{-\beta} - e^{-\beta(N_0+1)}} \ln\left(\frac{1}{\beta}\right). \tag{3.56}$$

Furthermore, if $\beta \ll 1$, then this can be further simplified to:

$$\frac{S_0}{N_0} \approx \beta \ln\left(\frac{1}{\beta}\right). \tag{3.57}$$

Similarly, the expression in equation (3.47) for Z can now be approximated by:

$$Z \approx \frac{S_0}{\lambda_2 N_0} \frac{e^{-\beta} - e^{-\beta(N_0+1)}}{1 - e^{-\beta}} \approx \frac{S_0}{\lambda_2 N_0} \frac{e^{-\beta}}{\beta} \approx \frac{\ln\left(\frac{1}{\beta}\right)}{\lambda_2}. \tag{3.58}$$

Although equation (3.58) cannot be solved for β analytically, it is much easier to solve numerically than is equation (3.44).

The above approximations allow us to greatly simplify the complicated equations for metrics like the species-abundance distribution $\Phi(n)$ and the energy distribution $\Psi(\varepsilon)$:

$$\Phi(n) \approx \frac{1}{\ln(\beta^{-1})} \frac{e^{-\beta n}}{n} \tag{3.59}$$

and

$$\Psi(\varepsilon) \approx \lambda_2 \beta \frac{e^{-\gamma}}{(1 - e^{-\gamma})^2}. \tag{3.60}$$

Equation (3.59) can be used to predict the number of very rare species from knowledge of the state variables S_0 and N_0. Suppose we ask: how many species in the community have no more than, say, 10 individuals? Calling that number $S(n \leqslant 10)$, we get from equation (3.35):

$$S(n \leqslant 10) \approx S_0 \frac{1}{\ln(\beta^{-1})} \sum_{n=1}^{10} \frac{e^{-\beta n}}{n} \approx \frac{S_0}{\ln(\beta^{-1})} \sum_{n=1}^{10} \frac{1}{n} \approx \frac{2.93 S_0}{\ln(\beta^{-1})},$$

and using equation (3.58) it becomes:

$$S(n \leqslant 10) \approx 2.93 \beta N_0 \tag{3.61}$$

It is known that the sum of the inverses of the first m integers is approximately given by $\gamma_{EM} + \ln(m)$, where $\gamma_{EM} = 0.577$ is the Euler–Mascheroni constant. Thus, we can further approximate and obtain:

$$S(n \leqslant m) \approx (0.577 + \ln(m)) \beta N_0, \tag{3.62}$$

which is a good approximation for $m \geqslant 10$. Notice that the predicted number of singleton species (species with $n = 1$) is just βN_0.

Contrasting Harte's METE against data

Harte's METE links together energetics, diversity, abundance, and spatial scaling to predict relatively accurately and without adjustable parameters the species-abundance distribution SAD (equation (3.49)), the spatial distributions of individuals within each of the species (equation (3.53)) and the abundance–energy relationship AER (Harte 2011). METE also does a decent job predicting the species–area relationship SAR (Harte 2011).

In particular, SAD $\Phi(n)$ is predicted to be a ***Fisher logseries distribution***: $\Phi(n) = (c/n)\exp(-\beta n)$; we will compare against measurements for different plant communities. Reliable prediction of the shape of this metric has been a central goal of macroecology. A general observation across a wide variety of ecosystems is their uneven abundance of species: some species are very common and others quite rare. This poses a difficulty concerning the graphical representation of such probability density functions. Namely, the shape of the distribution one obtains can depend on the way the binning intervals are chosen (Bulmer 1974). To avoid the ambiguity caused by having to select a binning interval, datasets providing empirical frequency distributions are often graphed in the form of rank-variable relationships. That is, one considers the list of abundances of the species in a plot in which there are S_0 species and N_0 individuals and arranges them in tabular form from most to least abundant. The most abundant species has a rank of 1, and the highest rank is the number of species S_0 (in cases with two species having the same abundance, each of them are ranked). Plotting abundance or frequency against rank is not particularly enlightening. But when log(abundance) or log(frequency) is plotted against rank, a simple pattern emerges: a straight line. The reason for this stems from the fact that often there are many species with quite low abundances and very few with very high abundance. Therefore, it is generally more instructive to avoid the ambiguity caused by having to select a binning interval, and instead plot the SAD in the format of a rank-abundance graph. Figure 3.4 shows observed versus predicted SADs for different plant communities (Harte 2011). Table 3.2 summarizes data for these sites.

Remark: In thermodynamics, temperature and pressure are termed 'intensive' variables in the sense that their values do not depend on the amount of the substance for which it is measured. For example, the temperature of a system in thermal equilibrium is the same as the temperature of any of its parts. In other words, in a combined system intensive variables are each the weighted averages of their values in the component systems. In contrast, volume and energy are termed 'extensive' variables because their values depend on the amount of substance and then are additive when systems are combined. In METE, A_0, N_0, and E_0 are extensive variables, but S_0 is intermediate between an intensive and an extensive variable; it neither adds linearly nor is it averaged when systems are adjoined and thus has no analogy in thermodynamics.

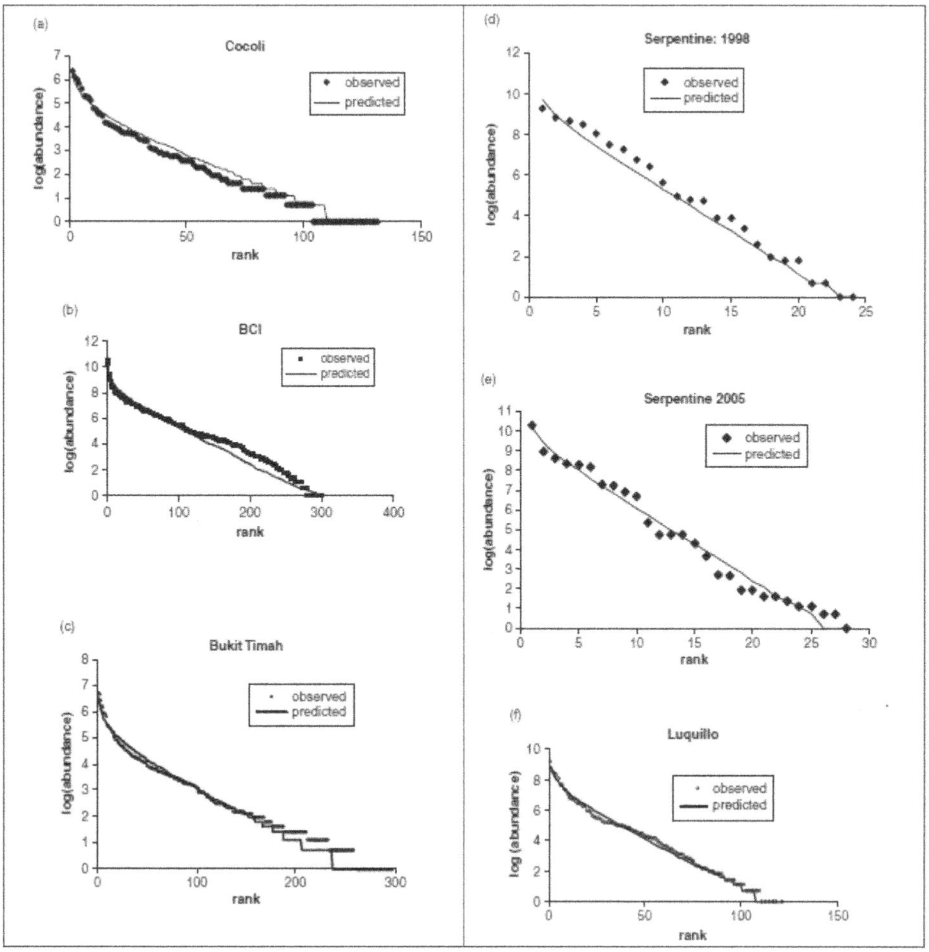

Figure 3.4. SAD: Observed versus predicted by METE for different plant communities. Reproduced from Harte 2011, copyright of OUP 2011.

Table 3.2. Properties of the six sites used to evaluate METE predictions.

	Location (year)					
Property	BCI (2005)	Bukit Timah	Coccoli	Luquillo	Serpentine (1998)	Serpentine (2005)
A_0	50 ha	2 ha	2 ha	16	0.0064 ha	0.0064 ha
S_0	302	296	130	137	24	28
N_0	213 791	11 843	4807	67 465	37 182	60 346
Variance in observed rank-log (abundance) explained	0.98	n.a.	0.97	0.99	n.a.	0.99

3.4 Neutral theories of ecology

A closely related theory to METE is Stephen Hubbell's ***Neutral Theory of Ecology*** (NTE) (Hubbell 2001). According to NTE the differences between members of an ecological community of trophically similar species are 'neutral', or irrelevant to their success since species do not interact with one another or with their environment in any explicit differentiated way. The different observed outcomes for the number of species and their differing abundances result only from stochastic birth, death, migration, and speciation events. So at the outset, species are not distinguished, but stochastic demographic rates result in differences in species' abundances. That is, biodiversity arises at random, as each species follows a random walk. The hypothesis has sparked controversy (McGill 2003, Wootton 2005, Clark 2008). Additionally, NTE can explain why some of the species of trees or birds in the forest are very abundant, while others are very rare, but because neutral theory is blind to traits, it cannot explain why colorful birds are rarer than drab birds (Harte 2011), etc.

The NTE has several fundamental tunable parameters: a speciation rate, a measure of the size of the community (a carrying capacity), and what is effectively a time constant representing an inverse birth rate. The NTE predicts SADs at two different scales. One is for the ***metacommunity***, i.e. the large collection of species and individuals that provide the pool of entrants into the plot being modelled. The other SAD is for the *local* community under dispersal limitation from the metacommunity, i.e. is for any particular plot that receives immigrants from the metacommunity. The predicted metacommunity SAD is the logseries distribution, also predicted by METE. On the other hand, the predicted plot SAD is the so-called ***zero-sum multinomial*** distribution. It describes quite well the SAD observed in the 50 ha BCI tropical forest plot (Hubbell 2001).

Harte (2011) explains very clearly the similarity between METE and NTE: 'METE is a neutral theory in the sense that, by specifying only state variables at the community level at the outset, no differences are assumed either between individuals within species or between species (Harte 2011). Differences between individuals and species do arise in METE because the abundance and metabolic rate distributions are not uniform, but these differences emerge from the theory; they are not assumed at the outset. Similarly, in the Hubbell NTE, differences between species arise from stochastic demographics; no *a priori* differences are assumed.' Besides, at a more technical level, additional points of similarity exist (see Harte 2011).

An interesting question is: why do theories that make such manifestly incorrect assumptions, like neutrality and no interactions, as do METE and NTE, work to the extent they do? A possible answer is that because there are so many ways in which species differ, the successes of neutral theories arise for reasons analogous to why statistical mechanics work: real molecules are not points objects that collide perfectly elastically, but the huge number of ways in which violations of those assumptions influence molecular movements average out somehow. In other words, neutral theories in ecology are a first approximation to reality. In the same way that ideal gases do not exist, neither do neutral communities. Similar to the kinetic theory of ideal gases in physics, neutral theories, like METE and NTE, provide the essential ingredients to further explore theories that involve more realistic assumptions. For instance, it is possible that the assertion that neutral theories are traitless needs to be revisited. Even though neutral theories assume no trait differences, they actually generate them. For example, no traits are assumed by METE but the theory predicts a distribution of metabolic rates across individuals within and across species, and thus trait differences, which can influence fitness, at the individual and species levels (Harte 2011).

3.5 Conclusion

Since Lotka proposed the idea of formulating ecology in terms of a statistical mechanics of populations almost a century ago (Lotka 1925) a growing number of ecological models based on statistical mechanics have been proposed (e.g. Kerner 1957, 1959, Leigh 1965, Hamann and Bianchi 1970, Goel *et al* 1971, Lurie and Wagensberg 1983, Matsuda *et al* 1992, Alexeyev and Levich 1997, Volkov *et al* 2004, Shipley *et al* 2006, Pueyo *et al* 2007, Harte *et al* 2008, Dewar and Port'e 2008, Volkov *et al* 2009, Banavar *et al* 2010, Fort 2013, Bertram and Dewar 2013). Indeed, these models differ considerably in both their approaches and areas of application, and Lotka's proposal remains largely unexplored (Bertram and Dewar 2015).

Statistical mechanics was originally developed to cope with non-living matter: gases, liquids, and solids. However, it has proven fruitful for the investigation of the collective dynamics of complex systems, including systems that lie beyond the scope of traditional physics. For example, what is nowadays called 'soft matter', which encompasses complex objects like colloids, membranes and biomolecules, objects which do not clearly fall into any one of the above classic categories.

Statistical mechanics has also found applications in other more distant disciplines such as economics (Lux 2007), finance (Ausloos *et al* 1999, Voit 2005) and social sciences (Castellano *et al* 2009). MaxEnt has been instrumental for this expansion outside physics. This is because MaxEnt has a wider applicability than either Boltzmann or Gibbs ensemble formulation of statistical mechanics since it neither requires the postulation of the existence of a Liouville equation, nor of a conserved quantity, nor of an ergodic hypothesis. Indeed, MaxEnt is a general inference method with a much broader scope.

In the next companion Application chapter we will use the MaxEnt derivations of subsection 3.2.3 to infer the parameters of a Lotka–Volterra model for describing the dynamics of tree species in tropical forests. In subsequent chapters we will revisit statistical mechanics approaches to model different ecosystems. For example, complex networks of interacting species (nodes) connected by interactions (edges) can be connected to Ising-like models we considered in this chapter. As another example of application of statistical mechanics to ecology we will show that the observed sudden onset of desertification, resulting from an increase in the harvesting pressure in arid ecosystems, shares many of the phenomenological features and signatures of a phase transition like boiling water. Indeed, the sudden increase of non-vegetated land cells is similar to the proliferation of bubbles in liquid water. This analogy between desertification and the onset of the liquid–gas phase transition allows us to resort to the wide panoply of quantitative techniques of statistical mechanics to explain and predict the dynamics of this phenomenon as well as to devise remedial actions to mitigate overstressed arid ecosystems.

Exercises

Exercise 3.1. An alternative way to obtain that the Lagrange multiplier of the Boltzmann distribution is $\lambda_1 = 1/k_B T$.

The entropy, energy and temperature are connected as $\partial S/\partial E = 1/T$, thus we can write

$$\frac{1}{T} = \frac{\partial(\mathcal{S}/N)}{\partial(E/N)} = \frac{\partial s}{\partial \bar{\varepsilon}},$$

where s is the entropy per molecule. Since both s and $\bar{\varepsilon}$ and are functions of $\lambda_1$we can use the chain rule for evaluating the derivative $\partial s/\partial\bar{\varepsilon}$ in such a way the above equation becomes:

$$\frac{1}{T} = \frac{\partial s}{\partial \lambda_1}\frac{\partial \lambda_1}{\partial \bar{\varepsilon}} = \frac{\frac{\partial s}{\partial \lambda_1}}{\frac{\partial \bar{\varepsilon}}{\partial \lambda_1}}.$$

Show that after some algebra we get $\lambda_1 = 1/k_B T$ and thus

$$p_i{*}=\frac{e^{-\frac{\varepsilon_i}{k_B T}}}{z}.$$

Exercise 3.2.

Use a binary tree to compute the entropy rate H/L for the 'alphabet' associated with the roll of a die {1,2,3,4,5,6}.

Exercise 3.3.

From equations (3.17), (3.24a) and (3.24b) show that the $P(\{x_i\})$ can be written as equation (3.25).

Exercise 3.4.

Compute the rates of the three messages of table 3.1 using (a) formula equation (3.17′) and (b) the average over the different branches of a binary tree.

Exercise 3.5.

Given the Lotka–Volterra linear equations:

$$\frac{dN_i}{dt} = r_i N_i \left(1 - \sum_{j=1}^{S} a_{ij} N_j\right) \qquad i = 1, \ldots, S \quad (a_{ij} \geqslant 0 \text{ for all } i \text{ and } j)$$

If $a_{ij} = -a_{ij}$ then show that the quantity $\Phi \equiv \sum_{i=1}^{S} [N_i(t) - N_i^* \ln N_i(t)]$ is conserved.

Exercise 3.6. A toy community

Consider a hypothetical community of $N = 12$ individuals distributed among $S = 6$ species as $n_1 = n_2 = n_3 = n_4 = n_5 = 1$ and $n_6 = 7$.

(A) Compare the multiplicity W and the entropy H of this partition $\{1,1,1,1,1,7\}$ with the multiplicity and entropy of an even partition $\{2,2,2,2,2,2\}$ (i.e. $n_i = 2$ for all species i).

(B) Now suppose we have partial knowledge (constraints) about the system and we want to incorporate this knowledge to infer the probability distribution $_{pi}$. We can imagine the six species community as a die with die numbers $k = \{1, 2, 3, 4, 5, 6\}$ corresponding to trait values. The average of this trait $\bar{k} = 3.5$. Use MaxEnt to obtain $_{pi}$. Could you have predicted the result you got?

Exercise 3.7.

Repeat part (B) of the previous exercise if we know that:

(A) The average $\bar{k} = 2.5$.

(B) The average $\bar{k} = 4.5$.

Plot and compare both distributions with the distribution for $\bar{k} = 3.5$.

Exercise 3.8.

Derive equations (3.40), (3.41) and (3.43).

Exercise 3.9.

Derive equations (3.44) and (3.45).

Exercise 3.10.

Derive the expression for the distribution of energies:

$$\Psi(\varepsilon) = \frac{S_0}{ZN_0}\left(-\frac{(N_0 + 1)e^{-\gamma(N_0+1)}}{1 - e^{-\gamma}} + \frac{e^{-\gamma}(1 - e^{-\gamma(N_0+1)})}{(1 - e^{-\gamma})^2}\right)$$

Exercise 3.11.

Verify equation (3.55).

Exercise 3.12.

Verify the approximations of equations (3.56), (3.57) and (3.58). (Hint: from Abramowitz and Stegun 1972 we have the following approximate formula if $\beta N_0 \gg 1$ and $\beta \ll 1$: $\sum_{n=1}^{N_0} \frac{e^{-\beta n}}{n} \approx \ln(\frac{1}{\beta})$.

References

Abramowitz M and Stegun I 1972 *Handbook of Mathematical Functions, with Formulas, Graphs, and Mathematical Tables* (New York: Dover) pp 228–37

Alexeyev V and Levich A 1997 A search for maximum species abundances in ecological communities under conditional diversity optimization *Bull. Math. Biol.* **59** 649–77

Ausloos M *et al* 1999 Applications of statistical physics to economic and financial topics *Phys.* A **274** 229–40

Banavar J R, Maritan A and Volkov I 2010 Applications of the principle of maximum entropy: from physics to ecology *J. Phys.: Condens. Matter* **22** 063101

Bertram J and Dewar R C 2013 Statistical patterns in tropical tree cover explained by the different water demand of individual trees and grasses *Ecology* **94** 2138–44

Bertram J and Dewar R C 2015 Combining mechanism and drift in community ecology: a novel statistical mechanics approach *Theor. Ecol.* **8** 419–35

Brown J H and Maurer B A 1989 Macroecology: the division of food and space among species on continents *Science* **243** 1145–50

Bulmer M 1974 On fitting the Poisson lognormal distribution to species-abundance data *Biometrics* **30** 101–10

Castellano C, Fortunato S and Loreto V 2009 Statistical physics of social dynamics *Rev. Mod. Phys.* **81** 591

Clark J S 2008 Beyond neutral science *Trends Ecol. Evol.* **24** 8–15

Cox R T 1961 *The Algebra of Probable Inference* (Baltimore, MD: Johns Hopkins Press)

Dewar R C and Port'e A 2008 Statistical mechanics unifies different ecological patterns *J. Theor. Biol.* **251** 389–403

Fort H 2013 Statistical mechanics ideas and techniques applied to selected problems in ecology *Entropy* **15** 5237–76

Gibbs J W 1902 *Elementary Principles in Statistical Mechanics* (New Haven, CT: Yale University)

Glazer M and Wark J 2006 *Statistical Mechanics. A Survival Guide* (New York: Oxford Univ. Press)

Goel N S, Maitra S C and Montroll E W 1971 On the Volterra and other nonlinear models of interacting populations *Rev. Mod. Phys* **43** 231

Hamann J R and Bianchi L M 1970 Stochastic population mechanics in the relational systems formalism: Volterra-Lotka ecological dynamics *J. Theor. Biol.* **28** 175–84

Harte J *et al* 2008 Maximum entropy and the state-variable approach to macroecology *Ecology* **89** 2700–11

Harte J 2011 *Maximum Entropy and Ecology: A Theory of Abundance, Distribution, and Energetics* (Oxford: Oxford University Press)

Hubbell S P 2001 *The Unified Neutral Theory of Biodiversity and Biogeography* (Princeton, NJ: Princeton University Press)

Jaynes E T 1957a Information theory and statistical mechanics I *Phys. Rev.* **106** 620–30

Jaynes E T 1957b Information theory and statistical mechanics II *Phys. Rev.* **108** 171–90

Kerner E H 1957 A statistical mechanics of interacting biological species *Bull. Math. Biophys.* **19** 57

Kerner E H 1959 Further considerations on the statistical mechanics of biological associations *Bull. Math. Biophys.* **21** 58

Kerner E H 1972 Gibbs ensemble: biological ensemble *The Application of Statistical Mechanics to Ecological, Neural and Biological Networks* (New York: Gordon and Breach) 167 pp

Landau L D and Lifshitz E M 1980 *Statistical Physics* 3rd revised and enlarged edn (Oxford: Pergamon)

Lotka A 1925 *Elements of Physical Biology* (London: Baillière, Tindall and Cox)

Leigh E G 1965 On the relation between the productivity, biomass, diversity of a community *Proc. Nat. Acad. Sci. U.S.A.* **53** 68

Lenz W 1920 Beiträge zum Verständnis der magnetischen Eigenschaften in festen Körpern *Phys. Z.* **21** 613–5

Lurie D and Wagensberg J 1983 On biomass diversity in ecology *Bull. Math. Biol.* **45**(2) 287–93

Lux T 2007 Applications of statistical physics in finance and economics *Economics Working Papers 2007-05* (Christian-Albrechts-University of Kiel, Department of Economics)

Matsuda H, Ogita N, Sasaki A and Sato K 1992 Statistical mechanics of population *Prog. Theor. Phys.* **88** 1035–49

Maxwell J C 1867 On the dynamical theory of gases *Phil. Trans. R. Soc. Lond.* **157** 49–88

Maynard Smith J 1978 *Models in Ecology* (Cambridge: Cambridge University Press)

McGill B J 2003 A test of the unified neutral theory of biodiversity *Nature* **422** 881–5

Pueyo S, He F and Zillio T 2007 The maximum entropy formalism and the idiosyncratic theory of biodiversity *Ecol. Lett.* **10** 1017–28

Schneier B 1996 *Applied Cryptography* 2nd edn (New York: Wiley)

Shannon C E 1948 A mathematical theory of communication *Bell Syst. Tech. J.* **27** 379–423

Shipley B *et al* 2006 From plant traits to plant communities: a statistical mechanistic approach to biodiversity *Science* **314** 812–4

Shore J E and Johnson R W 1980 Axiomatic derivation of the principle of maximum entropy and the principle of minimum cross-entropy *IEEE Trans. Inf. Theory* **26** 26–37

Voit J 2005 *The Statistical Mechanics of Financial Markets* (Berlin: Springer)

Volkov I, Banavar J R and Maritan A 2004 Organization of ecosystems in the vicinity of a novel phase transition *Phys. Rev. Lett.* **92** 218703

Volkov I, Banavar J R, Hubbell S P and Maritan A 2009 Inferring species interactions in tropical forests *Proc. Natl. Acad. Sci.* **106** 13854–9

Wootton J T 2005 Field parameterization and experimental test of the neutral theory of biodiversity *Nature* **433** 309–12

Chapter A3

Combining the generalized Lotka–Volterra model and MaxEnt method to predict changes of tree species composition in tropical forests

'The historical circumstance of interest is that the tropical rain forests have persisted over broad parts of the continents since their origins as stronghold of the flowering plants 150 million years ago.'

E O Wilson.

'In order to continue living, life must feed on order, but because there is no order—non highly organized—other than life, it is condemned to consume itself. It must destroy to live, must take its nourishment from systems that are nourishment to the extent that they can be ruined. Not ethics but physics determines this law.'

Stanisław Lem, His Master's Voice.

1. Tropical forests are mega-diverse communities for which it is difficult to obtain a sufficient sample of trees to provide an estimate of community diversity and structure, or determine the demographic parameters of individual species, in order to test ecological theory. One solution is to census large areas (i.e. tens of hectares) of forest, assuring that the sample comprises a reasonable number of individuals of rare species in addition to common species. Such large tropical Forest Dynamics Plots (FDPs), repeatedly censused over the long term play a significant role in developing theory to explain the species diversity and dynamics of tropical forests.

doi:10.1088/978-0-7503-2432-8cha3

2. Here we consider the 50 ha FDP located in Barro Colorado Island (BCI), in Panama. In this plot all trees and shrubs with diameter at breast height (dbh) of 1 cm or more have been identified and measured for diameter, and their locations mapped repeatedly across eight censuses: in 1981, 1985, and then at 5 year intervals until 2015. This painstaking effort has provided a goldmine of data on spatial and temporal dynamics critical to testing theoretical explanations for high species diversity in tropical forests: the total number of trees $N > 200\ 000$, covering $S > 300$ species, than were mapped and measured. We focus on the 20 most abundant tree species, that make up more than 60% of the tree community.
3. We divide the plot into quadrats of side l (e.g. $l = 10$ m). Thus, associated with each quadrat we have a vector of abundances $[n_1, n_2,\ldots,n_{20}]$, where n_i is the abundance of species i in this quadrat. Then we compute the first and second statistical moments (averages and covariances) and, using the results provided by MaxEnt in section 3.2.3, we obtain the corresponding Lotka–Volterra interaction matrix of tree species for the eight censuses.
4. We could use these matrices to compute the change in species abundances by means of the linear Lotka–Volterra generalized equations (LLVGE) of chapter 2. However, we find that the pairwise interspecific interaction strengths are much weaker than the intraspecific interactions. Therefore, following the parsimonious approach we have promoted throughout this book, we 'turn off' the interspecific interactions and treat each species as isolated and governed by a logistic equation.
5. To estimate the logistic parameters for each species we proceed as is customary in time series forecasting to assess the predictive performance of a model on new data: we partition the data into an earlier training period and a later validation period. We thus use the training period to estimate the parameters.
6. Once we have the logistic parameters for each species we substitute them into Lotka–Volterra finite difference equations to generate forecasts to be compared with the empirically measured species abundances corresponding to the validation period. It turns out that this method is able to predict the abundances of the 20 selected species with remarkable accuracy.
7. Finally, we check that taking into account the interspecific interactions we neglected does not substantially improve the accuracy of the predicted abundances.

A3.1 Background information

Tropical forests represent a small portion of the Earth's surface—two percent—but contain, together with reefs, most of the planet's biodiversity (Zimmerman *et al* 2008). Tropical forest communities are hyper-diverse. For example, a single hectare of Amazonian forest can support more than 280 tree species (Valencia *et al* 1994). In addition, in most of them only few species are common and many species are rare. In such communities it is difficult to obtain a sufficient sample of trees to provide an estimate of community diversity and structure, or determine the demographic parameters of individual species, in order to test ecological theory. One solution the ecologists have found is to census large areas (i.e. tens of hectares) of forest, assuring that the sample comprises a reasonable number of individuals of rare species in addition to common species. Such large tropical Forest Dynamics Plots (FDPs), repeatedly censused over the long term play a significant role in developing theory to explain the species diversity and dynamics of tropical forests.

In the FDPs belonging to the network coordinated by the Center for Tropical Forest Science (CTFS) all trees and shrubs with diameter at breast height of 1 cm or more are identified and measured for diameter, and their locations are mapped in plots that range in size from 2 to 52 ha (Zimmerman *et al* 2008). By virtue of their large size, the FDPs present a comprehensive picture of relative species abundances and species distribution in tropical forests, including the contribution of rare species. Interestingly, the plots have demonstrated that the shape of the species abundance curves is similar in a wide geographical and structural range of tropical forests. Mapped locations of trees and repeated censuses (usually at 5 year intervals) have provided a massive amount of data on spatial and temporal dynamics making it possible to parameterize multispecies models. Something that is critical to both testing theoretical explanations for high species diversity in tropical forests as well as to predict the fate of individual species.

An open question is whether or not tropical forests are in equilibrium. There are two major perspectives regarding species coexistence; equilibrium ***niche partitioning*** versus non-equilibrium stochastic ***dispersal-assembly*** (Clark and McLachlan 2003, Ishida *et al* 2003). According to the first hypothesis, in a community at equilibrium each species occupies a different niche that results from and reduces direct competition (Whittaker 1975). ***Stabilizing mechanisms***—like tradeoffs between species in terms of their capacities to disperse to sites where competition is weak, to exploit abundant resources effectively and to compete for scarce resources (Clark and McLachlan 2003)—play an important role. Alternatively, the dispersal-assembly hypothesis assumes that communities are open non-equilibrium assemblages of species that coexist only transiently though by chance, history, and random dispersal rather than by the stabilizing effects of niche differentiation, regarded as negligible (Hubbell 2008, 2009). This 'neutral model' thus emphasizes ***'equalizing' mechanisms*** (Chesson 2000), because competitive exclusion of similar species is slow. The relative importance of these two mechanisms is unknown. Therefore, two debates are intertwined: On the one hand, the competition versus neutral theory debate and, on the other hand, the debate over whether communities are or are not in equilibrium.

At any event, the BCI forest has exhibited considerable dynamism since the Forest Dynamics Project began (Hubbell 2011). A variety of population trajectories can be observed. Some species suffered steep population declines during the period 1982–1990 (probably as a result of severe droughts during that period), with most, but not all, then recovering (Condit *et al* 2017). So while the forest may have returned to a period of relative demographic stability after 1990, many species' abundances fluctuated over 35 years with no known cause (Condit *et al* 2017).

A3.2 Overview

What is our goal? The general goal is to develop quantitative tools for predicting the dynamics of multispecies communities of trees in tropical forests. This would be helpful to address a broad array of questions that are of particular importance to tropical systems. Including the mechanisms responsible for large-scale patterns of species abundance and distribution, species coexistence and the maintenance of the vast species diversity. These issues are not only important for the advancement of tropical ecology, but are crucial for our overall understanding of basic ecology in any system.

What do we want to know? We want to predict the dynamics of the tree species abundances in tropical forests. This is not merely of theoretical interest from the perspective of species coexistence, but rather is first and foremost essential for sustainable management, biodiversity conservation, and forest restoration.

What do we know? We have the species abundances of trees from eight censuses carried out from 1981 to 2015 in the 50 ha plot of BCI every five years.

What can we assume? We will assume that pairwise interactions are enough to describe the interspecific interactions among tree species.

How should we look at this model? Here we show that intraspecific competition allows us to explain the fluctuations of the abundances of the 20 most abundant species in the BCI plot. We partition the time series of eight available censuses into a training period, used to estimate the growth rates and carrying capacities of species, and a later validation period in which we check how well the estimated logistic parameters actually predict these later observed abundances. Our approach, which works independently of whether the ecosystem is in a steady state or not, is able to predict the abundances of the 20 selected species with remarkable accuracy. Additionally, using a MaxEnt approach like in Volkov *et al* (2009) we obtain the 20×20 interspecific interaction matrix for each census. We find that the pairwise interspecific interaction strengths are weak compared with the intraspecific interactions. Therefore, taking into account these interspecific interactions, through Lotka–Volterra equations, does not substantially improve the accuracy of the predicted abundances.

What will our model predict? It will predict the species abundances in future censuses.

Are the predictions valid? We will test our model quantitatively by comparing its predictions against empirical data.

How can we improve the model? At the end of this chapter we will discuss some caveats and possible further developments.

A3.3 Data for Barro Colorado Island (BCI) 50 ha tropical Forest Dynamics Plot

A3.3.1 Some facts about BCI

As we have seen, the BCI forest has exhibited considerable dynamism since the early eighties. However, some global quantities, defined for the whole population of trees, seem to have varied little along the 35 years across the eight censuses.

Conserved and non conserved quantities

Actually, which quantities remain steady or not depends on the threshold dbh we consider. A main threshold is dbh = 10 cm, which separates trees (⩾ 10 cm dbh) from saplings (1–10 cm dbh) (Condit *et al* 2017). For instance, for dbh ⩾ 1 cm the total number of trees, N, exhibits a drop of almost 15% from 1990 to 2005 and then the population seems to have more or less stabilized along the following three censuses (figure A3.1). On the other hand, for dbh ⩾ 10 cm (i.e. taking into account only trees) the total abundance N oscillates in a narrow band of less than 4% (figure A3.1).

Figure A3.1 also shows different quantities that seem to be conserved for dbh ⩾ 1 cm: Cover C_t = covering of trunks (discs of diameter = dbh), C_c = covering of crowns (discs of area = 0.234 dbh$^{1.43}$), total above-ground biomass (AGB) (proportional to the volume, which scales as dbh$^{8/3}$), Shannon equitability:

$$H = = -\sum_{i=1}^{S} p_i \ln(p_i)/ \ln(S),$$

with $_{pi} = N_i/N$.

The constancy of the total AGB is in agreement with calculations that estimated that the total AGB for the Barro Colorado forest is 281 ± 20 ton ha^{-1} with a 15 year

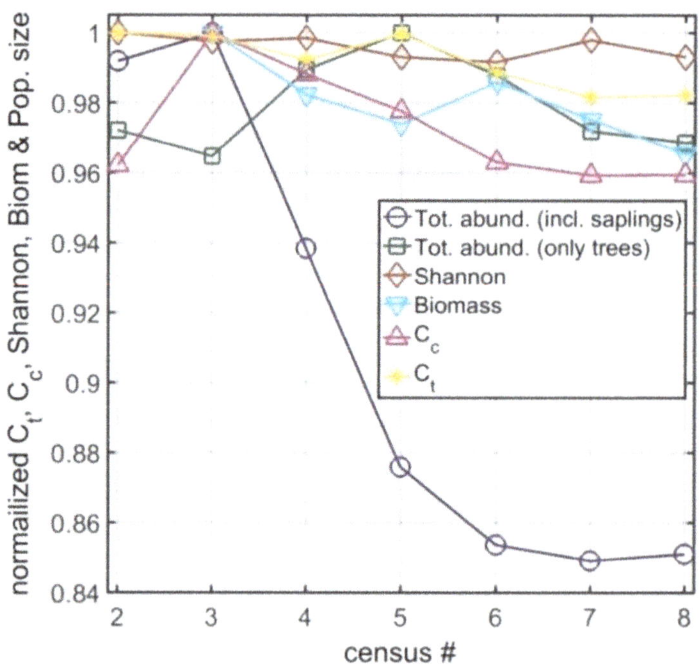

Figure A3.1. Empirical quantities for dbh ⩾ 1 cm for the eight censuses, except squares which correspond to dbh ⩾ 10 cm and includes only trees (no saplings). Census # 2 corresponds to 1985 and the following censuses are separated by 5 years, until census # 8 for 2015.

(1985–2000) average for the annual variation of +0.20 Mg ha^{-1} $year^{-1}$, but with a confidence interval that spanned zero (−0.68 to 0.63 Mg ha^{-1} $year^{-1}$) (Chave *et al* 2003).

Regarding species composition, a variety of population trajectories can be observed. Some species suffered steep population declines during the period 1982–1990 (probably as a result of severe droughts during that period), with most, but not all, recovering afterwards. Figure A3.2 shows these trajectories for the 20 most abundant species of trees with dbh $\geqslant$ 1 cm at the second census in 1985 (table A3.1).

Notice that figure A3.2(a) shows that for dbh $\geqslant$ 1 cm there are several species that did not reach yet a steady abundance, while for dbh $\geqslant$ 2.5 cm the steady assumption is more justified, still, at least seven species have clearly not yet reached a steady abundance in figure A3.2(b). Therefore, the BCI forest is not at equilibrium in terms of species composition.

A3.3.2 Covariance matrices and species interactions

Let us consider the 20 most abundant species for trees of dbh $\geqslant$ 1 cm at the second census (1985) listed in table A3.1. Next, we use the procedure of Volkov *et al* (2009) of dividing the 50 ha BCI plot into equally sized quadrats of side L, in such a way that one can assign to each quadrat a vector of 20 abundances $\mathbf{n} = [n_1, \ldots, n_{20}]$, and averaging over these quadrats, determine mean species abundances and their covariances.

The covariance is a measure of the joint variability of two random variables. In other words, this metric evaluates how much—to what extent—the variables change together. Unlike variance, which is non-negative, covariance can be negative or positive (or zero, of course). If the greater values of one variable mainly correspond with the greater values of the other variable, and the same holds for the lesser values, (i.e. the variables tend to show similar behavior), the covariance is positive. In the

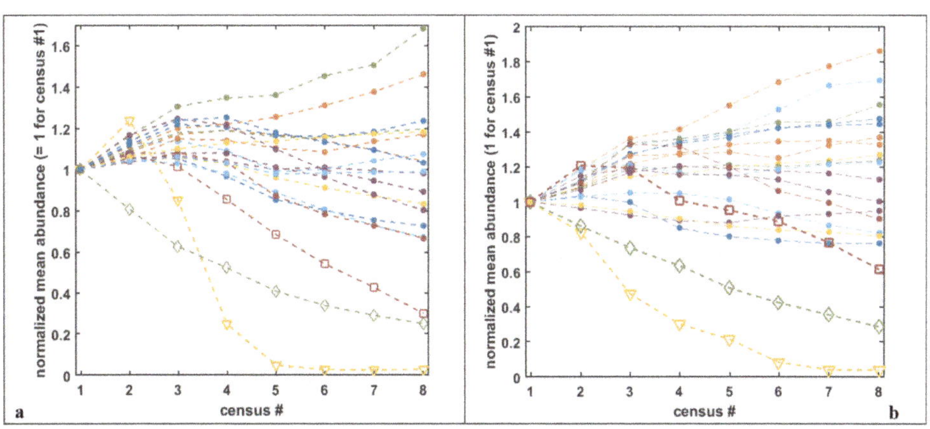

Figure A3.2. Normalized empirical mean abundances of the 20 most abundant species for the eight censuses. (a) dbh $\geqslant$ 1 cm. (b) dbh $\geqslant$ 2.5 cm.

Table A3.1. The 20 most abundant species at census # 2.

Species #	Name	Abbrev.	Abundance
1	*Hybanthus prunifolius*	hybapr	42 041
2	*Faramea occidentalis*	faraoc	25 464
3	*Trichilia tuberculata*	tri2tu	13 435
4	*Desmopsis panamensis*	des2pa	12 362
5	*Alseis blackiana*	alsebl	8328
6	*Mouriri myrtilloides*	mourmy	7788
7	*Psychotria horizontalis*	psycho	6620
8	*Hirtella triandra*	hirttr	4720
9	*Garcinia intermedia*	gar2in	4064
10	*Piper cordulatum*	pipeco	3898
11	*Capparis frondosa*	cappfr	3823
12	*Tetragastris panamensis*	tet2pa	3816
13	*Sorocea affinis*	soroaf	3453
14	*Tachigali versicolor*	tachve	3028
15	*Protium tenuifolium*	protte	2917
16	*Protium panamense*	protpa	2911
17	*Swartzia simplex*	swars2	2882
18	*Beilschmiedia pendula*	beilpe	2776
19	*Poulsenia armata*	poular	2766
20	*Rinorea sylvatica*	rinosy	2685

opposite case, when the greater values of one variable mainly correspond to the lesser values of the other, (i.e. the variables tend to show opposite behavior), the covariance is negative. The diagonal elements of the covariance of abundances matrix **c**, i.e. the intraspecific variances, are all large and positive, reflecting the clumping of single-species populations. The interspecific covariances are generally much smaller than the intraspecific variances, by approximately an order of magnitude, and can be both positive or negative (figure A3.3 and table A3.2 for $L = 20$ m).

A3.4 Modelling

A3.4.1 Inference of the effective interaction matrix from the covariance matrix via MaxEnt

From the covariance of abundances matrix, through equation (3.32), we obtain for each census c an effective interaction matrix, $\mathbf{J}^{(c)}$, given by:

$$\mathbf{J}^{(c)} = -\mathbf{C}^{(c)^{-1}}, \tag{A3.1}$$

whose diagonal elements are negative. To obtain the Lotka–Volterra interaction matrix $\mathbf{A}^{(c)}$, with -1 along the diagonal, we divide each row i of the $\mathbf{J}^{(c)}$ matrix by the absolute value of J_{ii}:

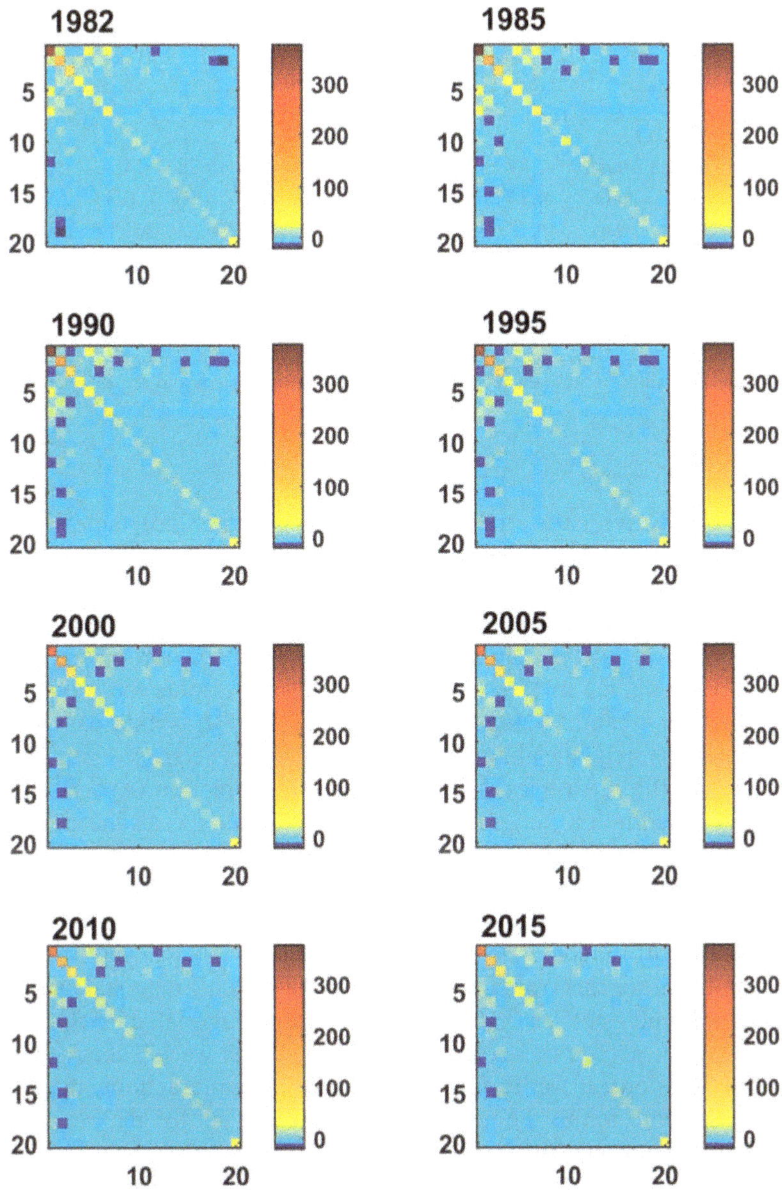

Figure A3.3. Covariance $\mathbf{C}^{(c)}$ of abundances matrices for the twenty most abundant species (dbh $\geqslant$ 1 cm) for each of the eight BCI censuses ($c = 1,\ldots,8$). Horizontal and vertical numbers correspond to the species number. The quadrats used to compute these covariance matrices were of side L = 20 m.

Table A3.2. Species with dbh ⩾ 1 cm in quadrats of $L = 20$ m. Mean, min and max, both for the intraspecific variance (diagonal terms of the covariance of abundances matrix **C**) and interspecific covariance (off-diagonal elements of **C**).

		census #							
		1	2	3	4	5	6	7	8
Intraspecific variances	**mean**	17.49	19.20	18.92	17.37	15.28	14.25	13.04	12.43
	min	4.74	5.17	5.09	1.47	0.16	0.09	0.09	0.10
	max	340.23	375.10	360.97	323.40	280.65	263.15	240.99	234.99
Absolute[a] value of interspecific covariances	**mean**	2.489	2.758	2.569	2.186	1.823	1.658	1.536	1.490
	min	0.003	0.006	0.005	0.002	0.001	0.004	0.006	0.003
	max	27.579	28.939	24.620	22.575	20.221	17.184	17.516	16.668

[a] We take the absolute value because interspecific covariances take both signs and are, on average, close to 0.

$$\alpha_{ij}^{(c)} \equiv J_{ij}^{(c)}/|\, J_{ii}^{(c)}\, |. \tag{A3.2}$$

We stress the effective nature of this interaction matrix since there is no simple relationship in this 'many-body' system between the covariance of abundances for a pair of species and their interaction. For example, two species could have a weak interaction and yet a very strong correlation because they are both strongly interacting with a third species. Similar emergent behavior occurs when one considers more complex interactions of multiple species plus the influence of the environment they inhabit. A negative interaction ($\alpha_{ij} < 0$) is akin to 'competition' between species, namely, the abundance of the ith species decreases with the increase of abundance of the jth species. Likewise, positive interaction ($\alpha_{ij} > 0$) coefficients correspond to 'facilitation', i.e. the abundance of the ith species increases with the increase of the jth species' abundance.

Similarly to what we observed for the covariance matrices, the diagonal elements of the $\mathbf{A}^{(c)}$ matrices, corresponding to intraspecific interactions strengths, are much greater than the off-diagonal elements, corresponding to interspecific interactions strengths. You can see this from figure A3.4: most of the off-diagonal elements have absolute values much smaller than the -1s along the diagonal. Therefore, as a first approximation, we neglect these off-diagonal terms (later we will check that this is a sound approximation). If we 'turn off' the interspecific interactions, the $\mathbf{A}^{(c)}$ matrix transforms itself into minus the identity matrix, i.e. we get for the evolution of the abundance of each species the logistic equation (1.9).

A3.4.2 Model equations

We want to predict the change in the abundances of the 20 most abundant tree species in the BCI plot. As we have just seen, neglecting the interspecific interactions, this implies considering a set of 20 logistic equations:

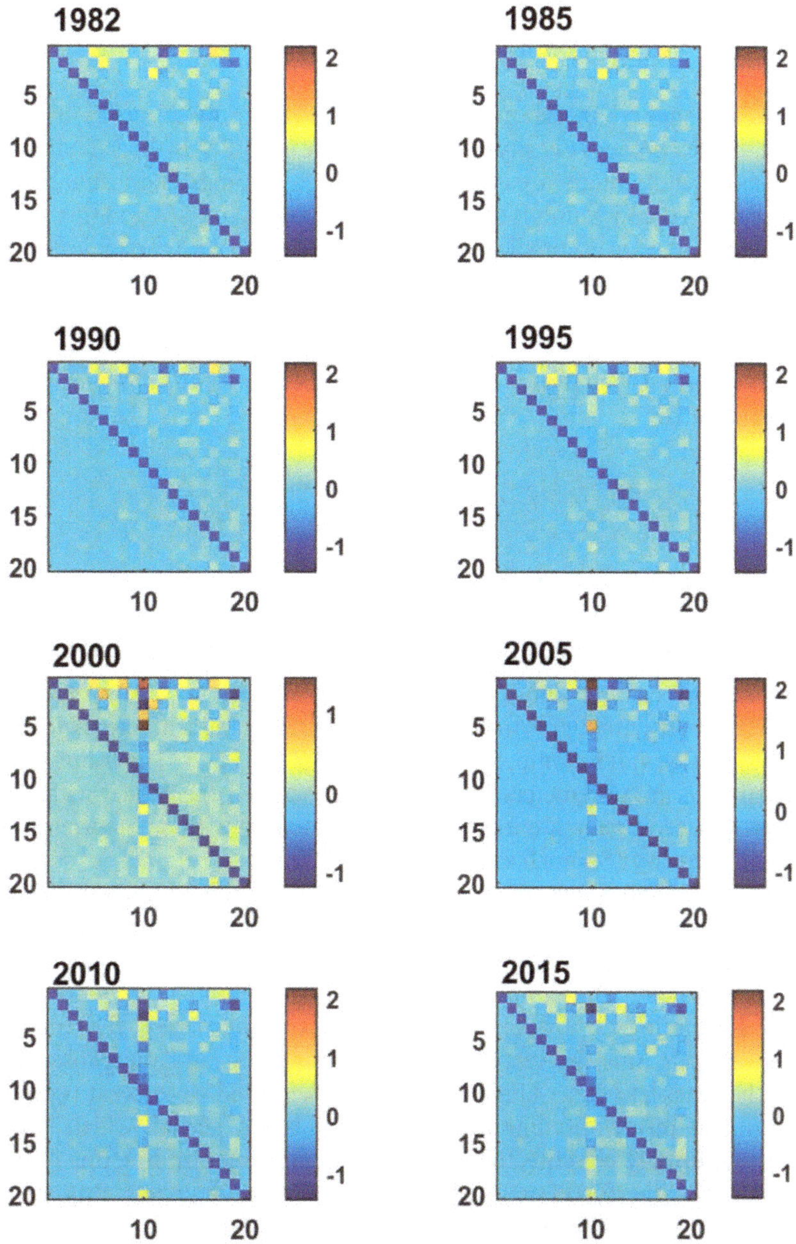

Figure A3.4. Interaction matrices $\mathbf{A}^{(c)}$ for the twenty most abundant species (dbh $\geqslant$ 1 cm) for each of the eight BCI censuses ($c = 1,\ldots,8$). These matrices were obtained as follows: first we compute $\mathbf{J}^{(c)} = -\mathbf{C}^{(c)^{-1}}$ and then, normalizing the diagonal elements of $\mathbf{J}^{(c)}$ to -1 we get $\mathbf{A}^{(c)}$. Horizontal and vertical numbers correspond to the species number. The quadrats used to compute the covariance matrices were of side L = 20 m.

$$\frac{dn_i}{dt} = r_i n_i \left(1 - \frac{n_i}{\mathcal{K}_i}\right) \qquad i = 1, \ldots, 20, \tag{A3.3}$$

where i denotes the tree species number, r_i is its intrinsic growth rate (with dimension time^{-1}), n_i is the species i density (individuals/area), $\mathcal{K}_i > 0$ is the carrying capacity (dimensions of individuals area^{-1}). We are not assuming equilibrium, rather we know the species composition is changing from census to census and this variation is precisely what we want to predict. Thus, besides the carrying capacities $\mathcal{K}_i$ we need to estimate the intrinsic growth rates r_i. In the next subsection we will show how to obtain all the model parameters.

Since the time interval for our data is five years, we will use instead of the differential equations equations (A3.3), the corresponding *finite difference* equations, which are given by:

$$n_i(c+1) - n_i(c) = r_i n_i(c)\left(1 - \frac{n_i(c)}{\mathcal{K}_i}\right), \tag{A3.4}$$

A3.5 Model validation using time series forecasting analysis

In time series forecasting to assess the predictive performance of a model on new data, it is customary first to partition the data into an earlier training period ($t = 1, 2, \ldots, T_{tr}$) and a later validation period ($t = T_{tr} + 1, T_{tr} + 2, \ldots$). The training period is used to estimate the parameters of the model, and then the model with estimated parameters is used to generate forecasts to be compared with data corresponding to the validation period (Shmueli and Lichtendahl 2016).

Varying the training period T_{tr}, i.e. the extents of the training and validation partitions, is generally used to develop multiple models. The validation partition is used to assess the performance of each model so that we can compare models and pick the best one.

Before attempting to forecast future values of the series, the training and validation periods must be recombined into one series, and the chosen model is rerun on the complete data. This final model is then used to forecast future values. The advantages in recombining are: (1) the validation period, which is the most recent period, usually contains the most valuable information in terms of being the closest in time to the forecasted period;. (2) with more data, some models can be estimated more accurately (Shmueli and Lichtendahl 2016).

A3.5.1 Estimation of intrinsic growth rates and carrying capacities using a training set of data

The recipe to obtain the intrinsic growth rates and the carrying capacities, arranged into 20×1 column vectors $\mathbf{r} = [r_i]$, $\mathbf{k} = [\mathcal{K}_i]$, is shown in box A3.1.

Box A3.1. The recipe to obtain the intrinsic growth rates and the carrying capacities.

1. We first rewrite equation (A3.4) as:

$$\frac{n_i(c+1) - n_i(c)}{n_i(c)} = -\frac{r_i}{\mathcal{K}_i} n_i(c) + r_i \qquad \text{(A3.4')}$$

2. Now, for each species i, we can use the empirical densities for the training period ($c = 1, 2, \ldots, T_{tr}$), which thereafter we will denote as n_i^e with the superscript e to distinguish them from the theoretical ones, to evaluate both sides of equation (A3.a'). Let us define:

$$y_i(c) \equiv \frac{n_i^e(c+1) - n_i^e(c)}{n_i^e(c)}, \qquad \text{(A3.5)}$$

and $x_i(c) \equiv n_i^e(c)$, in such a way that equation (A3.4') becomes the equation of a straight line with slope $r_i/\mathcal{K}_i$ and intercept r_i.

3. For each species i, we have eight values of $n_i^e(c)$ ($c = 1, \ldots, 8$). We use the first T_{tr} values, say $T_{tr} = 4$ (i.e. $c = 1, \ldots, 4$), as a training set to perform a least square fitting to obtain the parameters r_i and $\mathcal{K}_i$. With these four values we get three points in the x_i–y_i plane (remember that the numerator of y_i in equation (A3.5) involves a difference between consecutive censuses) as shown in figure A3.5. In this way we obtain the vector parameters **r**and **k**(see table A3.3).

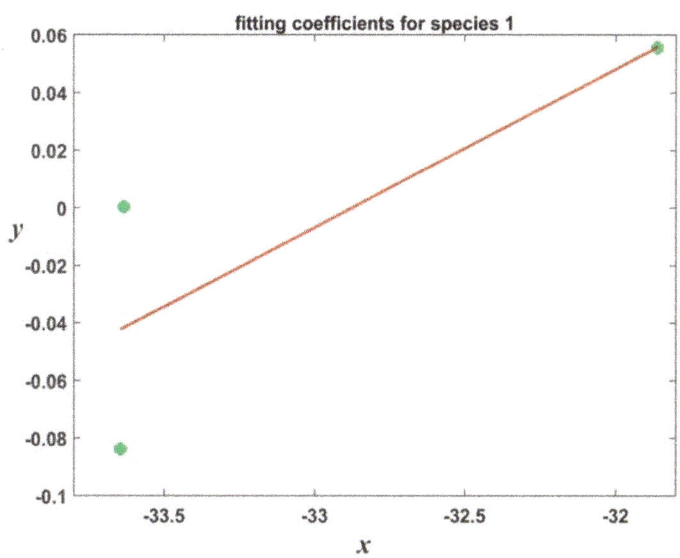

Figure A3.5. Illustration of the method for estimating logistic parameters for species 1, *Hybanthus prunifolius* (hybapr). The coordinates of green circles, x and y, were obtained from empirical data (see box A3.1) and the red line is the least squares fitting.

Table A3.3. Estimated values for intrinsic growth rates and carrying capacities when using $T_{tr} = 4$ censuses, $T_{tr} = 5$ censuses.

Species abrev.	$T_{tr} = 4$		$T_{tr} = 5$	
	r (y^{-1})	$\mathcal{K}$ (trees/ha)	r (y^{-1})	$\mathcal{K}$ (trees/ha)
hybapr	0.361	822	−0.078	896
faraoc	0.100	579	0.124	559
tri2tu	0.222	269	0.187	264
des2pa	0.225	247	0.221	244
alsebl	0.116	185	0.142	180
mourmy	0.238	155	0.243	151
psycho	0.147	119	−0.151	137
hirttr	0.133	106	0.168	102
gar2in	0.119	91	0.111	93
pipeco	−0.194	88	−0.194	88
cappfr	0.217	76	0.268	75
tet2pa	0.119	92	0.125	91
soroaf	0.323	68	−0.095	71
tachve	0.212	62	0.171	59
protte	0.143	64	0.164	62
protpa	0.114	64	0.146	62
swars2	0.086	63	0.094	63
beilpe	0.185	59	0.220	57
poular	−0.028	−144[a]	−0.040	7504
rinosy	0.210	54	0.206	53

[a] Warning: carrying capacities must be positive.

From table A3.3 we observe that most of the species estimated intrinsic growth rate coefficients r_i are positive (although this is not required). We can also see that all carrying capacities $\mathcal{K}_i$ estimated when using $T_{tr} = 4$ and 5 training censuses are very similar, with the exception of *Poulsenia armata* (poular). For this species the method for estimating parameters fails since it produces a negative carrying capacity when using $T_{tr} = 4$ training censuses, while the carrying capacity of a logistic equation must be positive by definition. An additional warning signal for this species is that for $T_{tr} = 5$ censuses the estimated $\mathcal{K}$ is very large and positive. Therefore, predictions of the abundance for this species are less trustable, although the relative errors are quite small (see table A3.4).

A3.5.2 Generating predictions to be contrasted against a validation set of data

Once we obtained the model parameters, vector parameters **r** and **k**, from the training set of data we can compare the theoretical abundances generated by the model equations against the abundances of the validation set. Starting with

Table A3.4. Relative errors ε as % for the predicted abundances using T_{tr} = 4 censuses, T_{tr} = 5 censuses, T_{tr} = 6 censuses and T_{tr} = 7 censuses for training.

Species abrev.	$T_{tr} = 4$				$T_{tr} = 5$			$T_{tr} = 6$		$T_{tr} = 7$
	c #5	c #6	c #7	c #8	c #6	c #7	c #8	c #7	c #8	c #8
hybapr	26.0	23.4	40.3	38.5	3.6	9.9	20.6	1.5	6.3	3.5
faraoc	3.5	7.3	12.1	18.2	3.7	8.3	14.2	5.8	12.0	8.9
tri2tu	8.7	14.3	19.4	25.0	11.3	17.0	22.5	1.4	2.7	0.1
des2pa	5.6	4.3	6.1	7.3	3.3	4.8	6.1	3.8	5.4	3.4
alsebl	4.0	5.2	3.0	1.5	1.9	0.1	1.5	1.1	2.3	1.8
mourmy	13.3	11.6	7.1	3.9	9.9	4.2	1.3	0.8	0.7	1.1
psycho	35.6	76.0	125.3	221.3	10.3	34.4	59.2	10.1	19.1	2.0
hirttr	7.4	12.4	16.5	23.6	7.1	11.5	18.3	8.5	16.1	12.3
gar2in	1.3	5.0	9.3	14.4	3.7	8.1	13.2	4.5	9.9	5.1
pipeco	0.0	8.9	10.0	10.0	8.9	10.0	10.0	6.1	8.8	4.9
cappfr	6.7	12.1	23.0	34.6	11.8	20.6	32.7	15.0	28.2	4.9
tet2pa	1.4	4.1	7.0	16.5	5.3	8.3	17.7	3.8	13.9	10.7
soroaf	17.5	24.9	41.6	51.6	1.2	0.8	9.0	4.0	15.4	6.3
tachve	20.9	34.3	44.2	57.7	23.7	36.7	50.3	3.2	6.5	0.2
protte	4.7	5.6	3.6	0.8	2.6	1.0	3.3	0.1	4.0	4.0
protpa	4.6	6.8	2.3	0.7	3.3	0.8	3.7	2.4	4.8	3.9
swars2	0.6	0.2	1.0	1.6	1.0	1.9	2.6	0.8	1.5	0.7
beilpe	11.8	23.5	30.3	38.2	19.3	25.7	33.4	17.3	28.3	12.4
poular	5.3	5.6	3.3	0.6	3.5	9.7	16.2	4.4	9.1	3.0
rinosy	7.0	6.5	6.1	5.8	4.7	4.4	4.1	1.6	2.2	0.9
Mean of $\|\varepsilon\|$	**9.3**	**14.6**	**20.6**	**28.6**	**7.0**	**10.9**	**17.0**	**4.8**	**9.9**	**4.5**

$n_i(T_{tr}) = n_i^e(T_{tr})$ the theoretical abundances are generated through equation (A3.4) as:

$$n_i(c+1) = n_i(c) + r_i n_i(c)\left(1 - \frac{n_i(c)}{\mathcal{K}_i}\right), \qquad c = T_{tr}, \ldots, 8.$$

Therefore, the relative percentage errors between $n_i(T_{tr}+1),\ldots,n_i(8)$ and the empirical ones are given by:

$$\varepsilon_i(c) = 100 \times (n_i(c)/n_i^e(c) - 1), \qquad c = T_{tr+1}, \ldots, 8. \qquad \text{(A3.6)}$$

A3.5.3 Verdict: model validated

Table A3.4 shows the relative errors of equation (A3.6) for the predicted abundances. We can see that increasing the training period from T_{tr} = 4 censuses (i.e.

Table A3.5. Predicted abundances for 2020.

Species	Predicted abund. 2020
hybapr	27 262
faraoc	25 683
tri2tu	10 221
des2pa	11 697
alsebl	9028
mourmy	7448
psycho	1228
hirttr	4587
gar2in	5434
pipeco	71
cappfr	2570
tet2pa	5564
soroaf	1952
tachve	1762
protte	3142
protpa	3154
swars2	3216
beilpe	2162
poular	729
rinosy	2573

predictions for censuses # 5 to 8) to $T_{tr} = 7$ censuses for training (i.e. predictions for census # 8) the mean relative errors decrease and converge to a value < 5% for five years ahead and < 10% for ten years ahead.

An increase of the average relative error of species abundance predictions of 1% per year seems quite good for a complex system like the tree community of a tropical forest, which is known to exhibit important variations in species composition. Thus the model passes the validation tests.

A3.6 Predictions

As we have seen, the final model to forecast future values requires recombining the training and validation periods into one series and estimating the model parameters for this complete dataset. Table A3.5 shows the resulting predictions for the species abundances for the future 2020 census. So we will have to wait until these data are published.

A3.7 Extensions, improvements and caveats

Remember that from the fact that the diagonal elements of the Lotka–Volterra interaction matrices $\mathbf{A}^{(c)}$(corresponding to intraspecific interactions strengths) were much greater than the off-diagonal elements (corresponding to interspecific

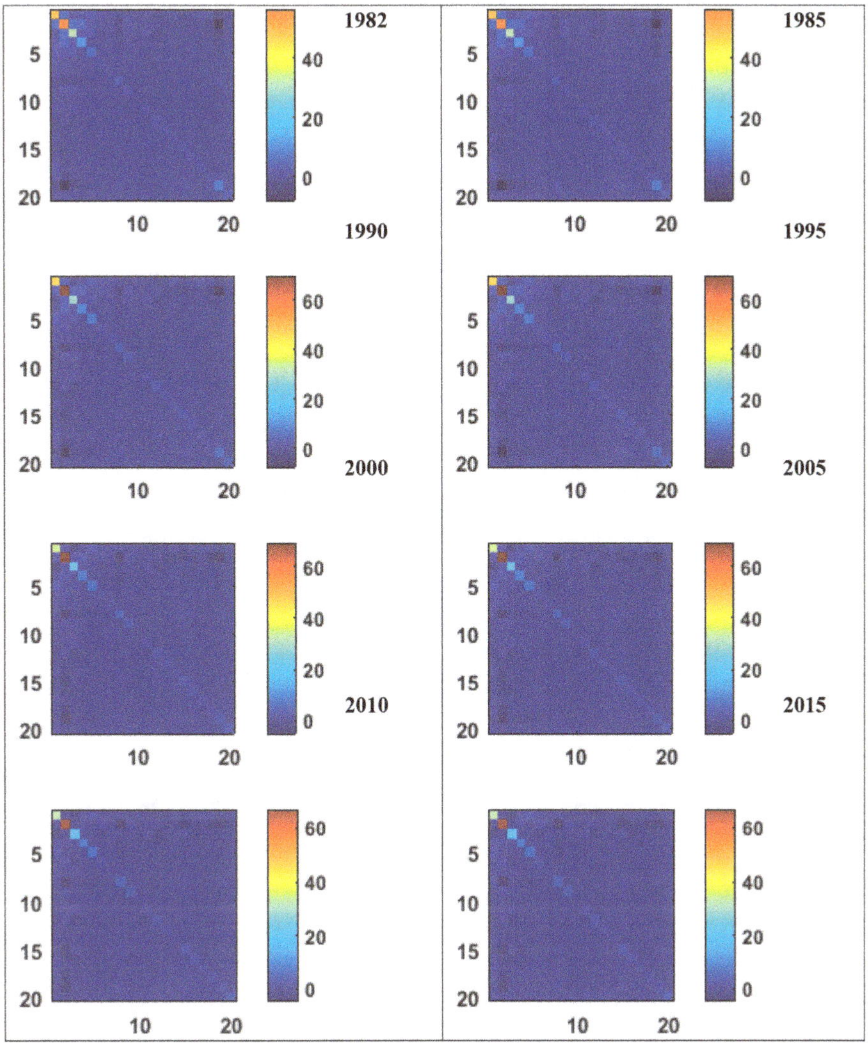

Figure A3.6. Covariance matrices for the eight censuses using dbh $\geqslant 2.5$ cm and $L = 10$ m.

interactions strengths), as a first approximation, we neglected these off-diagonal terms. Now we will check that considering the full matrix $\mathbf{A}^{(c)}$ rather than minus the identity matrix does not produce much better results. Another issue to explore when considering matrix $\mathbf{A}^{(c)}$ is the sensitivity of the quality of predictions to varying the length of the side L of the used quadrats and the dbh threshold. Thus, to kill two birds with one stone, we will repeat the calculations for $L = 10$ m and dbh $\geqslant 2.5$ cm. Figures A3.6 and A3.7 show, respectively, the covariance matrices and the Lotka–Volterra interaction matrices for the eight censuses using $L = 10$ m and dbh $\geqslant 2.5$ cm.

Once again we observe that the diagonal elements of both the covariance and the $\mathbf{A}^{(c)}$ matrices, have diagonal elements much greater than the off-diagonal elements.

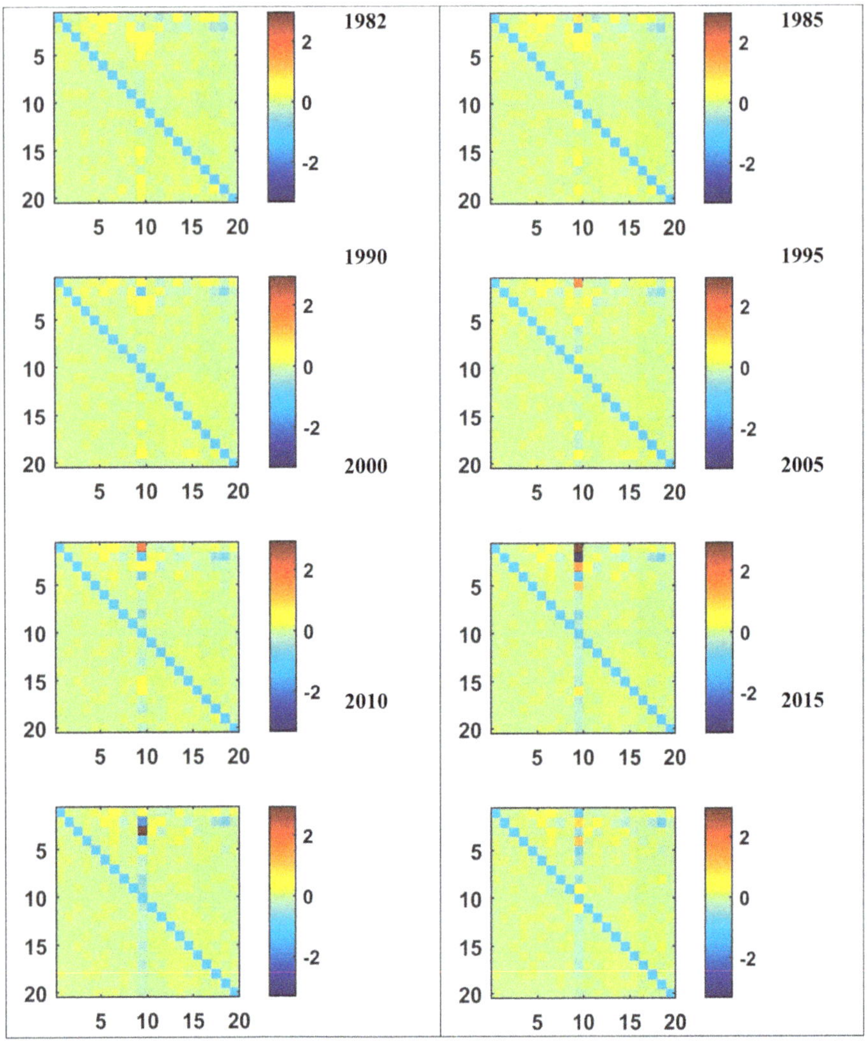

Figure A3.7. Interaction matrices for the eight censuses using dbh $\geqslant$ 2.5 cm and L = 10 m.

To make predictions that take into account the interspecific interactions we will use, instead of the Logistic equations (A3.4), the Lotka–Volterra interaction *finite difference* equations given by:

$$n_i(c+1) - n_i(c) = r_i n_i(c)\left(1 + \frac{\sum_{j=1}^{S} \alpha_{ij}(c) N_j(c)}{\mathcal{K}_i}\right). \tag{A3.7}$$

Table A3.6. Mean relative errors ε as % for the predicted abundances using $T_{tr} = 4$ censuses for training for $L = 20$ m versus $L = 10$ m and for full Lotka–Volterra interaction matrix versus only intraspecific interactions.

	Full matrix				Only intraspecific			
	c #5	c #6	c #7	c #8	c #5	c #6	c #7	c #8
$L = 20$ m	*9.4*	*23.4*	*24.3*	*31.4*	*8.0*	*16.7*	*26.7*	*98.1*
$L = 10$ m	*7.0*	*12.0*	*15.3*	*20.4*	*5.9*	*9.4*	*15.2*	*24.1*

Therefore, we have to change the $x_i(c)$ variable in the second step of the recipe of box A3.1 from $x_i(c) \equiv n_i^e(c)$, to $x_i(c) \equiv \sum_{j=1}^{S} \alpha_{ij}(c) n_j^e(c)$. The predicted abundances are then given by:

$$n_i(c+1) = n_i(c) + r_i n_i(c)\left(1 + \frac{\sum_{j=1}^{S} \alpha_{ij}(c) n_j(c)}{\mathcal{K}_i}\right), \qquad c = T_{tr}, \ldots, 8.$$

We can see from table A3.6 that there is no clear advantage regarding decreasing errors of predictions in taking into account the interspecific interactions through the full Lotka–Volterra interaction matrix.

A caveat of this method is that if we take the logistic growth seriously, the estimated carrying capacities must be positive and not vary much when changing the length of the training period T_{tr}. This is in general the case for most of the species. Nevertheless, we found for species *Poulsenia armata* the method we are using for estimating parameters yields, depending on the values of T_t, either a negative carrying capacity or a positive but absurdly large one. In these cases we should take the abundance predictions with a grain of salt.

Another observation regarding the $\mathbf{A}^{(c)}$ matrices by MaxEnt, is that positive and negative interactions occur simultaneously, so the observer sees only an integrated net effect of multiple interactions (Ewel and Hiremath 2005).

A3.8 Conclusion

The tree species composition of BCI forest is not stable and is changing for reasons which are not yet understood (Condit *et al* 2017); however, we can predict these changes as a result of self-regulation. We have shown that most of the trajectories of the 20 most abundant species are well described by logistic difference equations with parameters estimated by species forecasting techniques. This makes us confident in the proposed method to predict the future abundances of these species.

Additionally, we found the estimated species parameters, r_i and $\mathcal{K}_i$, vary strongly from one species to another. According to the dispersal-assembly hypothesis there are no important differences between species, thus our findings contradict the neutrality assumption. This is in agreement with the large ecological differences found among tree species in tropical forests indicating that the central assumption of neutral theory that species are equivalent is simply incorrect (Thomas 2003).

MATLAB code

All data on abundances and positions of trees from BCI forest can be obtained from Condit *et al* (2019). In particular, these data are necessary for computing the covariance matrices used in the MATLAB script below.

```
clear all
load('Selec20species8censuses.mat')      %Loads empiric abundances of the 20 most
                                          %abundant species for the 8 censuses
                                          %and the corresponding 8 Covariance
                                          %matrices CovNselM (N=1,2,…,8)
                                          %computed for quadrats of length Lq = 20.
%Computation of the 8 interaction matrices INM (N =1,2,…,8) in case the
% off-diagonal interaction coefficients are taken into account
I1M=-inv(Cov1selM);I2M=-inv(Cov2selM);I3M=-inv(Cov3selM);I4M=-inv(Cov4selM);
I5M=-inv(Cov5selM);I6M=-inv(Cov6selM);I7M=-inv(Cov7selM);I8M=-inv(Cov8selM);
Lq=20;      %quadrats of length Lq = 20.
Nquad=500*1000/Lq^2; % number of quadrats
neALL8M=[Ne1selV Ne2selV Ne3selV Ne4selV Ne5selV Ne6selV Ne7selV Ne8selV]/Nquad;
total_time=8; % There are 8 censuses for BCI
training_time=input('insert training time (it must be le. total_time; default= 5) = ')
validation_time=total_time-training_time

dneALL8M=diff(neALL8M')';
xM=-neALL8M(:,1:total_time-1);
yM=dneALL8M./neALL8M(:,1:total_time-1);
%In case off-diagonal interaction coeff. are taken into account
%xM=[I1M*neALL8M(:,1) I2M*neALL8M(:,2) I3M*neALL8M(:,3) I4M*neALL8M(:,4) …
%I5M*neALL8M(:,5) I6M*neALL8M(:,6) I7M*neALL8M(:,7)];
PolyM=zeros(20,2); % Polynomial fit (least squares)
for isp=1:20
  PolyM(isp,:)=polyfit(xM(isp,1:training_time-1), yM(isp,1:training_time-1),1);
  qV(isp)=PolyM(isp,1);
  rV(isp)=PolyM(isp,2); % in census -1
  KV(isp)=max(0.,PolyM(isp,2)/PolyM(isp,1)); % in trees per quadrat
end
NpredM=zeros(20,validation_time);
NpredM(:,1)=neALL8M(:,training_time).*(1+rV(:)-qV(:). *…
          (neALL8M(:,training_time)));
if validation_time>0
  for iter=2:validation_time
     NpredM(:,iter)= NpredM(:,iter-1).*(1+rV(:)-qV(:).*…
          (NpredM(:,iter-1)));
  end
errorM=NpredM-neALL8M(:,training_time+1:total_time);
error_percentM=100*(errorM./max(1,neALL8M(:,training_ time+1:total_time)));
```

```
pM=zeros(20,total_time-training_time);
for jc=1:total_time-training_time
 pM(:,jc)=neALL8M(:,training_time+jc)./sum(neALL8M (:, training_time+jc));
end
RMAE=sum(pM.*abs(error_percentM));
meanNe=mean(neALL8M(:,training_time+1:total_time))
denominatorE1=sum(abs(neALL8M(:,training_time+1:
total_time)-ones(20,1)*meanNe));
denominatorD1=sum(abs(NpredM-ones(20,1)*meanNe)+
abs(neALL8M(:,training_time+1:total_time)... -ones(20,1)*meanNe));
E1=1-sum(abs(errorM))./denominatorE1;
D1=1-sum(abs(errorM))./denominatorD1;
RMAE_E1_D1=[RMAE; E1; D1]
% Let's check that this logistic forecasting is better than adding to
%each abundance the average change over the training period
if total_time-training_time==1% 7 censuses for training
  Npred_mean_delta=[neALL8M(:,training_time)+...
mean(dneALL8M(:,1:training_time-1)')'.*ones(20,1)];
elseif total_time-training_time==2% 6 censuses for training
avdneALL8M_v=mean(dneALL8M(:,1:training_time-1)')'. *ones(20,1);
Npred_mean_delta=[neALL8M(:,training_time)+ avdneALL8M_v ...
        neALL8M(:,training_time)+2*avdneALL8M_v];
elseif total_time-training_time==3 % 5 censuses for training
avdneALL8M_v=mean(dneALL8M(:,1:training_time-1)')'. *ones(20,1);
Npred_mean_delta=[neALL8M(:,training_time)+ avdneALL8M_v ...
  neALL8M(:,training_time)+2*avdneALL8M_v ...
  neALL8M(:,training_time)+3*avdneALL8M_v];
elseiftotal_time-training_time==4 % 4 censuses for training

avdneALL8M_v=mean(dneALL8M(:,1:training_time-1)')'. *ones(20,1);
Npred_mean_delta=[neALL8M(:,training_time)+ avdneALL8M_v ...
         neALL8M(:,training_time)+2*avdneALL8M_v ...
   neALL8M(:,training_time)+3*avdneALL8M_v neALL8M(:,training_time)+...
         4*avdneALL8M_v];
end
error_percent_mean_delta=100*(Npred_mean_delta./...
   max(neALL8M(:,training_time+1:total_time),1)-1);
nro_of_times_the_error_of_predictions_of_addingDab_is_greater=...
 mean(abs(error_percent_mean_delta./error_ percentM))
 Npred_fin=NpredM(:,validation_time);
else
 Npred_fin=neALL8M(:,total_time);
end
r_years_minus_1=rV'/5
K_trees_per_ha=10000*KV'/Lq^2
```

References

Chave J *et al* 2003 Spatial and temporal variation of biomass in a tropical forest: results from a large census plot in Panama *J. Ecol.* **91** 240–52

Clark J S and McLachlan J S 2003 Stability of forest biodiversity *Nature* **423** 635–8

Chesson P 2000 Mechanisms of maintenance of species diversity *Annu. Rev. Ecol. Syst.* **31** 343–66

Condit R *et al* 2017 Demographic trends and climate over 35 years in the Barro Colorado 50 ha plot *Forest Ecosyst.* **4** 17

Condit R, Perez R, Aguilar S, Lao S, Foster R and Hubbell S P 2019 *Complete Data from the Barro Colorado 50-ha Plot: 423617 Trees, 35 Years, 2019 Version* https://doi.org/10.15146/5xcp-0d46

Ewel J J and Hiremath A J 2005 Plant–plant interactions in tropical forests *Biotic Interactions in the Tropics* ed D F R P Burslem, M A Pinard and S E Hartley (Cambridge: Cambridge University Press)

Hubbell S P 2008 Approaching tropical forest complexity, and ecological complexity in general, from the perspective of symmetric neutral theory *Tropical Forest Community Ecology* ed W Carson and S Schnitzer (New York: Wiley)

Hubbell S P 2009 Neutral theory and the theory of island biogeography *The Theory of Island Biogeography at 40: Impacts and Prospects* ed J Losos and R E Ricklefs (Princeton, NJ: Princeton University Press)

Hubbell S P 2011 To know a tropical forest. What mechanisms maintain high tree diversity on Barro Colorado island, Panama? *The Ecology of Place* ed I Billick and M V Price (Chicago, IL: University of Chicago Press)

Ishida A *et al* 2003 Leaf physiological adjustments to changing lights: partitioning the heterogeneous resources across tree species *Pasoh: Ecology of a Lowland Rain Forest in Southeast Asia* ed T Okuda, N Manokaran, Y Matsumoto, K Niiyama, S C Thomas and P S Ashton (Tokyo: Springer)

Shmueli G and Lichtendahl K C 2016 *Practical Time Series Forecasting with R: A Hands-on Guide* 2nd edn (Green Cove Springs: Axelrod Schnall)

Thomas S C 2003 Comparative biology of tropical trees: a perspective from Pasoh *Pasoh: Ecology of a Lowland Rain Forest in Southeast Asia* ed T Okuda, N Manokaran, Y Matsumoto, K Niiyama, S C Thomas and P S Ashton (Tokyo: Springer)

Volkov I, Banavar J R, Hubbell S P and Maritan A 2009 Inferring species interactions in tropical forests *Proc. Natl. Acad. Sci.* **106** 13854–9

Valencia R, Balslev H and Paz y Miño G 1994 High tree alphadiversity in Amazonian Ecuador *Biodivers. Conserv.* **3** 21–8

Whittaker R H 1975 *Communities and Ecosystems* (New York: MacMillan)

Zimmerman J K, Thompson J and Brokaw N 2008 Large tropical forest dynamics plots: Testing explanations for the maintenance of species diversity *Tropical Forest Community Ecology* ed W P Carson and S A Schnitzer (Singapore: Wiley)

Chapter 4

Catastrophic shifts in ecology, early warnings and the phenomenology of phase transitions

'There is an hour of the afternoon when the plain [*llanura*, flatness] is about to say something; it never says it, or perhaps it says it infinitely and we do not understand it, or we understand it but it is untranslatable like a music...'

—J L Borges *The End.*

'Human history becomes more and more a race between education and catastrophe.'

—H G Wells.

Summary

Catastrophe theory is a branch of mathematics whose origins were intimately tied up with biology. It originated with the work of the French mathematician René Thom in the 1960s to study and classify phenomena characterized by sudden shifts in behavior arising from small changes in circumstances. Catastrophes are bifurcations between different equilibria, or alternative stable states.

Catastrophes have characteristic fingerprints or 'wave flags' that may serve as early warnings of unnoticed ongoing catastrophic shifts. Some of the standard catastrophe flags are: modality, sudden jumps, hysteresis and a large or anomalous variance. We illustrate these early warning signals using a simple ecological model, representing a species subject to exploitation (grazing, harvesting or predation), that exhibits alternative stable states. It turns out that overgrazing is known to be one of the greatest causes of a common catastrophic shift: desertification.

Neglecting all spatial heterogeneities, we first consider the simpler mean field (MF) or non-spatially explicit version of the model. MF models are easy to analyze

doi:10.1088/978-0-7503-2432-8ch4

and in cases without significant heterogeneity their predictions are not very different from those of spatial models.

However, often spatial dimensions profoundly alter the dynamics and outcomes in the real world. Thus, in the next section, we address the development of *spatial* early warnings of catastrophic shifts.

Finally, we consider the similarities and differences between the grazing model, describing the desertification catastrophic shift, and the van der Waals model used to qualitatively describe the boiling of water. The formal analogy between these two models provides valuable insights regarding the kinetic of the transition between alternative stable states and different early warning signals.

4.1 Catastrophes

4.1.1 Catastrophic shifts and bifurcations

External conditions to ecosystems such as climate, inputs of nutrients or toxic chemicals, groundwater reduction, harvest or loss of species diversity often change gradually with time (Vitousek *et al* 1997, Tilman *et al* 2001). The state of some ecosystems may respond in a smooth, continuous way to such trends. Other ecosystems may be quite inert over certain ranges of conditions, and suddenly respond very drastically when conditions approach a certain critical level (Scheffer *et al* 2001). This is typical of many nonlinear systems in which a variation of certain parameters can cause different types of ***bifurcations***. Equilibria can appear or disappear, or change from attracting to repelling and vice versa, leading to a shift from one steady state to a radically different one, from which recovery is exceedingly difficult. That is, for certain environmental conditions, the ecosystem has two alternative ***stable states*** (we will use the terms ***attractors*** or equilibria as synonyms of stable states), separated by an unstable equilibrium. Each attractor has its own ***basin of attraction***, i.e. the sets of points from which a dynamical system spontaneously moves to this particular attractor. The unstable equilibrium marks the border between the basins of attraction of both stable states. Examples of such ***catastrophic shifts*** (hereafter we will use the term ***catastrophes*** as a synonym) range from massive extinctions to desertification processes.

Before entering into the mathematics let us illustrate the catastrophic shift phenomenon qualitatively by means of the example of desertification of drylands subject to overgrazing (Adeel *et al* 2005). Overgrazing is regarded as one of the major causes of desertification (Dregne 1986, Wiesmeier 2015). Figure 4.1 shows how the vegetation density decreases as a result of the grazing pressure in drylands. At first, increasing the grazing pressure from point A has no dramatic consequences. When we reach point F_1, in addition to the stable 'high' vegetation equilibrium two additional equilibria appear out of the blue: an unstable equilibrium (dashed line) and another stable 'low' vegetation equilibrium. However, the system remains in its upper equilibrium branch without providing any significant warning signal (we only observe a gradual decrease in the vegetation density). If we continue increasing the grazing pressure, when we reach point F_2 the 'high' vegetation equilibrium disappears. Thus, at this point, a sudden steep drop of the vegetation density

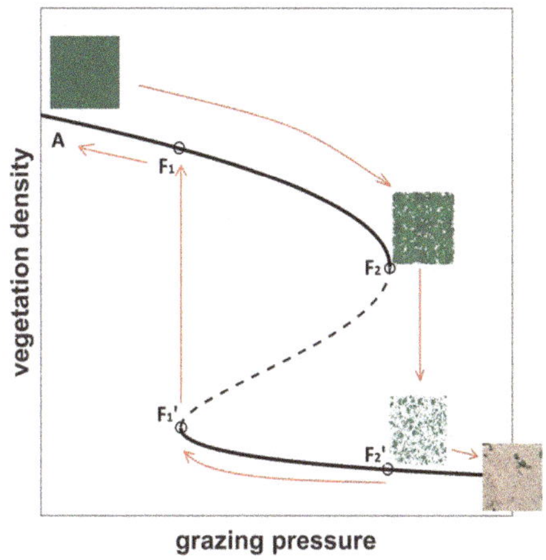

Figure 4.1. A schematic evolution of the vegetation density as a function of the grazing pressure. The vegetation density decreases more or less linearly with the grazing pressure from A to F_1. Increasing the grazing pressure beyond F_1 has no apparent dramatic consequences. However, a new stable much lower density equilibrium appeared for the ecosystem. If one continues to increase the grazing pressure we finally arrive at a point F_2 where the higher vegetation density equilibrium disappears and the ecosystem experiences a steep drop to the lower vegetation density equilibrium. Further increasing the grazing pressure inevitably leads to desertification.

towards the only remaining stable 'low' vegetation equilibrium takes place (from F_2 to F_2'). In fact, we will see later in this chapter that the transition from one attractor to the other can occur *before* we reach the 'point of no return' F_2, somewhere between F_1 to F_2.

At any event, a necessary condition for a catastrophic shift to occur is that, under the same external conditions, the system can be in two or more ***alternative stable states*** (ASS) (Scheffer *et al* 2001, Carpenter 2001). The appearance of ASS is equivalent to saying that the system experiences a bifurcation. Hence, when subjected to a slowly changing external factor (such as climate or human activities), an ecosystem may show little change until it reaches a critical point where a sudden shift to an alternative contrasting state occurs.

In mathematics, *Catastrophe Theory* (Thom 1989) is the branch of *Bifurcation Theory* in the study of dynamical systems, which deals with such sudden collapses resulting as a consequence of small changes in environmental conditions. Bifurcation theory studies and classifies phenomena characterized by sudden shifts in behavior arising from small changes in circumstances, analyzing how the qualitative nature of the solutions of an equation depends on the parameters that appear in the equation. This may lead to sudden and dramatic changes, for example the unpredictable timing and magnitude of a landslide or a pest outbreak. Bifurcations occur in both continuous systems—described by ordinary differential equations (ODEs) or partial differential equations (PDEs)—and discrete systems—described by difference equations or maps.

To make concrete the program of *Elementary* Catastrophe Theory (ECT) (Gilmore 1981), we look for solutions $\psi_i\ (t;\mathscr{c}_\mu)$ of ***autonomous*** dynamical systems of the form[1]:

$$\frac{d\psi_i}{dt} = f_i\,(\psi_j;\ \mathscr{c}_\mu), \qquad i = 1,\ \ldots,\ N;\ \mu = 1,\ \ldots,\ P, \tag{4.1}$$

where t is the time and $\mathscr{c}_\mu$ denotes a set of P parameters. Since the parameters may control the qualitative properties of the solutions ψ_j, the $\mathscr{c}_\mu$ are called ***control parameters***. Notice that we are using a ***mean field*** (MF) ***description*** which, neglecting all spatial heterogeneities, describes the change over time of some average variable that characterizes the state of the ecosystem.

Of particular interest are the equilibria $d\psi_i/dt = 0$ of the above systems. Since these equilibria are by definition time independent we denote them simply as $\psi_i(\mathscr{c}_\mu)$. In summary:

Elementary catastrophe theory is the study of how the equilibria $\psi_i(\mathscr{c}_\mu)$ change as the control parameters $\mathscr{c}_\mu$ change.

4.1.2 A simple population (mean field) model with a catastrophe

Worldwide, desertification is making approximately 12 million hectares useless for cultivation every year (World Resources Institute 1988). The Sahara desert appears to be advancing downward into some places of the Sahel, the semiarid region of 5000 km extending across Africa south of the Sahara desert, at the rate of 18 feet per hour due to overgrazing (Wade 1974). Thus our starting point is a population equation which is commonly used for grazing systems (Noy-Meir 1975) and that was already introduced to describe the grass dynamics in the Application 1 chapter of this book. This same model has been later used to describe several ecosystems (May 1977) and in particular for the case of the spruce budworm (Ludwig *et al* 1978, Murray 1993). It involves the total grass biomass density X which evolves in time according to:

$$\frac{dX}{dt} = \mathscr{r}X\left(1 - \frac{X}{\mathscr{K}}\right) - \mathscr{c}\frac{X^2}{X^2 + \mathscr{h}^2}, \tag{4.2}$$

where $\mathscr{r}$ is the intrinsic per capita growth rate, $\mathscr{K}$ is the carrying capacity, i.e. the biomass density which can be supported in a given area within natural resource limits, $\mathscr{c}$ is the maximum consumption rate or ***grazing pressure*** (the stress on plant populations due to the grazing of animals) and $\mathscr{h}$ is a half-saturation constant, i.e. it corresponds to the value of X such that the effective consumption is half of the maximum consumption rate. We emphasize that this is a mean field model since X does not depend explicitly on space coordinates, rather it represents a global mean (over space) vegetation density. We can rewrite equation (4.2) in terms of non-dimensional quantities: $t = rt,' = X/\mathscr{h}$, $K = \mathscr{K}/\mathscr{h}$ and $c = \mathscr{c}/(\mathscr{h}\mathscr{r})$, as:

---

[1] Actually, more rigorously, the autonomous systems must be also ***gradient***, i.e. f_i must be derivable from a potential function $V(\psi_i\ (t;\mathscr{c}_\mu))$: $f_i = -\dfrac{\partial V(\psi_j;\ \mathscr{c}_\mu)}{\partial \psi_i}$.

$$\frac{dX}{dt} = X\left(1 - \frac{X}{K}\right) - c\frac{X^2}{X^2 + 1}. \tag{4.3}$$

The right-hand side of equation (4.3) may be thought of as the derivative of a potential V associated with the problem:

$$\frac{dX(t)}{dt} = -\frac{dV(X)}{dX} \text{ with } V = -\frac{X^2}{2} + \frac{X^3}{3K} + c(X - \arctan X). \tag{4.4}$$

The equilibria X^* of equation (4.4) correspond to the roots of the first derivative of V:

$$X^*\left(1 - \frac{X^*}{K}\right) - c\frac{X^{*^2}}{X^{*^2} + 1} = 0. \tag{4.5}$$

This equation has one or three real roots (besides the trivial unstable solution $X^* = 0$), corresponding to one stable equilibrium state or two alternative stable states (separated by an unstable one). Figure 4.2 shows a three-dimensional plot with the equilibrium values of the state variable, X^*, on the vertical axis, for each pair of

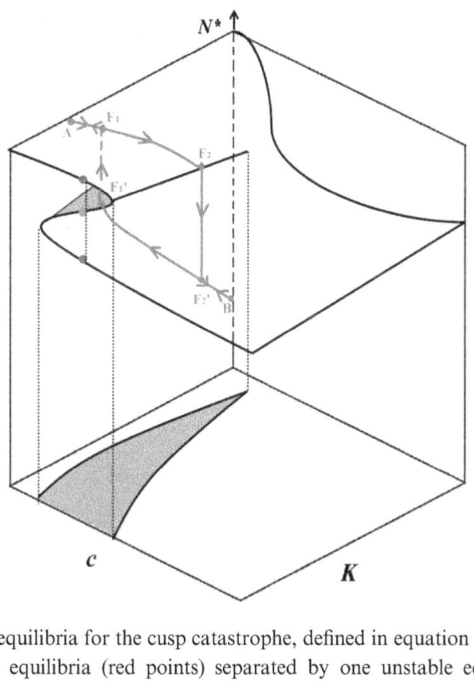

Figure 4.2. The surface of equilibria for the cusp catastrophe, defined in equation (4.5), showing a fold which corresponds to two stable equilibria (red points) separated by one unstable equilibrium (blue point). A hysteresis loop (green curve) describes the evolution of the grass biomass density from the vegetation density state in the upper equilibrium branch (point 'A') to the vegetation density state in the lower equilibrium branch (point 'B'), by increasing the grazing pressure (section A–F_1–F_2–F_2'–B), and then back to the initial state, by decreasing the grassing pressure (section B–F_2'–F_1'–F_1–A). The gray wedge-shaped region on the c–K plane corresponds to the projection of the fold onto this plane and delimits the region in parameter space with alternative stable states.

values of the parameters c and K. When the surface generated $X^*(c,K)$ has a fold there are ASS (i.e. for a fixed value of the parameters c and K there are two alternative stable equilibrium states and one unstable between them). Projecting X^* onto the plane (c,K) we get the solid lines which are the edges of the folds and their point of intersection is called the ***cusp***, and this is why this catastrophe is called a ***cusp catastrophe***. There are three equilibrium points inside the wedge region, two stable points and an unstable one. Outside of the wedge, on the plane of control parameters (c,K), there is only one equilibrium point, which is stable. It is worth noting that the presence of ASS is linked to the functional form assumed for the density dependent consumption. As we have seen (subsection 1.3.3) this can be modelled by different consumption functions, being the most popular ones: linear (or Holling type I), hyperbolic (or Holling type II) and sigmoidal (or Holling type III) (Holling 1959). Only for the sigmoidal consumption of equation (4.5) there occur two stable equilibria separated by an unstable one and therefore do we have ASS.

In figure 4.3 the response curves of X^* for different fixed values of K (i.e. iso-K curves) and varying the consumption rate parameter c are depicted (exercise 4.1). These curves can be obtained as different slices of the surface of equilibria in figure 4.2. For $K \leqslant K_c = 3^{3/2} \cong 5.196$ only one stable solution exists for each c. As long as we consider quasi-stationary evolution for increasing c, the system will exhibit a smooth response. On the other hand, for $K > K_c$, the response curve is folded backwards at two ***saddle-node*** bifurcation points, i.e. it has the same S-shaped form as the curve depicted in figure 4.1. For certain values of c the system can be found either in the upper or the lower stable branch. If the system starts on the upper branch and c increases slowly, X will vary smoothly until a threshold value is found, where a catastrophic transition to the lower branch occurs. If at this point we want to reverse this transition by decreasing c, the system will not be able to recover its

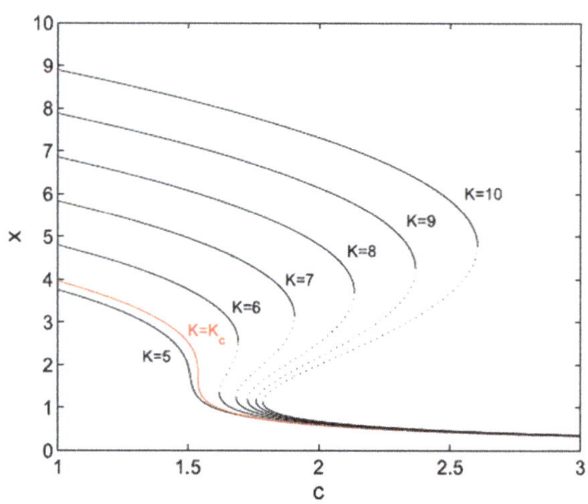

Figure 4.3. Folding curves for different values of K. Filled (dotted) iso-K lines correspond to stable (unstable) equilibria. Reprinted from Fernández and Fort 2009.

original state. Instead, the system will remain on the lower branch, until we decrease c enough to reach another threshold value and 'jump' to the upper branch. This dependency of the state of a system not only on the current values of the control parameters but also on its history is called ***hysteresis*** (see below and figure 4.3). From an ecological management viewpoint, it would be desirable to anticipate these transitions in order to avoid them.

4.2 When does a catastrophic shift take place? Maxwell versus delay conventions

A problem regarding the dynamics of catastrophic shifts is how the qualitative properties of a system change when its control parameters change. This is a crucial issue to assess how close or how far an ecosystem is from experiencing a drastic change.

As we have seen, for systems that can be described by potentials depending on control parameters, the results of ECT are immediately applicable. However, such systems must of necessity be static, because we have removed all dynamical considerations (all explicit dependence on time). Therefore, the time evolution of the equilibria of a gradient system is beyond the scope of ECT (Gilmore 1981).

A change of a control parameter, reflecting changes of the external conditions, modifies the form of the potential. Thus, as the shape of the potential changes, an original global minimum in which the system sits may become a ***metastable*** local minimum because another minimum assumes a lower value, or it may even disappear. In this case the system must jump from the original global minimum to the new one. ECT does not tell us when, and to which minimum, the jump occurs. Indeed, to discuss dynamics, ECT must be extended by incorporating some external assumptions called *conventions*. Before discussing conventions we need to introduce two important sets of points in parameter space which control structural changes of the potential.

The first such set of points is the ***bifurcation set*** S_B (Gilmore 1981), the locus of the points (c,K) such that the second derivative of the potential, d^2V/dX^2, vanishes and then an attractor pops out or in (the first derivative of V also vanishes, since this is the condition to have an attractor). S_B is a ***separatrix***, i.e. a boundary separating two modes of behavior in a differential equation. It is shown in figure 4.4 as filled lines dividing the phase space into two regions corresponding either to single stability or bistability of the system (the shadowed wedge-shaped region in figure 4.2).

The second set of points is called the ***Maxwell set*** S_M (Gilmore 1981), marked as a dot-dashed line in figure 4.4. On the Maxwell set the values of V at both stable equilibria are equal (see the inset of V for $K = 7.5$, $c = 1.91$ in figure 4.4). Each point of S_M is a ***Maxwell point***.

S_B and S_M are connected, respectively, to two commonly applied criteria or conventions. Systems which remain in an equilibrium state until this state disappears are said to obey the *Delay* convention. On the other hand, systems which always seek a global minimum of V are said to obey the *Maxwell* or the *absolute minimum* convention (since it requires that the variable ψ remains always at the absolute

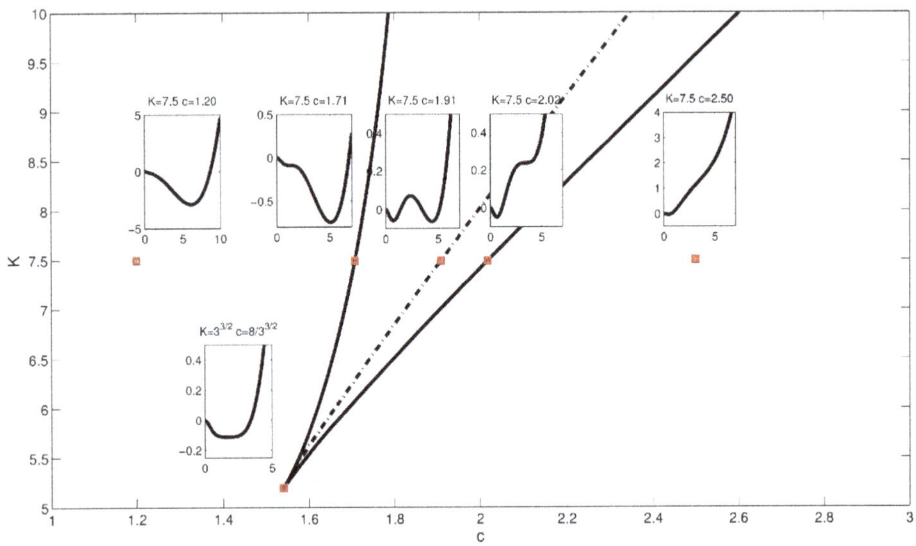

Figure 4.4. The control parameter plane of the *spruce budworm* model and its cusp catastrophe. The bifurcation set S_B, is a separatrix for the cusp catastrophe, i.e. the lines separate the region in which there are three alternative stable states—two of them stable, and one unstable—from a single stable state (solid line), with a cusp point at $c_c = 8/3^{3/2}$, $Kc = 3^{3/2}$. The Maxwell set S_M (dot-dashed line) is the locus of points such that the potential V for both stable equilibria is equal. The potential V is shown for selected values of c and K along an horizontal trajectory at $K = 7.5$ and at the cusp. Reprinted from Fernández and Fort 2009.

minimum of the potential $V(\psi\ ;\ c)$ and therefore ψ jumps as soon as a lower minimum appears. Figure 4.5 shows schematically the difference between these two conventions.

Mean-field models, like equation (4.3), obey the Delay convention when their control parameters are varied. Suppose we start at an equilibrium X belonging to the upper equilibrium branch (point ‘A’ in figure 4.2), by increasing the grassing pressure the point representing the system will describe the section A–F_1–F_2–F_2'–B of the green hysteresis cycle. That is, it will remain in the upper equilibrium branch until this branch disappears at point F_2, when it jumps to the lower equilibrium branch at F_2'. Similarly, if we start at an equilibrium X belonging to the lower equilibrium branch (point ‘B’) if we then decrease the grassing pressure the system will describe the section B–F_2'–F_1'–F_1–A (exercise 4.4).

What about real systems? Indeed the delay and Maxwell conventions correspond to two extremes in a continuum of possibilities. In fact, real systems may obey either of these two conventions depending on combinations of several factors, like the spatial heterogeneity, the rate of change of the control parameters or on other external environmental conditions, etc. For instance, when the control parameters, and thus the shape of V, change very slowly the system tends to follow the delay convention. In contrast, when control parameters change more quickly due to drastic perturbations of the environment (e.g. a severe drought), the Maxwell convention may describe the dynamics better. In particular, close to the folds, small

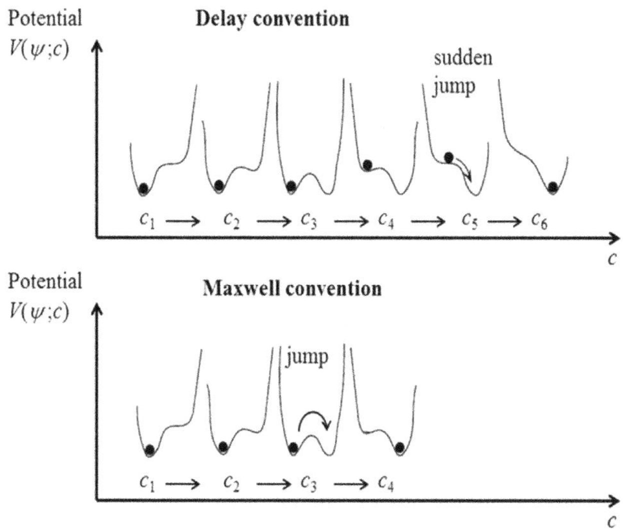

Figure 4.5. Scheme of delay convention versus Maxwell convention. The potential function $V(\psi;c)$ depends on the variable x and the parameter c. As parameter c increases from c_1 to c_2, etc, the form of the potential changes. Initially $V(\psi;c_1)$ has only one minimum, then a second local minimum appears for $c = c_2$ to the right of the absolute minimum, for $c = c_3$ the depth of both minima become equal. If c increases above this value, the right minimum becomes the absolute minimum while the left minimum becomes the local minimum. If c further increases to $c = c_4$ the left minimum disappears. In the case of delay convention here is when a sudden jump to the new equilibrium state (right minimum) takes place, because the system has no other possible equilibrium state. On the other hand, according to Maxwell convention a jump to the new equilibrium state starts at $c = c_3$ when the depth of the right minimum becomes equal to the depth of the left minimum.

perturbations in vegetative biomass that are unrelated to stocking levels may tip the system over the breakpoint curve and cause the vegetative biomass to shift abruptly to the other stable point (Petraitis 2013). We refer the reader interested in learning more about the issue of which convention governs in catastrophic shifts to the enlightening discussion in chapter 6 of Petraitis (2013).

For example, satellite data show that forests and savannas tend to coexist roughly equally throughout the bistability range, then this would be consistent with the delay convention (van Nes *et al* 2014). Additionally, in a recent study it was found that forests and savannas in South America and Africa tend to coexist around a Maxwell point (Staal *et al* 2016). This would indicate that coexistence or bistability occurs at a wider range of conditions than the Maxwell convention would imply for shifts between forests and savannas. Indeed, ecologists usually assume the delay convention to be the norm. However, Petraitis (2013) discusses a theoretical example of a population with an Allee effect (i.e. a positive density dependence, or the positive correlation between population density and individual fitness to grow) and shows that the Maxwell convention applies across a range of model parameters.

We will show that for phase transitions in physics, like boiling water, it is the other way around: under normal conditions the Maxwell convention applies; only under much more stringent experimental conditions is it possible to observe the delay convention.

We will also show that taking into account spatial heterogeneity together with the locality of interactions can affect the type of convention a dynamical system obeys. But before this, in the next section, we will review a series of signals displayed by systems when approaching a catastrophic shift that serve as early warnings of this shift.

4.3 Early warnings of catastrophic shifts[2]

The presence of ASS implies that if a system has gone through such a state shift, it tends to remain in the new state until the control parameter is changed back to a much lower level. This hysteresis phenomenon of 'history dependent' alternative equilibrium states, which is well known in physics, makes it very difficult to restore ecosystems to their original 'good' equilibrium states.

The identification of early warning signals that allow us to predict such catastrophic events before they happen is therefore invaluable for making management decisions to avoid or mitigate them.

Catastrophic shifts have characteristic wave flags. This section is devoted to a description of some of the standard catastrophe flags. The first three flags—***modality, sudden jumps, hysteresis***—occur when the physical control parameters can move into a region of control parameter space in which the corresponding potential has more than one minimum, i.e. there are ASS. They all occur together, with the possible exception of hysteresis, which does not occur if the Maxwell convention is obeyed. The remaining two flags, ***critical slowing down*** and ***large anomalous variance***, can occur even when the potential does not have multiple minima (Gilmore 1981).

I. *Modality.*

   This means that the system has two or more distinct ASS in which it may occur (e.g. vegetation versus desert, rain forest versus savanna, etc). This implies that the potential describing the system has more than one local minimum for some range of the external control parameters.

II. *Sudden Jumps.*

   If either the Maxwell convention or the delay convention is adopted, then a small change in the value of the control parameter may result in a large change ('sudden jump') in the value of the state variable as the system jumps from one local minimum to another. Under the delay convention, jumps occur at the folds of S-shaped response curves, where small continuous changes in parameters cause a sudden discontinuous change in the equilibrium values of the state variables (figure 4.1). Under Maxwell's convention, the jump occurs at the Maxwell point. We remark that the occurrence of sudden jumps is independent of the convention adopted.

[2] This section is mostly a summary of the material presented in chapter 9 of Gilmore (1981).

III. *Hysteresis.*

Hysteresis occurs whenever a physical process is not strictly reversible. That is, the jump from local minimum 1 to local minimum 2 does not occur for the same value of the control parameter as the reciprocal jump from local minimum 2 to local minimum 1. Hysteresis fails to occur only when the Maxwell convention is obeyed. For example, hysteresis occurs for the cusp catastrophe as illustrated by the green curve in figure 4.2. We have adopted the delay convention and assumed that by first increasing the control parameter c the state of the state of the system moves from point 'A' to point 'B', and then attempting to undo the change by decreasing c the point sweeps out the characteristic hysteresis loop depicted in green.

IV. *Critical slowing down.*

Critical slowing down is the phenomenon of a system relaxing to equilibrium much more slowly than usual. It occurs near some bifurcation points, due to the loss of linear stability. In statistical mechanics, this phenomenon is a signature of a second-order phase transition (see next section and exercise 4.5).

It turns out that for many state variables $X_1, X_2, \ldots$, equation (4.4) can be generalized to

$$\frac{dX_i}{dt} = -\frac{\partial V}{\partial X_i}. \tag{4.4'}$$

We can expand the potential V around an equilibrium point X_i^* so that equation (4.4′) becomes (remember that, by definition, an equilibrium verifies that $V(X_i^*) = 0$):

$$\frac{dX_i}{dt} = -\left(\frac{\partial^2 V}{\partial X_i \partial X_j}\right)_{X_i^*} \delta X_j + \mathcal{O}(2). \tag{4.4''}$$

By neglecting the terms of order 2 and higher, the dynamical equations reduce to a simple system of linear equations. The equilibrium at X_i^* is stable if all the eigenvalues of the stability matrix $V_{ij}(X_i^*) = \left(\frac{\partial^2 V}{\partial X_i \partial X_j}\right)_{X_i^*}$ are positive. The normal modes have time dependences of the form $\exp(-\lambda_i t)$, where $\lambda_1, \lambda_2, \ldots$, are the eigenvalues of V_{ij}, so that $1/\lambda_i$ is the characteristic relaxation time of the ith eigenmode.

We have seen that the second derivative of the potential vanishes at bifurcation set S_B. Similarly, as the bifurcation set (delay convention) is approached, one or more of the eigenvalues of V_{ij} approaches zero. The relaxation time of the corresponding mode becomes larger and larger.

V. *Anomalous variance.*

Suppose we have an ensemble of replicas of a physical system. For example, in the case of our grazing system these replicas can correspond to different external conditions (like climate, rain, etc) or different times or places (with differences in soil properties) and we focus on the state variable X. Therefore, the system is specified by a probability distribution $P(X)$ defined over the space of state variables rather than by an isolated point in the space of the state variable. Such a system is characterized by the moments of the probability distribution function

$$\begin{aligned} 1 &= \int P(X)dX \\ \overline{X} &= \int X\, P(X)dX \\ \overline{X^2} &= \int X^2 P(X)dX \end{aligned} \tag{4.6}$$

The variance is defined by:

$$\overline{(X-\bar{X})^2} = \overline{X^2} - \overline{X}^2 \tag{4.7}$$

Assuming that $P(X)$ is a decreasing function of $V(X)$ (in such a way it is maximum at the equilibrium points of the potential), we can write (exercise 4.6):

$$P(X) \approx Ne^{-a\lambda(X-X^*)^2}, \tag{4.8}$$

where N is a normalization constant, a is a positive constant and $\lambda = \left(\frac{\partial^2 V}{\partial X^2}\right)_{X^*}$. We can compute the variance using equation (4.8) and get

$$\overline{(X-\bar{X})^2} = \frac{1}{2}\frac{1}{a\lambda} \tag{4.9}$$

This suggests that the variance may experience an anomalous growth when we approach a point of S_B which, by definition, has $\lambda = \left(\frac{\partial^2 V}{\partial X^2}\right)_{X^*} = 0$.

4.4 Beyond the mean field approximation

Alternative stable states in ecology have been well studied in isolated, well-mixed systems. Nevertheless, most ecosystems in nature exist on spatially extended landscapes. Besides, organisms in nature are discrete entities that interact only within their immediate neighborhood and therefore are neither distributed uniformly nor at random. Rather, they are prone to form characteristic spatial patterns like patchy structures or gradients. This spatial variance in the environment creates diversity in communities of organisms, as well as affecting their stability, dynamics and pattern generation (May 1974, Tilman *et al* 1997). However, these spatial effects have been long ignored by most ecologists because of the difficulties they pose for modelling. Instead, a common assumption in ecological modelling is the **well mixing hypothesis** or, in physics parlance, the mean-field (MF) approximation.

The MF assumption, that spatial heterogeneity of population densities is negligible, is a good approximation when the physical environment is homogeneous and physical forces exist that cause strong mixing of organisms, or when organisms themselves are highly mobile, etc. As conditions depart from those above, the MF assumption becomes less and less appropriate. A lack of mixing generates neighborhoods around individuals that deviate from the spatial averages. This is particularly the case when organisms only interact over short distances and thus heterogeneity in local environmental conditions becomes especially important.

A breakthrough in community theory was the recognition of the patchy nature of most systems, and the importance of spatially localized disturbances in maintaining diversity. From the seminal paper by Watt (1947) and later work in the marine intertidal zone (Levin and Paine 1974), forests (Pickett and White 1985), and in other systems this has been a very active area of research in ecology. Often spatial dimensions profoundly alter the dynamics and outcomes in the real world (Steinberg and Kareiva 1997). In fact, the oversimplification of MF models casts doubt on whether the occurrence of an alternative stable state could be an artifact. Besides, the analysis of spatially explicit models is relevant for other reasons—for example, to understand phenomena like aggregation and spatial segregation in plant communities (Levin and Pacala 1997). Indeed, vegetation patches, which have been extensively studied for arid lands (Aguiar and Sala 1999), can be approached as a pattern formation phenomenon (Klausmeier 1999, Von Hardenberg *et al* 2001, von Hardenberg *et al* 2010). Interestingly, vegetation patchiness seems to provide a signature of imminent catastrophic shifts between alternative states (Rietkerk *et al* 2004). Shortly after, evidence was found that vegetation patches in arid Mediterranean ecosystems were present over a wide range of size scales, so that the patch-size distribution of vegetation follows a power law (Kéfi *et al* 2007). Additionally, it was also found that with increasing grazing pressure, the field data revealed deviations from power laws. Hence, it seems that this power law behavior may be used as an ***early warning*** signal for the onset of desertification. This and other spatial early warnings, e.g. the spatial variance (Fernández and Fort 2009, Dakos *et al* 2010, Donangelo *et al* 2010) complement temporal ones like the variance of time series introduced to detect trophic shifts in lakes (Carpenter and Brock 2006) or the impact of pollutants (Brock and Carpenter 2006).

Thus in this section we address the subject of *spatial* early warnings of catastrophic shifts applying tools and ideas borrowed from statistical mechanics to a model of overgrazing leading to desertification.

Let us start by introducing the spatial version of the grazing model given by equation (4.2) which we have shown has ASS. In order to take into account the spatial heterogeneity of the landscape, we will take one of the two parameters, the ***local*** parameter, as dependent on the position and constant in time. The other parameter, the ***global control*** parameter, is taken as uniform throughout the system and changing slowly with time.

Our goal is to use this framework to address in this section the following questions:

(i) How do the spatial heterogeneity of the environment and the diffusion of matter and organisms affect the existence of alternative stable states?
(ii) Are emergent characteristic spatial patterns really useful as early warning signals?
(iii) How are these early warnings connected with temporal signs of catastrophic shifts?
(iv) Does the patch-size distribution obey a power law for some value of the control parameter close to the catastrophic shift?

4.4.1 Spatial model: cellular automaton

A straightforward two-dimensional continuum spatial version of the MF model equation (4.3) can be written as

$$\begin{aligned}\frac{dX(x, y; t)}{dt} = X\left(1 - \frac{X(x, y; t)}{K(x, y)}\right) - c(t)\frac{X(x, y; t)^2}{1 + X(x, y; t)^2} \\ + D\left(\frac{\partial^2 X(x, y; t)}{\partial x^2} + \frac{\partial^2 X(x, y; t)}{\partial y^2}\right),\end{aligned} \tag{4.10}$$

where the carrying capacity $K(x, y)$ is a local, i.e. spatially heterogeneous, parameter that varies from point to point, the parameter $c(t)$ is taken as uniform but time-dependent, and D is a diffusion coefficient for X (with units of the inverse of the intrinsic growth rate $1/r$ of equation (4.2)).

In order to simulate equation (4.10) we will use a cellular automaton (CA) (Wolfram 1994). A CA is a collection of 'colored' cells on a grid of specified shape that evolves through a number of discrete time steps according to a set of rules based on the states of neighboring cells. The rules are then applied iteratively for as many time steps as desired. Cellular automata were studied in the early 1950s as a possible model for biological systems (Wolfram 2002). Thus we will represent space by a $L \times L$ regular square lattice so that each cell of side a, centered at coordinates ($x = a{\cdot}i$, $y = a{\cdot}j$) with i and j integer numbers, can be associated with an elementary quadrat, representing the resolution at which we want to describe a land region (figure 4.6). For example, in a recent study of the African Sahel region the resolution imposed by the remote sensing limitations of satellite images was 30 m × 30 m (Weissmann *et al* 2017). Thus, depending on the size we want to cover, we can use a = 30 m, 60 m, 300 m, etc. Likewise, when working with real data, commonly used vegetation indices as proxies of the vegetation density $X(i,j)$ are NDVI (normalized difference vegetation index) or the more reliable EVI (enhanced vegetation index), that incorporates corrections to both soil reflectance and atmospheric disturbances (Weissmann *et al* 2017). In figure 4.6, the greener the cell the higher the vegetation density in the corresponding quadrat.

In addition to the grid on which a cellular automaton lives and the colors its cells may assume, the neighborhood over which cells affect one another must also be specified. The simplest choice is 'nearest neighbors,' in which only cells directly adjacent to a given cell may be affected at each time step. A common neighborhood

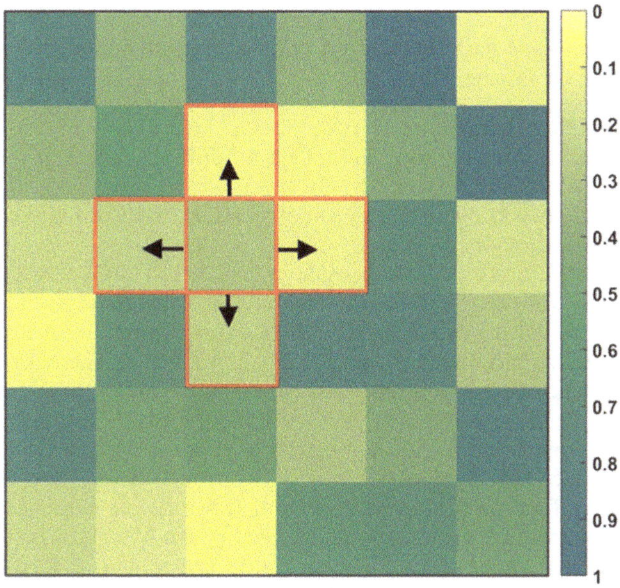

Figure 4.6. A portion of a cellular automaton. In this plot each cell has a normalized vegetation density varying between 0 to 1. The cell marked with 'x' interacts with its von Neumann neighborhood (highlighted in red).

in the case of a two-dimensional cellular automaton on a square grid is the so-called von Neumann neighborhood (the four cells highlighted in red surrounding the cell marked with an 'x' in figure 4.6). Box 4.1 summarizes the properties defining the CA.

The discretized form of equation (4.10) reads:

$$\begin{aligned} X(i, j; t+1) = X(i, j; t) &+ X(i, j; t)\left(1 - \frac{X(i, j; t)}{K(i, j)}\right) \\ &- c(t)\frac{X(i, j; t)^2}{1 + X(i, j; t)^2} \\ &+ d(X(i+1, j; t) + X(i-1, j; t) \\ &+ X(i, j+1; t) + X(i, j+1; t) - 4X(x, y; t)), \end{aligned} \tag{4.11}$$

where d is a reduced diffusion coefficient related to D and the lattice spacing a by $d = 4D/a^2$. This equation provides the CA 'updating rule' (box 4.1).

4.4.2 Early warning signals

Several quantities can be measured from the time series produced by the model:

- The *spatial mean* at fixed time t, $\langle X(t) \rangle$:

$$\langle X(t) \rangle = \frac{1}{L^2}\sum_{i,j} X(i, j; t) \tag{4.12}$$

Box 4.1. Properties and parameters of the cellular automaton used in the explicitly spatial version of the grazing model.

- Shape of lattice: square.
- Size L: ranging from 100 to 1000 (in fact, for different values of L in this range, no important differences were found).
- Neighborhood: von Neumann neighborhood, i.e. each cell is connected to its four nearest neighbors.
- Boundary conditions: periodic boundary conditions (PBC). That is, the cells at the border ($x = 1$ or L or/and $y = 1$ or L) only have two neighbors instead of four. One way to avoid artifacts due to border effects is to assume that cells at $i = 1$ and $i = L$ as well as those at $j = 1$ and $j = L$ are neighbors. These are the so-called PBC; it is equivalent to curling the square into a cylinder, and then joining the cylinder's circular ends to each other, yielding a torus or donut-shape form.
- Number of time steps T: The number of iterations is typically 1000.
- Updating rule: equation (4.11).
- The ranges of values for the parameters that we use are chosen to contain the region of alternative stable states determined by the MF equations: the carrying capacity $K(i, j)$ varies randomly from cell to cell around a fixed spatial mean $\langle K \rangle = 7.5$ in the interval $[-\delta K, \delta K]$ where $\delta K = 1.0$–2.5. Typical values for the consumption rate c are between 1 and 3 and for d between 0.1 and 0.5.

- The *spatial variance* $\sigma^2{}_X$:

$$\sigma_X^2 = \langle X(t)^2 \rangle - \langle X(t) \rangle^2 \tag{4.13}$$

- The *temporal variance* σ_t^2, computed from mean values of X at different times, which is defined as:

$$\sigma_t^2 = \frac{1}{\tau} \sum_{t'=t-\tau}^{t} \langle X(t') \rangle^2 - \left(\frac{1}{\tau} \sum_{t'=t-\tau}^{t} \langle X(t') \rangle \right)^2 \tag{4.14}$$

for temporal bins of size τ (typical values for τ are from 50 to 150).
- The *patchiness* or *cluster structure*. Clusters of high (low) X are defined as connected regions of cells with $X(i, j, t) > X_m$ ($X(i, j, t) < X_m$ where X_m is a threshold value. There are different criteria for defining X_m (see below).
- The *two-point correlation function* for pairs of cells at (i_1, j_1) and (i_2, j_2), separated by a given distance R, which is given by:

$$G_2(R) = \langle X(i_1, j_1) X(i_2, j_2) \rangle - \langle X(i_1, j_1) \rangle \langle X(i_2, j_2) \rangle \tag{4.15}$$

Let us now study the effect of gradually increasing stress on the system, varying c from 1 to 3 in 1000 steps of 0.002, i.e. each 'measure' is performed for a different value of the control parameter c.

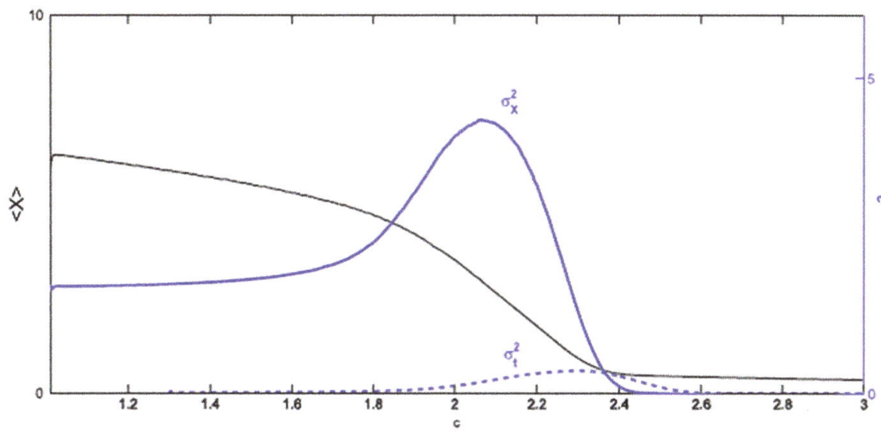

Figure 4.7. $\langle X \rangle$, $\sigma^2{}_X$ and $\sigma^2{}_t$ for $d = 0.1$, $\langle K \rangle = 7.5$. The peak of $\sigma^2{}_X$ occurs at $c_m \cong 2.08$ and the peak of $\sigma^2{}_t$ at $c \cong 2.30$. Reprinted from Fernández and Fort 2009.

Let us see that some characteristics of the spatial structure may serve as early warnings of catastrophic shifts of the system (Fort 2013).

Mean and variances

In figure 4.7 we plot $\langle X \rangle$, $\sigma^2{}_X$ and $\sigma^2{}_t$ in terms of an increasing c for a fixed average capacity $\langle K \rangle = 7.5$, $d = 0.1$ and the initial condition for each $X(i, j)$ in the interval $[0, \langle K \rangle]$. Notice that the spatial variance has a peak at $c_m \cong 2.08$. This peak occurs nearly 110 time steps earlier than the peak for the temporal variance. So $\sigma^2{}_X$ works better than $\sigma^2{}_t$ as a warning signal for the upcoming catastrophic shift. This is because when estimating the temporal variance one must consider past values in the time series, which correspond to situations where the ecosystem is far from undergoing a transition. The spatial variance instead only considers the present values; hence a signal announcing the shift is not blurred by averages including situations where these indications are smaller, or even absent.

It is interesting to look at the process inverse to desertification or **vegetation restoration**. For this we need to analyze what happens when c is decreased. Figure 4.8 shows the hysteresis cycles, for the mean vegetation (black curves) when we first increase c so that desertification takes place and then reverse the trend and decrease c, for different values of d. Notice two remarkable things. First, the peak in $\sigma^2{}_X$ is always narrower for the backward transition than for the forward transition. Second, the width of the hysteresis loop decreases with d, so diffusion tends to make the shift steeper.

Correlation function

It is easy to check that the spatial variance is a particular case (distance $R = 0$) of the two-point correlation function defined by equation (4.15). In figure 4.9 the two-point correlation is depicted for distances $R = 0, 1, 2, 3$. Notice that the peak of the correlation for any R occurs at nearly the same value of the control parameter $c \approx c_m = 2.08$.

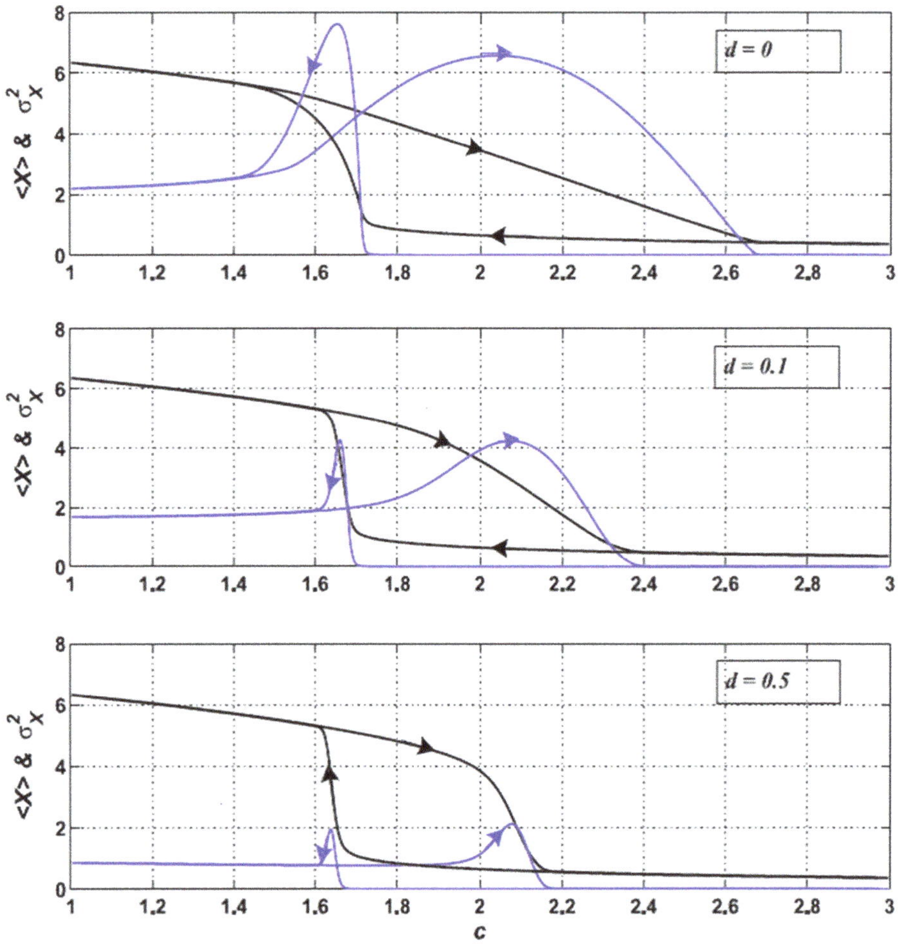

Figure 4.8. $\langle X \rangle$ (black curves) and $\sigma^2{}_X$ (blue curves) for $\langle K \rangle = 7.5$ and $\delta K = 2.5$, computed for forward and backward changes of the control parameter c. Results for $d = 0$ (above), $d = 0.1$ (middle) and $d = 0.5$ (below). Reprinted from Fernández and Fort 2009.

Cluster structure: patchiness

To study the cluster structure first we have to define a threshold X_m which separates low from high density vegetation cells. For $\langle K \rangle = 7.5$ and $d = 0.1$ the maximum in $\sigma^2{}_X$ is given at $c_m \cong 2.08$ (figure 4.6). One possible choice is to take X_m coinciding with the average value of $X(i,j)$ at $c = c_m$, i.e. $X_m = \langle X \rangle(c_m) \cong 2.89$. This is the threshold we will take. In the first column of figure 4.10 we include snapshots of typical patch configurations for $c = c_m - 0.1$, $c = c_m$ and $c = c_m + 0.1$. Densely vegetated patches correspond to clusters of cells which are all between yellow and red. Conversely, bare patches appear in blue. The second column a binary representation, i.e. dark red (blue) cells correspond to cells for which $X > \langle X \rangle\ c_m$ ($X < \langle X \rangle\ c_m$). The plots in the third column are the corresponding cluster distributions. At $c = cm$ the patch-size distribution follows a power law over two

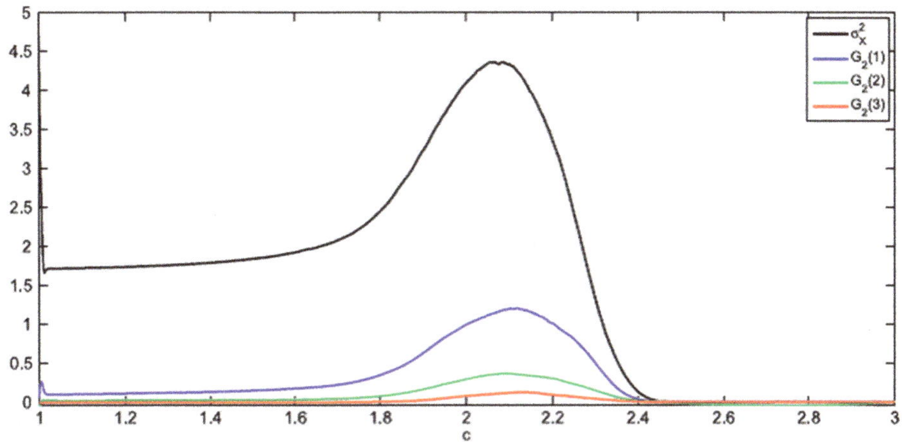

Figure 4.9. Two-point correlation function for different lengths R, d = 0.1, $\langle K \rangle$ = 7.5. Reprinted from Fernández and Fort 2009.

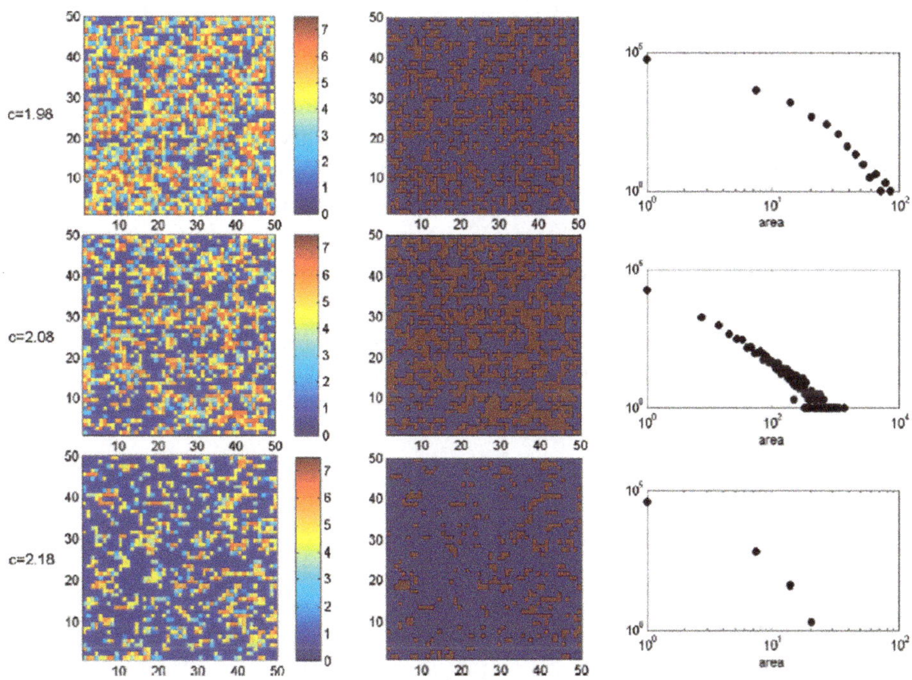

Figure 4.10. *First column*: a portion of 50 × 50 cells from the original 800 × 800 lattice is shown, grids representing the value taken by $X(i, j)$ at each cell for $\langle K \rangle$ = 7.5, d = 0.1. The rows correspond to c = 1.98, c = 2.08 and c = 2.18. *Second column*: same as the first, for binarized data (blue: cells with low vegetation density, red: cells with high vegetation density). *Third column*: number of clusters versus area on a logarithmic scale. Reprinted from Fernández and Fort 2009.

decades—with exponent $\gamma \approx -1.1$ for $d = 0.1$ and $\gamma \approx -0.9$ for $d = 0.5$—which disappears for smaller or greater values of c. Therefore, the power law distribution may be considered as a signature of an ongoing desertification.

4.5 A comparison with the phenomenology of the liquid–vapor phase transition

In this section we show how the desertification transition can be connected with one of the most widely analyzed phase transitions in physics: the boiling of a liquid. We will do this by mapping the grazing model into the simplest state equation able to account for the liquid–vapor phase transition, the ***van der Waals equation of state***. This correspondence allows further insights on the phenomenology of catastrophic shifts as well as to connect the behavior of typical observables, like the spatial variance, the two-point correlation function and the patchiness.

4.5.1 Beyond the ideal gas: the van der Waals equation of state for a fluid and its formal correspondence with the grazing model

Water, like many simple substances, has sharply defined sublimation, melting, and boiling temperatures which change as a function of the pressure. The solid–gas (sublimation), solid–liquid (melting) and liquid–gas (boiling) coexistence curves are commonly summarized in a ***phase diagram***, like the one depicted in figure 4.11 in pressure–temperature axes. Across these three curves we have ***first-order phase transitions***, which are characterized by two distinguishable phases coexisting during the transition. There is a point of particular interest in this diagram: the ***critical point*** C. At the critical point the transition becomes second-order, the difference between the liquid and gas phases ceases to exist. Hence, beyond the critical point C the substance is said to be in a 'fluid' state. The temperature and pressure of this critical point are the ***critical temperature***, T_c, and ***critical pressure***, P_c.

A signature of a second-order phase transition is ***scale invariance***. Intuitively speaking, scale invariance means that if part of a system is magnified until it is as large as the original system; one would not be able to tell the difference between the magnified part and the original system. Whereas the only length scale that defined

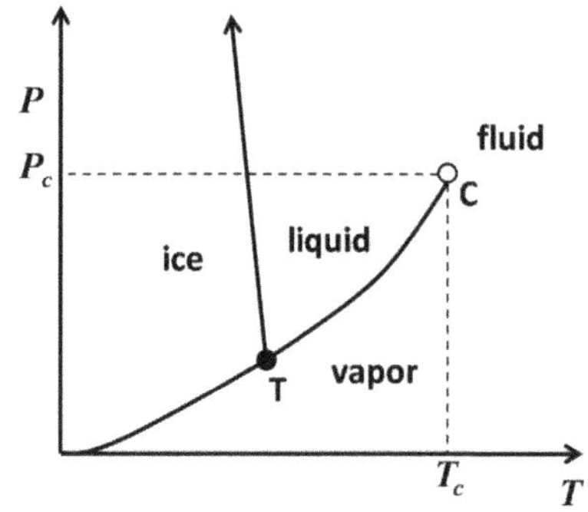

Figure 4.11. The phase diagram of water in P–T axes. Coexistence curves separate the three phases of water: solid (ice), liquid and gas (vapor). At the triple point T the three coexistence curves meet. At the critical point C, the liquid–gas coexistence curve ends. And, above T_c or P_c we can no longer distinguish between these two phases that we collectively call 'fluid'.

the system before the critical point was the atomic spacing, at the critical point all length scales become important and the material takes on fractal character. Second-order phase transitions are characterized by the growth of fluctuations on all length-scales. This means that 'critical fluctuations' are scale invariant since they extend from the atomic scale to over the entire system. As a consequence, pairs of measurable quantities depend upon each other in a power-law fashion, with well-defined ***critical exponents*** (Callen 1985). For instance, the size of liquid droplets is governed by a power law of $(T–T_c)$.

The line TC that goes from the triple point T to the critical point, separating the liquid and gas phases, is the ***liquid–gas coexistence line***. Suppose we want to describe the boiling phase transition occurring across this line. Once more we could try the simplest model we have for a gas, our old friend the ideal gas model. Its equation of state (the equation relating the three main thermodynamic variables, namely the pressure P, the molar volume v and the temperature T) is given by:

$$Pv = RT, \tag{4.16}$$

where R is the gas constant. We have seen that this model allows us to understand many properties of gases. However, the ideal gas model cannot describe the condensation of a gas into a liquid. This is because the ideal gas model neglects the intermolecular interactions, which are crucial to explain this phase transition in which the gas condenses into a liquid. Thus we need to incorporate these intermolecular forces into the gas equation of state.

The van der Waals equation of state; named after Johannes Diderik van der Waals (1873) generalizes the ideal gas law in a simple way. Firstly, to account for the volume that a real gas molecule occupies, the van der Waals equation replaces v in the ideal gas law with $(v - b)$, where b is the volume that is occupied by one mole of the molecules. This leads to:

$$P(v - b) = RT, \tag{4.16'}$$

The second modification made to the ideal gas law accounts for the fact that gas molecules do in fact attract each other and that real gases are therefore more compressible than ideal gases. Van der Waals reasoned that in order for a molecule to feel the attraction from another molecule both must be close enough. The chance for a molecule to occupy a small region around a point is proportional to the gas density at this point, or equivalently to the inverse of volume v. The chance of two molecules to be in this small region is then proportional to $1/v^2$. Therefore, van der Waals provided for intermolecular attraction by adding to the observed pressure P in the equation of state a term a/v^2, where a is a constant whose value depends on the gas. The van der Waals equation is therefore written as:

$$(P + a/v^2)(v - b) = RT. \tag{4.17}$$

The isotherm curves (constant temperature curves in the volume–pressure plane) of the van der Waals equation of state, are sketched in figure 4.12. For many gases the

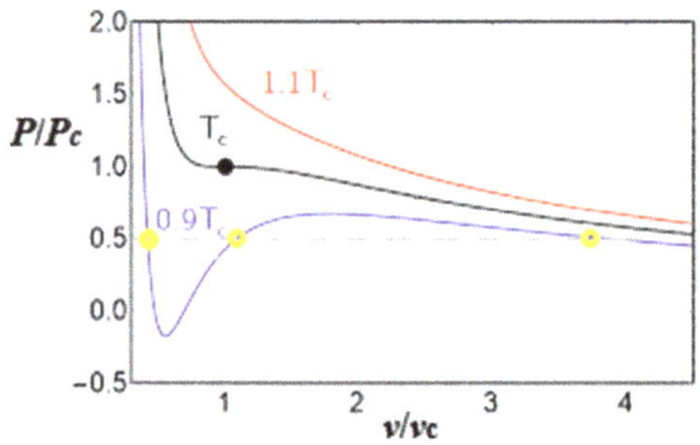

Figure 4.12. Sketch of the isotherms of the van der Waals gas for temperatures below the critical temperature T_c (blue curve), at the critical temperature T_c (black curve), and above the critical temperature T_c (red curve).

shape of the isotherm curves (constant temperature curves in the volume–pressure plane) is well represented by equation (4.17), at least semi-quantitatively (Callen 1985). As we can see, for temperatures below the critical temperature, T_c, the isotherm $v(P)$ (blue curve) exhibits a 'wiggle', similar to the one of the curves and is triple valued in P (three yellow disks). For the isotherm corresponding to the critical temperature $T = T_c$, called the ***critical isotherm*** (in black on the sketch), the folded part disappears (black curve). The isotherms above T_c (red curve) tend to be nearly hyperboloidal like the ones of the ideal gas, with the volume decreasing monotonically as the pressure is increased, according to Boyle's law (P inversely proportional to the volume at constant temperature).

The van der Waals equation can be rewritten in terms of the gas density, $n = 1/v$, in a form equivalent to the equation (4.5):

$$(P/a + n^2)(1/n - b) = RT/a. \tag{4.18}$$

Box 4.2 summarizes this formal correspondence between the two equations (Fort 2013). Therefore, the biomass density X would correspond to the fluid density, the liquid to the high biomass density attractor and the vapor to the low biomass density attractor. In consequence, 'islands' of vegetation on a desert background correspond to droplets of liquid in vapor. Likewise, bare patches on vegetation regions can be interpreted as bubbles of vapor in the bulk liquid.

It is enlightening to analyze in a P–v diagram the liquid–gas transition. The coexistence line for these two phases now becomes in figure 4.13 the ***binodal curve***, which denotes the condition at which the liquid and gas phases may coexist (dot-dashed curve). Figure 4.13 also includes the ***spinodal curve***, denoting the boundary of instability of pure phases (liquid or gas) to decomposition into a liquid–gas mixture. That is, the isotherms or lines of constant temperature in the P–v plane (blue curves in figure 4.13) produced by the van der Waals equation have a non-physical piece, called 'kink', in which the volume V grows when the pressure is simultaneously P growing. The spinodal lines (red dotted lines) are the locus of the end points (big red dots) delimiting kinks in the van der Waals isotherms. From a mathematical point of view spinodals are the locus of bifurcations. Hence three types of regions can be distinguished: the region outside the coexistence curve, which is absolutely stable,

Box 4.2. Mapping from the grazing model to the van der Waals equation of state (from Fort 2013).

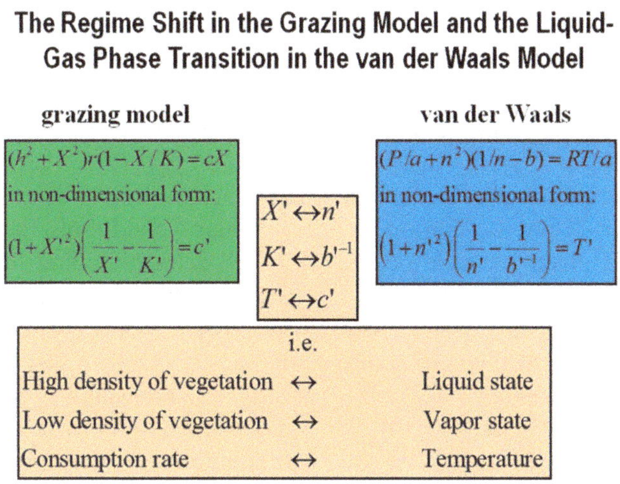

$$n' = \sqrt{a/P}\,n, \quad b' = \sqrt{P/a}\,b, \quad T' = RT/\sqrt{aP}.$$

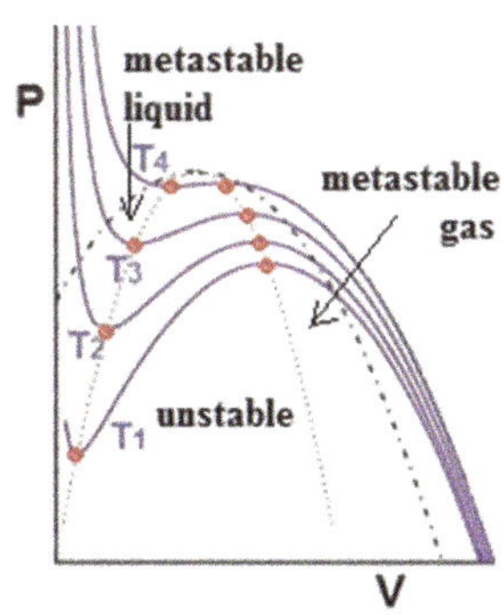

Figure 4.13. Filled blue curves are isotherms (constant temperature Pv curves) of the van der Waals equation of state for four different temperatures $T_1 < T_2 < T < T_4$. The dotted curve is the spinodal, denoting the limit of local stability with respect to small fluctuations and is mathematically defined by the condition that the derivative of the isotherms is zero. The dot-dashed curve is the ***coexistence curve*** or ***binodal curve***, denoting the condition at which liquid and gas may coexist. Equivalently, it is the boundary between the set of conditions in which it is thermodynamically favorable for the system to be fully mixed and the set of conditions in which it is thermodynamically favorable for it to phase separate. The extremum of a binodal curve in temperature coincides with that of the spinodal curve and is known as a critical point. Reprinted from Fort 2013.

the spinodal region (below the spinodal curve), which is absolutely unstable, and a region between these two, which is ***metastable***. In physics, a metastable state is a stable state of a dynamical system other than the system's state of least energy. An important difference is that a metastable state is not eternal (as the global minimum is). Being an excited state—i.e. with an energy above this global minimum or ***ground state***—it will eventually decay to a more stable state, releasing energy. Below the spinodal curve, infinitesimally small fluctuations in composition and density will lead to rapid phase separation via the mechanism known as ***spinodal decomposition*** (Goldenfeld 1992).

Which convention applies to the liquid–gas phase transition? To answer this let us return to the P–T phase diagram. Under ordinary conditions the liquid to gas transition for a given pressure occurs at a well-defined boiling temperature T_b, corresponding to the abscissa of the intersection between the horizontal fixed pressure line and the coexistence line (point 3 in figure 4.14). For example, at $P = 1$ atm, $T_b = 100$ °C. Thus, the coexistence line can be identified with S_M and the phase transition occurs according to the Maxwell convention. Surrounding the coexistence line we have in figure 4.13 the ***gas spinodal*** and ***liquid spinodal*** lines. Each spinodal line delimits a metastable state: the ***superheated*** water and ***supersaturated*** vapor, respectively. A superheated liquid is a liquid heated to a temperature higher than its boiling point without boiling, while a supersaturated vapor is a vapor cooled below its condensation point without condensing. Such superheated water and supersaturated vapor are metastable states where transitions to the stable phase are induced either by external or internal effects, and hence these two spinodal lines can be identified with S_B. For instance, superheating is achieved by heating a homogeneous sample of a substance in a clean container, free of nucleation sites (e.g. impurities or spots on the heating surface with lower wetting properties), while

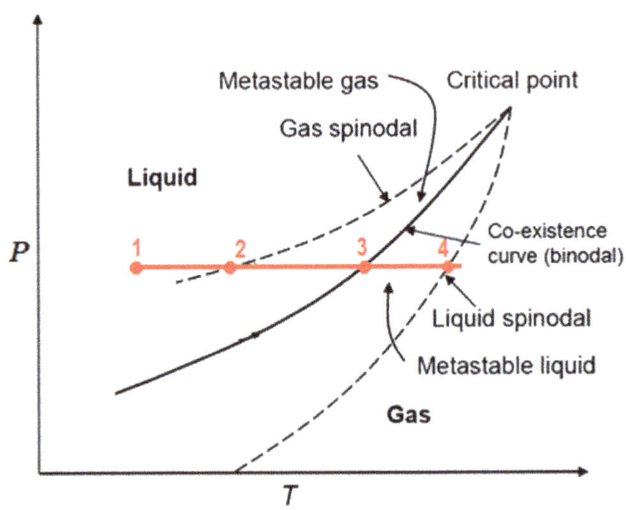

Figure 4.14. At fixed pressure the liquid water can be superheated above the boiling temperature defined by the binodal line (point 3), and corresponding to S_M, but not above the high-temperature liquid spinodal line (point 4). The spinodal lines play the role of S_B. Reprinted from Fort 2013.

taking care not to disturb the liquid. In this way, the transition from liquid to gas can also occur according to the delay convention, in some point between the binodal curve and the liquid spinodal (for example, in a process at constant pressure along the red horizontal line in figure 4.14, between points 3 and 4).

4.5.2 Similarities and differences between desertification and the liquid–vapor transition

The catastrophic shift we are considering occurs for example when c moves across the bifurcation set. To analyze similarities and differences between desertification and the liquid–vapor transition we compare the catastrophe flags introduced in section 4.2 for the fluid versus the ecosystem.

- Modality: the fluid is bimodal in the neighborhood of the liquid–gas coexistence curve, having well-defined liquid and gas states. These two states correspond to high and low vegetation density, respectively. Thus in this aspect both systems are similar.
- Sudden jumps: in the case of the fluid it is certainly true that sudden jumps occur, since there is an abrupt increase in volume when a liquid transforms into vapor. However, this large change in volume occurs when a slight change in the temperature and pressure moves the fluid from one side of the coexistence curve (identified with S_M) to the other. Hence, the liquid–vapor phase transition obeys under normal conditions the Maxwell convention. On the other hand, the shift in the MF grazing model always obeys the delay convention: the ecosystem remains in its upper attractor (higher vegetation density) until the bifurcation set is completely traversed. However, we have seen in section 4.2 that when perturbations are big enough to allow the switching between equilibria on different stability branches, the system might follow the Maxwell convention. Hence, to check this, let us consider the effect of a sudden environmental shift, like a severe drought. We can represent this environmental disturbance by a sharp decrease of the carrying capacity K followed by a recovery to its original value. Figure 4.15 illustrates this from a non-spatially explicit MF point of view: K is initially equal to 7.5, and for a value of the control parameter $c_1 = 1.68$ suddenly decreases to $K = 6$. Afterwards, K increases more slowly (at a comparable pace as c is increasing) until it reaches its original value just when the system crosses S_M at $c_M = 1.915$. It turns out that X returns to the upper attractor branch for a very close value of the grazing pressure, $c = 1.923$. We can check that varying the value of c_1 when the external environmental shift takes place in general only changes the shape of the wedge-like section of the curve $X(c)$ (exercise 4.7).

What happens in the case of the spatially heterogeneous and diffusive model? Figure 4.16 shows the evolution of the system for a completely similar change but in the average carrying capacity $\langle K \rangle$ rather than K (with $d = 0.1$).

Therefore, there is a dramatic difference between the MF picture (figure 4.15) and the spatially explicit picture (figure 4.16): in both cases the vegetation density drops

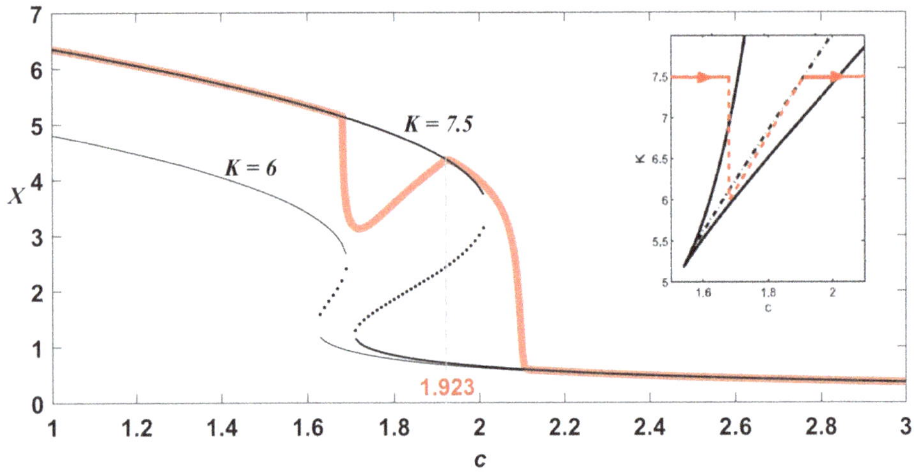

Figure 4.15. The effect on X of a global perturbation on K which suddenly decreases it from 7.5 to 6 and slowly recovers later (iso-K' curves for K = 7.5 and K = 6.0 are depicted in black). The red curve is the 'trajectory' followed by X, which returns to the upper attractor branch for c = 1.923. Inset: path followed by the system under perturbation in the c–K phase space.

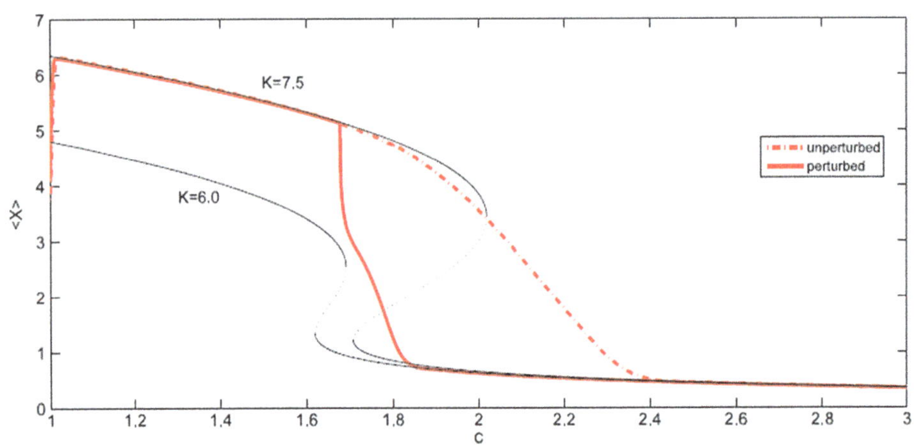

Figure 4.16. The effect on $\langle X \rangle$ of a global perturbation on $\langle K \rangle$ which suddenly decreases from $\langle K \rangle = 7.5$ to 6 and slowly recovers later (iso-K' curves for $K = 7.5$ and $K = 6.0$ are depicted in black) for $d = 0.1$. Red curves correspond to the 'trajectory' followed by $\langle X \rangle$ in the unperturbed (dot-dashed) and perturbed case (filled).

quickly when the carrying capacity instantaneously reduces from 7.5 to 6. However, in the first case the vegetation density recovers almost as fast as K and returns to the upper branch of the iso-K 7.5 curve, for c = 1.923 i.e. after the Maxwell point $c_M = 1.915$ but before this upper branch curve ends at $c = 2.01$. Conversely, for the spatial case, the drop continues until the system reaches the lower branch of the iso-K 6 curve. Thus, we conclude that, **when spatial heterogeneity and the locality of**

interactions are taken into account, perturbing the system this way allows us to change from the delay to the Maxwell convention (Fernández and Fort 2009).

- Hysteresis: in everyday situations one does not observe hysteresis in the liquid–gas phase transition of water—the liquid usually boils at the same temperature as that at which the vapor condenses. In other words, water's changes of state obey, under ordinary conditions, the Maxwell convention. Nevertheless, as we have seen, a careful experimentalist can obtain a hysteresis cycle by first raising the temperature and superheating the liquid above its boiling point without boiling, and after evaporation, cooling the gas below the condensation point (supersaturated vapor). Remember the coexistence curve is surrounded by the two spinodal lines (figure 4.13), which determine the limits to superheating and supersaturation, and can then be identified with S_B. Thus, even though hysteresis is not observed in water boiling under ordinary conditions, this does not represent a fundamental difference between both systems.
- Anomalous variance: when a fluid condenses (boils) from its gas (liquid) to its liquid (gas) state, small liquid droplets (bubbles) are formed. As a consequence, the variance of the volume may become large, which is similar to what happens for vegetation under grazing pressure.
- Cluster structure: the cluster structure and the kinetics of the transition are related with the mechanism behind the transition. The liquid–gas phase transition is first-order, the two distinguishable phases coexist during the transition inside the spinodal curve (in complete similitude to the desertification shift). The transformation from one phase to another in a first-order phase transition usually occurs via a ***nucleation*** process. Nucleation is the first step in the formation of a new thermodynamic phase, and it is usually how first-order phase transitions start. In general nucleation occurs around impurities that serve as seeds for nucleation of the new phase or on the heating surface, at *nucleation sites*. However, ***homogeneous nucleation*** is much simpler and easier to understand than this much more frequent ***heterogeneous nucleation***. For the former we have the ***classical nucleation theory*** (**CNT**) (Abraham 1974) to quantitatively study the kinetics of droplet formation. The classical nucleation theory assumes that even for a microscopic nucleus of the new phase or a 'droplet' (which can mean either a region of liquid phase embedded in the vapor phase or a bubble in the liquid phase), we can express the energy of such droplet as the sum of a bulk term that is proportional to its volume, and a surface term that is proportional to its surface area. The volume term is the difference in energy between the thermodynamic phase nucleation is occurring in, and the phase that is nucleating. For example, imaging the nucleation of a bubble of vapor occurring in the bulk of liquid water when the pressure is reduced so that the liquid becomes superheated with respect to the pressure-dependent boiling point. Since the vapor phase has smaller free energy this volume term is negative. On the other hand, the surface term corresponds to the work required to create the surface of the bubble against the ***surface tension***, which is always positive. That is, at liquid–gas interfaces, surface tension results from the greater attraction of liquid molecules

> to each other, due to cohesion, than to the gas molecules, due to adhesion. The net effect is an inward force at the surface of the bubble as if this surface were covered with a stretched elastic membrane. When the radius r of the bubble is small, the surface term (proportional to r^2) 'wins' over the volume term (proportional to r^3). Conversely, above a critical radius r^* the volume term becomes larger than the surface term and the bubble can expand spontaneously (i.e. it can jump the energy barrier preventing the nucleation of a bubble). In a completely similar way, for the nucleation of a droplet in the bulk of the vapor phase there is a critical radius beyond which liquid drop can grow spontaneously. In the case of desertification, the analogs of bubbles and liquid droplets are, respectively, bare and vegetation patches. For instance, small patches, for which the area of the surface region is large with respect to their 'volume', are under stronger stress from their neighborhood, while the effect of surface stress is vanishingly small for large patches (Weissmann and Shnerb 2016). This phenomenon, which is analogous to the opposing effects of surface tension and bulk free energy that govern the physics of nucleation in first-order transitions, explains the main characteristic of bistable systems that allows for catastrophic transitions, is positive feedback. In a local patch, small populations go extinct and large populations are self-sustained. This outcome is in agreement with empirical measurements using satellite images of the African Sahel region which reveal that small patches tend to shrink but large patches tend to grow (Weissmann *et al* 2017).

An important remark is that this power law behavior for vegetation domains or patches exhibited as early-warning indicators in regime shifts has nothing to do with *critical point* phenomena. As we have seen, a critical point in statistical mechanics or thermodynamics occurs under conditions, such as specific values of temperature and pressure, at which no phase boundaries exist (e.g. at the critical point of water the properties of liquid and vapor become indistinguishable) and is associated with second-order phase transitions. Nucleation, on the other hand, is associated with first-order transitions in which both *distinguishable* phases coexist. The power law behavior, synonym of scale invariance, can be explained from the fact that nucleation near a spinodal appears to be very different from classical nucleation. The surface tension of a droplet vanishes as the spinodal is approached, and hence the assumption of classical nucleation theory that the nucleating droplet is compact is no longer applicable (Gould and Klein 1993). That is, near the spinodal, the nucleating droplet is a diffuse, fractal-like (Klein and Unger 1983), and the process of nucleation is due to the coalescence of these droplets, rather than the growth of a single one (Monette and Klein 1992). As a consequence the domain size grows with a power law for spinodal decomposition (Goldenfeld 1992).

4.6 Final comments

We have analyzed first a non-spatially explicit or mean-field (MF) ecological model with alternative stable states or attractors, that has been widely used to describe

Glossary of Thermodynamic terms

Droplet: can mean either a region of liquid phase embedded in the vapor phase or a bubble in the liquid phase.

Hysteresis: is the dependence of a system not only on its current state but also on its history.

This dependence arises because the system has alternative stable states. Therefore, to predict its future development, in addition to its present state, the knowledge of its history is also required. A paradigmatic example of hysteresis in physics is ferromagnetism. When an external magnetic field is applied to a ferromagnetic material such as iron, the atomic dipoles align themselves with this field. Even when the external magnetic field is removed, part of the alignment will be retained: the material has become magnetized. Once magnetized, the magnet will stay magnetized indefinitely. To demagnetize it requires heat or a magnetic field in the opposite direction. Hysteresis implies that the forward 'path' described by the succession of states of a system when a control parameter (the external magnetic field in the case of ferromagnetism or the grazing pressure in desertification) increases is different from the backward path when this parameter is decreased, giving rise to a *hysteresis loop*.

Nucleation process: is the start of the process of forming a new thermodynamic phase in first-order phase transitions by formation of 'droplets' or nuclei of new thermodynamic phase, in the bulk of a different phase, via self-assembly. Typical examples of nucleation are the formation of gaseous bubbles in liquids (boiling) or liquid droplets in saturated vapor (condensation).

Phase transition: is a change of certain properties of a given medium, often discontinuously, as a result of the change of external conditions, such as temperature, pressure, etc. For example, when heating a liquid to its boiling point it becomes gas, resulting in an abrupt change in volume.

Order of a phase transition: phase transitions can be divided into two large groups. First-order transitions, in which distinguishable phases coexist and the medium absorbs or releases a fixed amount of energy per volume known as **latent heat**. For example, when boiling water in a pan on the stove provides this latent heat the water absorbs and we can observe coexistence of liquid and gas bubbles (both phases are distinguishable because they have very different densities). In second-order transitions the phases are no longer distinguishable and no latent heat must be provided or released. This is why second-order transitions are also called 'continuous' phase transitions. The indistinguishability of both phases implies power law behavior for the size of droplets of the new phase embedded in the bulk of the original phase (however, the reciprocal is not true). This scale invariance is an important phenomenon of second-order phase transitions.

Spinodal lines: are lines denoting the limits of metastability, i.e. the boundary of instability of pure phases, liquid or gas, to decomposition into a liquid–gas mixture.

Superheating or ***boiling delay***: is the phenomenon in which a liquid is heated to a temperature higher than its boiling point, without boiling. This is a metastable state, where boiling might occur at any time, induced by external or internal effects.

Supersaturation: is the phenomenon in which vapor is cooled below its condensation point, without condensing. This is a metastable state.

van der Waals equation of state: the van der Waals equation represents a step further than the simpler ideal gas equation of state because it takes into account the molecular interaction forces (neglected by the ideal gas approximation). These intermolecular forces are crucial to yield a phase transition from gas to a liquid state, and this is why the van der Waals equation is able to model the liquid–gas phase transition, e.g. water boiling in a pan (the ideal gas equation of state only serves to describe the gas phase).

different relevant processes, ranging from pest outbreaks to habitat desertification and harvesting of aquatic plants. We used this model to introduce several early warning signals of ongoing catastrophic shifts. Providing early warning signals is central both for management and recovery strategies for ecosystems.

Next, we considered a spatially explicit version of this model with local interactions and subject to random spatial dispersion. For large enough values of the diffusion coefficient d, the system self-organizes, producing characteristic spatial patterns. We have considered several spatial quantities both to characterize the state of the ecosystem and to use them as early warnings. One such observable is the spatial variance σ^2_X, obtained by measuring samples of X on a grid of points. In the next Application chapter we will show that a grid containing relatively few points might be sufficient for the purpose of extracting an appropriate signal, and that a significant growth in σ^2_X could serve as an early warning of an imminent transition. The spatial variance shows an advantage over the temporal one, as σ^2_X rises before σ^2_t. The explanation for this is simple: it occurs since the former corresponds to a snapshot of the present state of the system while the latter includes in its computation data for previous times where the fluctuations were still small. The origin of the rise in σ_X is tied to the emergence of spatial patterns, in the form of patches of high/low concentration of X. We then conclude that the visualization of the onset of those patches, for example by aerial or satellite imaging, may be another indicator of the imminence of a catastrophic shift and an effective way of anticipating this transition. Furthermore, we have shown that at the very maximum of σ_X the distribution of sizes of patches becomes power law, so this particular distribution could serve as an early warning. Power law distribution has also been found in other systems as a signature of self-organization (Pascual *et al* 2002, Vandermeer *et al* 2008).

Another spatially based observable of interest is the two-point correlation. We found that as long as the diffusion coefficient d increases, the peak in σ^2_X decreases and the correlation increases. This dependence on d connected to the spatial patterns that emerge as d increases is the result of two factors that point in opposite directions: the intrinsic spatial heterogeneity of the ecosystem versus the dispersion or diffusion. Therefore, for low diffusion, σ^2_X is the most appropriate of these two indicators for detecting catastrophic shifts while the correlation works less well. On the other hand, for high diffusion, the correlation may become a more useful quantity to analyze.

We have shown that the grazing model can be mapped on the very popular van der Waals thermodynamic model, used to describe the liquid–gas phase transition.

This formal correspondence serves to provide useful insights on different aspects of the issue of desertification by overgrazing in particular as well as catastrophic shifts in general. For example, when changing the control parameter c of the grazing model, the transition from one attractor to the other is according to the delay convention, both for the mean field and the spatially explicit model. Nevertheless, we showed that for the spatial model this transition occurs according to the Maxwell convention when a large enough shift in the environmental conditions is considered. This is similar to what happens with the liquid–gas transition, which occurs, under ordinary conditions, according to the Maxwell convention: water either boils or condenses when it reaches the saturation temperature. However, with sufficient care, it is possible to superheat water or supercool vapor, in such a way that the transition follows the delay convention (although any disturbance will produce an immediate change of phase). Another benefit from linking ecological catastrophic shifts with phase transitions in physical systems is that in ecology the studies on early warnings in general focus either on spatial early warnings (for instance see Rietkerk *et al* 2004, Kéfi *et al* 2007) or on temporal signals (e.g. Brock and Carpenter 2006, Carpenter and Brock 2006, Dakos *et al* 2008). On the other hand, in statistical thermodynamics, the connection between spatial and temporal phenomena for systems near a phase transition—like hysteresis, critical slowing down, long range order, etc—is well known. We used this in our attempt to link both types of early warning signals.

Extending the analogy we can visualize the occurrence of different processes in desertification inspired by the phenomenology of the liquid–gas transition. For instance, superheating a liquid causes irregular boiling or 'bumping' (Chang 2007) until a loss of thermodynamic stability of the liquid phase occurs upon intersecting the spinodal line. At this point, the existence of the metastable state becomes impossible, and it goes over into the two-phase state by means of an explosive transition (Martynyuk 1977). Such a phase explosion near the spinodal may be particularly important when heating foods and drinks in a microwave oven due to the explosive production of steam from the superheated water. Hence, superheating allows a change of the phase transition convention from Maxwell to delay. On the other hand, ecological resilience can be generally defined as the maximum disturbance that a system can tolerate without switching to the alternative stable state (Holling 1973). In the case of a spatially explicit model of an ecosystem, the above definition can be extended to local disturbances and ecological resilience can be defined as the size of the area of a strong local disturbance needed to trigger a shift (van de Leemput *et al* 2015). As environmental conditions change, a sudden drop in resilience occurs at the Maxwell point, where one state becomes more resilient than the other, rather than the classical gradual decrease towards the fold bifurcation, characteristic of the delay convention. In summary, the delay (Maxwell) convention seems to be the norm in ecological catastrophic shifts (thermodynamics phase transitions). However, a change of convention is possible provided the metastable states are impeded (promoted).

A remark is in order; while the theory of phase transitions is well established in thermodynamic equilibrium, its use in nonequilibrium systems is less understood. However, in nonequilibrium systems a (dynamical) scaling of variables may occur even in first-order transitions, when the order parameter, for example the difference between the volume of liquid and gas, jumps at the transition. This is exactly the kind of phenomenon that we are observing for the spatial ecological model we studied in this chapter.

To conclude this chapter: how helpful are all these warning signals in designing effective management protocols? Leaving aside economic considerations (which are beyond the scope of this chapter) this depends on different factors. For example, on the degree of diffusion (the size of *d*): as we will see in the next Application chapter, the larger the diffusion, the earlier the corrective action should be taken.

Exercises

Exercise 4.1. *Spruce budworm* model

(A) Obtain graphically and numerically the equilibria for the *spruce budworm* model (equation (4.2)) with $\mathscr{r} = 1$ and $\mathscr{K} = 1$, testing different values of $\mathscr{c}$ for $\mathscr{h} = 0.1$ and 0.2.

(B) For the model with dimensionless magnitudes (equivalent to taking $\mathscr{r} = \mathscr{h} = 1$), plot X^* versus K for a fixed value of c leading to four equilibria or alternative stable states.

(C) Reproduce the plot in figure 4.2.

Exercise 4.2. The 'fold curve' of the *Spruce budworm* model in the control parameter space *c–K*

Obtain the equations of the curves S_{B} that delimit the region of alternative stable states which are the solutions of equation (4.5) in the phase diagram for the non-dimensioned model ($r = h = 1$) and graph them for K in the interval [5,10] and c in the interval [1, 3].

Exercise 4.3. A simple model with a fold bifurcation

(A) Show that the first-order differential equation:

$$\frac{dx}{dt} = r - x - e^{-x},$$

undergoes a fold bifurcation, like the two bifurcations in the spruce budworm model, by varying the parameter r and find the r_B value of this parameter for which the bifurcation occurs.

(B) Plot the equilibrium value x^* versus r for $r < r_B$ and $r > r_B$.

Exercise 4.4.

Write a script that solves equation (4.3) by transforming it into a finite difference equation.

(A) For $K = 5$ and $c = 1$, let the system start from different arbitrary initial values of the vegetation density X_0 and check that the vegetation density reaches an equilibrium X^* after a number of iterations.

(B) Keeping K fixed to 5, start from the above equilibrium value X^* as initial condition and varying the parameter c from 1 to 3 in steps of 0.01 (300 steps) compute the corresponding curve followed by X.

(C) Repeating (A) and (B) for $K = 6, 7, \ldots, 10$, check that you obtain the iso-K curves of figure 4.3.

Exercise 4.5. Critical slowing down

Near some bifurcation points, due to the loss of linear stability, the rates of decay of perturbations become extremely long.

(A) Obtain the analytical solution to the differential equation $dx/dt = -x^3$ for an arbitrary initial condition. Show that $x(t) \to 0$ as $t \to \infty$, but that the decay is not exponential but rather is according to a much slower varying algebraic function of t.

(B) To get some intuition about the slowness of the decay, make a numerically accurate plot of the solution for the initial condition $x_0 = 10$, for $0 \leqslant t \leqslant 10$. Then, on the same graph, plot the solution to $dx/dt = -x$ for the same initial condition.

Exercise 4.6. Anomalous variance

Suppose we have an ensemble of replicas of a system specified by a probability distribution $P(X)$ defined over the space of state variable X. Assuming that $P(X)$ is a decreasing function of the potential $V(X)$, i.e. $P(X) = P(V(X))$. Therefore, since $V(X)$ is minimum for an equilibrium state X^*, $P(V(X^*))$.

(A) By Taylor expanding $\log(P(V(X)))$ around an equilibrium point, show that $P(X)$ can be approximated by

$$P(X) \approx Ne^{-a\lambda(X-X^*)^2},$$

where N is a normalization constant, a is a positive constant and $\lambda = \left(\frac{\partial^2 V}{\partial X^2}\right)_{X^*}$.

(B) Compute the variance using equation (4.8) and check that we obtain:

$$\overline{(X - \bar{X})^2} = \frac{1}{2}\frac{1}{a\lambda}$$

Exercise 4.7. Simulating an environmental shift

(A) Modify the script you wrote in exercise 4.4 to include a sudden drop of the carrying capacity, from $K = 7.5$ to $K = 6$, at a value of the grazing pressure parameter $c_1 = 1.68$, and recovering to $K = 7.5$ for $c_2 = 1.915$ ($d = 0.1$), so you can reproduce figure 4.15.

(B) Change the values of c_1 and c_2 and plot the curve of X versus c.

(C) Repeat for $d = 0.5$.

Exercise 4.8.

(A) Write a script in MATLAB or C or R or FORTRAN, etc that implements the simulation of equation (4.11) according to the specifications of box 4.1.
(B) Use this script to reproduce figures (4.6) and (4.7).
(C) Consider a sudden drop of the mean carrying capacity, from $\langle K \rangle = 7.5$ to $\langle K \rangle = 6$, at a value of the grazing pressure parameter $c_1 = 1.68$, and recovering to $K = 7.5$ for $c_2 = 1.915$, so you can reproduce figure 4.16.

References

Abraham F F 1974 *Homogeneous Nucleation Theory* (New York: Academic)

Adeel Z, Safriel U, Niemeijer D and White R 2005 Ecosystems and human well-being: desertification synthesis *A Report of the Millennium Ecosystem Assessment* (Washington DC: World Resources Institute)

Aguiar M R and Sala O E 1999 Patch structure, dynamics and implications for the functioning of arid ecosystems *Tree* **14** 273–7

Brock W A and Carpenter S R 2006 Variance as a leading indicator of regime shift in ecosystem services *Ecol. Soc.* **11** 9

Callen H B 1985 *Thermodynamics and an Introduction to Thermostatistics* (New York: Wiley)

Carpenter S 2001 Alternate states of ecosystems: evidence and some implications *Ecology: Achievement and Challenge* ed N Huntly, M C Press and S Levin (Oxford: Blackwell) pp 357–83

Carpenter S R and Brock W A 2006 Rising variance: A leading indicator of ecological transition *Ecol. Lett.* **9** 311–8

Chang H 2007 When water does not boil at the boiling point *Endeavour* **31** 7–11

Dakos V, Scheffer M, van Nes E H, Brovkin V, Petoukhov V and Held H 2008 Slowing down as an early warning signal for abrupt climate change *Proc. Natl. Acad. Sci.* **105** 663–9

Dakos V, van Nes E H, Donangelo R, Fort H and Scheffer M 2010 Spatial correlation as leading indicator of catastrophic shifts *Theor. Ecol.* **3** 163–74

Donangelo R, Fort H, Dakos V, Scheffer M and van Nes E H 2010 Early warnings of catastrophic shifts in ecosystems: Comparison between spatial and temporal indicators *Int. J. Bifurcat. Chaos* **20** 315–21

Dregne H E 1986 Desertification of arid lands *Physics of Desertification* ed F El-Baz and M H A Hassan (Dordrecht: Martinus Nijhoff)

Fernández A and Fort H 2009 Catastrophic phase transitions and early warnings in a spatial ecological model *J. Stat. Mech.* **2009** P09014

Fort H 2013 Statistical mechanics ideas and techniques applied to selected problems in ecology *Entropy* **15** 5237–76

Gilmore R 1981 *Catastrophe Theory for Scientists and Engineers* (New York: Dover)

Goldenfeld N 1992 *Lecture on Phase Transitions and the Renormalization Group* (Reading, MA: Perseus Books) pp 219–22

Gould H and Klein W 1993 Spinodal nucleation effects in systems with long-range interactions *Physica* D **66** 61–70

Holling C S 1959 The components of predation as revealed by a study of small mammal predation of the European pine sawfly *Can. Entomol.* **91** 293–320

Holling C S 1973 Resilience and stability of ecological systems *Annu. Rev. Ecol. Syst.* **4** 1–23

Klausmeier C A 1999 Regular and irregular patterns in semiarid vegetation *Science* **284** 1826–8

Kéfi S, Rietkerk M, Alados C L, Pueyo Y, Papanastasis V P, ElAich A and de Ruiter P C 2007 Spatial vegetation patterns and imminent desertification in Mediterranean arid ecosystems *Nature* **449** 213–7

Klein W and Unger C 1983 Pseudospinodals, spinodals, and nucleation *Phys. Rev.* B **28** 445

Levin S A and Paine R T 1974 Disturbance, patch formation, and community structure *Proc. Natl. Acad. Sci.* **71** 2744–7

Levin S A and Pacala S W 1997 Theories of simplification and scaling of spatially distributed processes *Spatial Ecology: The Role of Space in Population Dynamics and Interspecific Interaction* ed D Tilman and P M Kareiva (Princeton, NJ: Princeton University Press) pp 271–96

Ludwig D, Jones D D and Holling C S 1978 Qualitative analysis of insect outbreak systems: The spruce budworm and forest *J. Anim. Ecol.* **47** 315–22

May R M 1977 Thresholds and breakpoints in ecosystems with a multiplicity of stable states *Nature* **269** 471–7

May R M 1974 *Stability and Complexity in Model Ecosystems* (Princeton, NJ: Princeton University Press)

Martynyuk M M 1977 Phase explosion of a metastable fluid *Combust. Explos. Shock Waves* **13** 178–91

Monette L and Klein W 1992 Spinodal decomposition as a coalescence process *Phys. Rev. Lett.* **68** 2336–9

Murray J D 1993 *Mathematical Biology* (Berlin, Germany: Springer)

Noy-Meir I 1975 Stability of grazing systems: An application of predator–prey graphs *J. Ecol.* **63** 459–81

Pascual M, Roy M, Guichard F and Flierl G 2002 Cluster size distributions: signatures of self-organization in spatial ecologies *Philos. Trans. R. Soc. Lond.* B **357** 657–66

Petraitis P 2013 *Multiple Stable States in Natural Ecosystems* (Oxford: Oxford University Press)

Pickett S T A and White P S 1985 *The Ecology of Natural Disturbance and Patch Dynamics* (Orlando, FL: Academic)

Rietkerk M, Dekker S C, de Ruiter P C and van de Koppel J 2004 Self-organized patchiness and catastrophic shifts in ecosystems *Science* **305** 1926–8

Scheffer M *et al* 2001 Catastrophic shifts in ecosystems *Nature* **413** 591–6

Staal A *et al* 2016 Bistability, spatial interaction, and the distribution of tropical forests and savannas *Ecosystems* **19** 1080–91

Steinberg E K and Kareiva P 1997 Challenges and opportunities for empirical evaluation of spatial theory *Spatial Ecology: The Role of Space in Population Dynamics and Interspecific Interaction* ed D Tilman and P M Kareiva (Princeton, NJ: Princeton University Press) pp 318–32

Thom R 1989 *Structural Stability and Morphogenesis: An Outline of a General Theory of Models* (Reading, MA: Addison-Wesley)

Tilman D, Lehman C L and Kareiva P 1997 Population dynamics in spatial habitats *Spatial Ecology: The Role of Space in Population Dynamics and Interspecific Interaction* ed D Tilman and P M Kareiva (Princeton, NJ: Princeton University Press) pp 3–20

Tilman D *et al* 2001 Forecasting agriculturally driven global environmental change *Science* **292** 281–4 van de Leemput I A, van Nes E H and Scheffer M 2015 Resilience of alternative states in spatially extended ecosystems *PLoS One* **10** e0116859

van der Waals J D 1873 Over de Continuiteit van den Gas- en Vloeistoftoestand *PhD Thesis* University of Leiden, Netherlands

van Nes E H, Hirota M, Holmgren M and Scheffer M 2014 Tipping points in tropical tree cover: linking theory to data *Glob. Change Biol.* **20** 1016–21

Vandermeer J, Perfecto I and Philpott S M 2008 Clusters of ant colonies and robust criticality in a tropical agroecosystem *Nature* **451** 457–9

Vitousek P M, Mooney H A, Lubchenco J and Melillo J M 1997 Human domination of Earth'secosystems *Science* **277** 494–9

Von Hardenberg J, Meron E, Shachak M and Zarmi Y 2001 Diversity of vegetation patterns and desertification *Phys. Rev. Lett.* **87** 1981011

Von Hardenberg J, Kletter A Y, Yizhaq H, Nathan J and Meron E 2010 Periodic versus scale-free patterns in dryland vegetation *Proc. R. Soc.* B **277** 1771–6

Wade N 1974 Sahelian drought: No victory for western aid *Science* **185** 234–7

Watt A S 1947 Pattern and process in the plant community *J. Ecol.* **35** 1–22

Weissmann H and Shnerb N M 2016 Predicting catastrophic shifts *J. Theor. Biol.* **397** 128–32

Weissmann H, Kent R, Michael Y and Shnerb N M 2017 Empirical analysis of vegetation dynamics and the possibility of a catastrophic desertification transition *PLoS ONE* **12** e0189058

Wiesmeier M 2015 Environmental indicators of dryland degradation and desertification *Environmental Indicators* ed R H Armon and O Hänninen (Berlin: Springer) ch 14

Wolfram S 1994 *Cellular Automata and Complexity, Collected Papers* (Reading, MA: Addison-Wesley)

Wolfram S 2002 *A New Kind of Science* (Champaign, IL: Wolfram Media)

World Resources Institute 1988 *World Resources 1988–89* (New York: Basic Books)

IOP Publishing

Ecological Modelling and Ecophysics
Agricultural and environmental applications
Hugo Fort

Chapter A4

Modelling eutrophication, early warnings and remedial actions in a lake

… by foreseeing them at a distance, which is only done by men of wisdom, the evils which might arise from them are soon cured; but when, from want of foresight, they are suffered to increase to such a degree they are perceptible to everyone, there is no longer a remedy.

—Machiavelli, *The Prince* (1513).

1. *Eutrophication*, is the overenrichment of water with minerals and nutrients which induces excessive growth of algae. This process may result in oxygen depletion of the water body. One example is an 'algal bloom' or great increase of phytoplankton in a water body as a response to increased levels of nutrients. Eutrophication is often induced by the discharge of nitrate or phosphate-containing detergents, fertilizers, or sewage into an aquatic system. Eutrophication is a serious environmental problem since it results in a general deterioration of water quality and other effects that reduce and preclude its use.
2. Historical records in several lakes show that for a given level of nutrients in water (for instance phosphorus) two alternative stable states can occur: clear and turbid water. Indeed, the transition from clear to turbid water when blooms of noxious blue–green algae take place is very rapid and difficult to forecast. In addition, some lakes have recovered after sources of nutrients were reduced. In others, recycling of phosphorus from sediments enriched by years of high nutrient inputs causes lakes to remain eutrophic even after external inputs of phosphorus are decreased thus displaying hysteresis. As we have seen in chapter 4, modality, sudden jumps and hysteresis are all wave flags of ongoing catastrophic shifts.

doi:10.1088/978-0-7503-2432-8cha4

3. In a pioneering work, Steve Carpenter (2005) developed a population (mean-field) model to describe phosphorus dynamics and the roles of internal recycling and slow dynamics of soil phosphorus in lakes and their watersheds. Values of the parameters were estimated for the watershed of Lake Mendota in Madison, Wisconsin. Lake Mendota is probably one of the most studied lakes in the world. Nevertheless, it has unpredictable and undesirable behavior as evidenced by irruptive blooms of noxious blue–green algae, explosive colonization by exotic species, and depletion of fish population.
4. In this application our goal is to get spatio-temporal early warnings for the catastrophic shift from clear to turbid water. Thus we extend Carpenter's model for Lake Mendota to have a spatially explicit description of phosphorous dynamics. To do this, we represent the lake as a bi-dimensional grid. We use the parameters estimated for Mendota Lake (external nutrient loading, rates of recycling, sedimentation and outflow of phosphorus) to feed a cellular automaton (CA) model whose cell variables are the phosphorus densities in surface sediment and in water.
5. We run a simulation across 1500 years. The first 250 years correspond to a simplified history of Lake Mendota from settlement to our days. The next 1250 years are used to predict the future evolution of this lake, including a drastic eutrophication shift. We compute both the temporal and spatial variances of phosphorous density in water. Since the spatial variance considers only the present values, it provides an earlier signal announcing the shift which is not blurred by averages including situations where these indications are absent.
6. We use the cellular automaton model to explore the feasibility as well as the effectiveness of a remedial action designed to reduce and control phosphorus loading into the lake.

A4.1 Background information

There is increasing evidence that ecosystems can pass thresholds and go through *catastrophic regime shifts* (Scheffer and Carpenter 2003) where sudden and large changes in their functions take place. Most importantly, these changes are very difficult and costly to reverse. Thus, an important problem in environmental sciences is to get early warning signals of these impending catastrophic regime shifts in ecosystems to allow addressing currently intractable problems in ecosystem management, such as the avoidance of ecological surprises, and the maintenance of systems in desired states.

An example of such a regime shift is provided by many shallow (thermally unstratified) lakes which display alternate stable states. One is a clear water state, with low algae population but abundant rooted aquatic plants, whereas the other is a turbid state where shading by abundant algae suppresses rooted plants (Scheffer 1997). Some lakes change between these states from time to time, whereas others persist for years in either the clear water or turbid state. Quite often these lakes suddenly switch from clear to turbid water due to algae blooms. These blooms are connected to *eutrophication*, i.e. the overenrichment of aquatic ecosystems with nutrients, principally phosphorus (Carpenter 2008). The role of phosphorus is firmly established. Blue–green blooms can be triggered by adding phosphorus to whole lakes (Schindler 1988). An aquatic environment with a limited availability of phosphorus and nitrogen is described as *oligotrophic* while one with high availability of these elements is called *eutrophic*. When the eutrophication phenomenon becomes particularly intense, undesirable effects and environmental imbalances are generated.

The two most acute phenomena of eutrophication are *hypoxia* (lack of oxygen) in the deep part of the lake and algal blooms that produce harmful toxins, processes that can destroy aquatic life in the affected areas. Indeed, blooms of blue–green algae are a serious water quality problem in many productive lakes (Paerl and Sandgren 1988), because when they occur, many of the ecosystem services which humans derive from these systems, such as potable water, fisheries and places for recreation, can be lost or deteriorated. The estimated cost of damage mediated by eutrophication in the US alone is approximately $2.2 billion annually (Dodds *et al* 2009).

Therefore, reducing eutrophication should be a key concern when considering future policy, and a sustainable solution for everyone, including farmers and ranchers, seems feasible. Today, the main control mechanism of the eutrophic process is based on removal of the nutrients that are introduced into water bodies (Chislock *et al* 2013). Water quality can often be improved by reducing the concentrations of one of the two main nutrients, nitrogen and phosphorus. In particular phosphorus, which is considered to be the limiting factor for the growth of algae (Carpenter 2005), by acting on localized loads (associated with waste water) and widespread loads from diffuse sources (such as land and rain).

However, nutrient reduction can be difficult (and expensive) to control, especially in agricultural areas where the algal nutrients come from nonpoint sources. Furthermore, in lakes where external loading of nutrients has been reduced, internal loading of nutrients from sediments may prevent improvements in water quality (Søndergaard *et al* 2003). Likewise, it is often difficult, costly and/or impossible to reverse these changes once a certain threshold has been crossed. Lake eutrophication has proven to be a stubborn environmental problem (National Research Council 1992, Carpenter 2005). Instead of alternating regimes, many lakes remain eutrophic for extended periods of time. This is why early warnings of these shifts are so important to ecosystem management.

A general problem is that our world is changing at an unprecedented rate and we need to understand the nature of the change and to forecast the way in which it might affect ecosystems of interest. In order to address questions about the impact of environmental change, and to understand what might be done to mitigate the predicted negative effects, ecologists need to develop the ability to project models into novel, future conditions. This will require the development of models based on understanding the (dynamical) processes that result in a system behaving the way it does (Fort 2012). A main difficulty with the majority of current ecological models is that they are excellent at describing the way in which a system has behaved, but they are poor at predicting its future state, especially under novel conditions (Evans 2012).

In chapter 4, we have mentioned that a common simplification of these mathematical ecological models is that they rely on the well-mixed assumption or, in the physics parlance, the mean-field (MF) approximation. Using such a mean-field, i.e. non-spatially explicit, model it was shown that the rising of the temporal variance for the nutrient concentration works as an early warning signal (Carpenter and Brock 2006).

Later on it was shown that, if one takes into account explicitly the space, the spatial variance provides an even earlier early warning (Fernández and Fort 2009, Donangelo *et al* 2010, Dakos *et al* 2010). This is an interesting motivation for relaxing the MF assumption and formulating a spatially explicit model, whose straightforward implementation is by means of a cellular automaton. Another more compelling reason to take space into account is that, as we have shown in chapter 4, the MF assumption breaks down when there is spatial heterogeneity, produced for instance by a gradient in some parameter that controls the dynamics. This is the situation in the problem that we will analyze here. Thus, we will present a cellular automaton, the *Mendota Lake cellular automaton* (MLCA) (Fort 2012), that models a lake as a square lattice with the phosphorous concentration as the dynamical variable defined on lattice cells.

A4.2 Overview

What is our goal? Our goal is twofold: to forecast sudden switches of lakes from clear to turbid water, as well as to simulate remedial actions to assess their usefulness to keep/restore the clear water state.

What do we want to know? We want to know what variable to monitor in lakes to anticipate the shift from clear to turbid water, as well as to estimate safety thresholds for phosphorus (P) in water.

What do we know? Nutrient loadings and lake water quality are inextricably linked. Excessive loadings of nitrogen and, particularly, phosphorus cause lakes to become eutrophic and often to exhibit noxious algal blooms (Schindler 1977). We know long-term nutrient and water clarity data for Lake Mendota (Lathrop 1992). In particular, there are estimates of external nutrient loading, rates of recycling, sedimentation and outflow of P per unit lake area.

How should we look at this model? We should regard this model as a general model of phosphorus dynamics in stratified lakes and their watersheds that uses a highly simplified representation of both the spatial structure and the history of Lake Mendota.

What will our model predict? It will predict, with some anticipation, an ongoing shift from clear to turbid water due to eutrophication.

Are the predictions valid? Results suggest that dynamics of soil phosphorus may control alternate stable states, potentially causing eutrophication to last for centuries.

Are the predictions useful? We will analyze the practicality and usefulness of early warnings for designing remedial actions.

How can we improve the model? At the end of this chapter we will discuss some caveats and possible further developments.

A4.3 Data for Lake Mendota

Lake Mendota is the northernmost and largest of the four lakes in Madison, Wisconsin (figure A4.1). Although the lake is comparatively small (surface area 40 km^2) and shallow (mean depth 12.2 m), it is significant as the birthplace of modern limnology and an early instance of artificially accelerated eutrophication (World Lake Database 2019).

In Lake Mendota, blue–green algal blooms have occurred since at least the 1880s (Lathrop 1992). In fact, owing to the influx of domestic wastewater, the lake has suffered from eutrophication since the beginning of the 20th century. From 1912 to 1958, copper sulfate was applied over the lake to reduce the overgrowth of algae. The algal bloom was effectively suppressed but the lake became polluted with copper, which still remains accumulated in the bottom sediments as insoluble copper carbonate (World Lake Database 2019).

External loadings of P to Lake Mendota prior to the mid-1950s are impossible to quantify. Nutrient loadings most likely increased during the 1920s when sewage discharges from upstream communities first entered Lake Mendota. During the 1940s, increases in both in-lake and stream dissolved reactive P concentrations, indicated that P loadings from the sewage effluents increased substantially (Lathrop 1992). Loadings probably continued to increase as agricultural fertilizers were used extensively after World War II. In the ensuing years, loadings increased because of additional urban storm water runoff as Madison grew. P loadings to Lake Mendota remained high until the early 1970s. Then, since the late 1970s, loadings have been

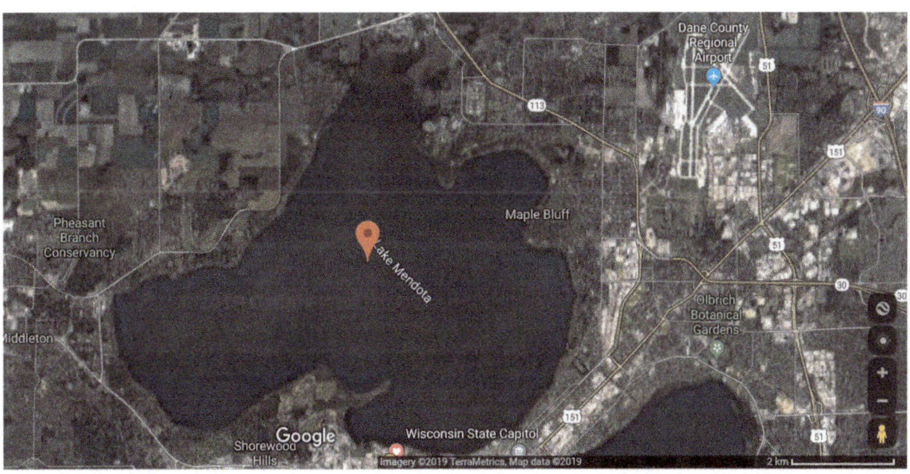

Figure A4.1. Google earth photograph of Lake Mendota. Imagery © 2019 TerraMetrics, Map data © 2019.

Table A4.1. Seasonal water clarity responses to relative levels of nutrients in Lake Mendota since the early 1900s (adapted from Lathrop 1992).

	Period						
Lake condition	Late 1800s–1920s	1940s	1960s–1977	1978–summer 1983	Fall 1983–summer 1987	Fall 1987–1988	1989
Nutrients'	Low	Medium	High	Medium–high	Medium	Medium	High
Water clarity							
Spring	Low	Medium	Medium–high	Low	Low	High	High
Summer	Medium	High	Medium	Low	Low–medium	High	Medium
Fall	Medium–high	Medium	Medium–high	Low	Medium–high	High	Medium–high

lower (mainly due to the suppression of sewage inputs). In-lake phosphorous concentrations during the 1980s period experienced a downward trend.

An improvement in water clarity throughout the 1980s mirrored the decrease in lake P concentrations (Lathrop 1992). While it is impossible to quantify the absolute levels of nutrients (either loadings or in-lake concentrations) since the early 1900s, it is possible to assign relative levels (low, medium, or high) for different time periods that reflect generally stable conditions in the historical record (table A4.1).

Notice that the same overall water clarity condition can be produced by different nutrient regimes. The disturbance of aquatic equilibria may be more or less evident according to the enrichment of water by nutrients (phosphorus and nitrogen).

A4.4 Modelling[1]

A4.4.1 The *Mendota Lake* cellular automaton

Our starting point is the MF model introduced by Carpenter (2005), to evaluate the roles of internal recycling and slow dynamics of soil phosphorus, consisting of three differential equations for phosphorus densities in soil (U), in surface sediment (M) and in water (P). Variables P and M are attached to the lake while U describes the surroundings of the lake. They obey the equations:

$$dU(t)/dt = W + F - H - cU(t), \tag{A4.0a}$$

$$dP(t)/dt = cU(t) - (s + h)P(t) + rM(t)f(P), \tag{A4.0b}$$

[1] Excerpts of this and the next section originally appeared in Fort (2012).

$$dM(t)/dt = sP(t) - bM(t) - rM(t)f(P), \tag{A4.0c}$$

where

$$f(P) = P^q/(P^q + m^q). \tag{A4.0d}$$

Values of the parameters were estimated for the watershed of Lake Mendota, Wisconsin (Carpenter 2003, 2005) and are defined in table A4.2. These parameters are consistent with estimates of presettlement phosphorus inputs and modern measurements of phosphorus inputs.

Using what we learned in chapter 4, here we transform the above model into the *Mendota Lake cellular automaton* (MLCA) (Fort 2012) to analyze the catastrophic transition from clear to turbid water and discuss possible early warning signals to this shift. Thus we take P and M as local spatial variables, while we take U as a global (non-spatial) variable since this variable is not attached to the lake itself but to its surroundings. The evolution equations for $U(t)$, $P(x,y;t)$ and $M(x,y;t)$ are:

$$dU(t)/dt = W + F - H - cU(t), \tag{A4.1}$$

$$dP(x, y; t)/dt = cU(t) - (s + h)P(x, y; t) + rM(x, y; t)f(P) + D\nabla^2 P(x, y; t), \tag{A4.2}$$

$$dM(x, y; t)/dt = sP(x, y; t)) - bM(x, y; t) - rM(x, y; t)f(P). \tag{A4.3}$$

Table A4.2. Physical model parameters.

Symbol	Definition	Units	Value[2]
b	Permanent burial rate of sediment P	y^{-1}	0.001
c	P runoff coefficient	y^{-1}	0.001 15
F	Annual agricultural import of P to the watershed g m^{-2} y^{-1}	31.6 per unit lake area	
H	Annual agricultural export of P from the watershed g m^{-2} y^{-1}	31.6 per unit lake area	
h	Outflow rate of P	y^{-1}	0.15
m	P density in the lake when recycling is 0.5 r	g m^{-2}	2.4
r	Maximum recycling rate of P	g m^{-2} y^{-1}	0.019
q	Parameter for steepness of f (P) near m	Unitless	8
s	Sedimentation rate of P	y^{-1}	0.7
W_0	Nonagricultural inputs of P to the watershed before disturbance, per unit lake area	g m^{-2} y^{-1}	
W_D	Nonagricultural inputs of P to the watershed	after disturbance, per unit lake area	g m^{-2} y^{-1}

[2] From Carpenter (2005).

In addition to the empirical parameters our MLCA model includes two *free parameters* that were not estimated from empirical data. First, there is a diffusion coefficient D to model diffusion effects. Another modification, in order to incorporate the effect of mechanical stirring of the lake waters (wind, currents, animals) is that we consider that at each time t, $a(t) \equiv cU(t)$ results from a spatio-temporal variable $a(x,y;t)$ that fluctuates locally, from point to point, around its average global value $a(t)$ in the interval $[a(t) - \Delta, a(t) + \Delta]$, where Δ is the second free parameter of this model.

The lake is represented by a square lattice of $L \times L$ cells each one identified by its integer coordinates (i,j). That is, on each cell there are two local variables assigned: $P(i,j)$ and $M(i,j)$. Therefore, equations (A4.1)–(A4.3) lead to the following synchronous update rules in discrete time, where now t represent the time measured in years:

$$U(t+1) = U(t) + W(t) + F(t) - H(t) - cU(t), \tag{A4.1$'$}$$

$$\begin{aligned} P(i,j;t+1) = {} & P(i,j;t) + a(i,j) - (s+h)P(i,j;t) \\ & + rM(i,j;t)P(i,j;t)^q/(P(i,j;t)^q + h^q) \\ & + 0.25D(P(i-1,j;t) + P(i+1,j;t) + P(i,j-1;t) \\ & + P(i,j+1;t) - 4P(i,j;t)) \end{aligned} \tag{A4.2$'$}$$

$$\begin{aligned} M(i,j;t+1) = {} & M(i,j;t) + sP(i,j;t) - bM(i,j;t) + rM(i,j;t) \\ & P(i,j;t)^q/(P(i,j;t)^q + h^q). \end{aligned} \tag{A4.3$'$}$$

Remarks.

- An important approximation is that the system is two-dimensional, i.e. depth is ignored, something that works for shallow lakes which are thermally unstratified (but see section A4.7 on improvements and caveats).
- Of course, lakes of arbitrary shape could be studied by embedding them into a square lattice like the one above, with appropriate boundary conditions.

A4.4.2 Catastrophic shifts in lakes and their spatial early warnings

As we have seen in chapter 4, catastrophic shifts (Thom 1989) have characteristic fingerprints or 'wave flags'. Some of the standard catastrophe flags are: modality (at least two well defined attractors), sudden jumps and a large or anomalous variance (Gilmore 1981). Therefore, we will calculate the following two observable quantities for the phosphorous density in water P, which is the relevant variable in our case.

1. The ***spatial variance*** of $P(i,j;t)$, σ_s^2, defined as

$$\sigma_s^2 \equiv \langle P^2 \rangle - \langle P \rangle^2 = \frac{\sum_{i,j=1}^{L} P(i,j;t)^2 - \left(\sum_{i,j=1}^{L} P(i,j;t) \right)^2}{L^2}, \tag{A4.4}$$

where $\langle X \rangle$ denotes the *spatial mean* of a spatial variable $X(i,j)$,

2. The patchiness or cluster structure.
   Remember that the rise of σ_s is produced by the onset of fluctuations in the spatial dependence of $P(x,y;t)$, leading to patchiness.
   In addition, even if we neglect all spatial heterogeneity, in a completely similar way as we do with mean-field models, we can compute:
3. The temporal variance σ_t^2.

This variable has been suggested by Brock and Carpenter (Brock and Carpenter 2006, Carpenter and Brock 2006) as an early warning signal. At an arbitrary point, say $(x,y) = (0,0)$, is defined as

$$\sigma_t^2 = \frac{\sum_{t'=t-\tau+1}^{t} P(0,0;\, t')^2 - \left(\sum_{t'=t-\tau+1}^{t} P(0,0;\, t') \right)^2}{\tau}, \qquad \text{(A4.5)}$$

for temporal bins of size τ.

A4.5 Model validation

A4.5.1 Simulations and results

The properties and parameters of the MLCA are summarized in box A4.1.

We ran simulations for 1500 years to illustrate changes of Lake Mendota over time. The first 250 years of each simulation are a highly simplified representation of the history of Lake Mendota until now. The remaining years of the simulation serve to explore a possible future evolution, including a shift towards eutrophication and the corresponding early warning signals.

Following Carpenter (2005), we initiate the simulation at stable equilibrium values calculated with $F = H = 0$ and W for undisturbed conditions. These represent

Box A4.1. Properties and parameters of the cellular automaton used in the explicitly spatial version of the Lake Mendota model.

- Shape of lattice: square.
- Size L: ranging from 20 to 500 (in fact, for $L > 100$, no important differences were found).
- Neighborhood: von Neumann neighborhood, i.e. each cell is connected to its four nearest neighbors.
- Free boundary conditions: That is, cells at the border ($i = 1$ or L or/and $j = 1$ or L) only have two neighbors instead of four.
- Updating rule: set of equations (A4.1′)–(A4.3′).
- Values for the parameters: $D = 0.1$, $\Delta = 0.125$ (the reader can verify that the results do not depend much on the choice of these values).

presettlement conditions, and were maintained for years 0–100. For years 100–200, W was changed to the value for disturbed conditions (W_D, table A4.2), representing the advent of agriculture in the region. For years 200–250, F and H were increased to the values estimated for a period of intensive industrialized agriculture in the Lake Mendota watershed, and W was maintained at W_D (table A4.2). After year 250 we assume the application of management policies focused on reducing loads from diffuse agricultural sources, so that the phosphorus budget of agriculture is balanced but other inputs remained at the post disturbance rate. Certainly, low flux of phosphorus from over-fertilized soils may be even more important for maintaining eutrophication of lakes in agricultural regions.

That is, we consider four different stages:

(I) Years 0–100 $W = W_0 = 0.147$, $F = H = 0$.
(II) Years 101–200 $W = W_D = 1.55$, $F = H = 0$.
(III) Years 201–250 $W = W_D = 1.55$, $F–H = 31.6–18.6 = 13$.
(IV) After year 250 $W = W_D = 1.55$, $F–H = 0$.

Before using the MLCA let us see what we get when we integrate Carpenter's (2005) mean-field equations (A.40). Hence, the corresponding evolutions of the three phosphorus density variables are shown in figure A4.2. In particular, for P, we can see a gradual first shift that start after year 200 and stabilizes at year 250 and a second much more drastic and larger jump after year 500 that quickly stabilizes at around 7.5 g m^{-2}. Indeed, this type of eutrophication is not reversible unless there are substantial changes in soil management. Actually, results suggest that dynamics of soil phosphorus may control alternate stable states, potentially causing eutrophication to last for centuries.

Figure A4.3 displays the 1500 years history generated by the MLCA for P. As we can see, the curve for $\langle P \rangle$ is very similar to the one for P produced by Carpenter's (2005) non-spatial model depicted in figure A4.2 (green curve): Two jumps, the first one smaller and less steep than the more drastic second one occurring between years 440 and 485. Notice that the σ_s produced by the MLCA reaches at year 425 three times the constant value it had along the first 400 years. Therefore, it provides an early warning of the coming shift in around 60 years. σ_s reaches its maximum value at year 467 which coincides with the *Maxwell point* (Fernández and Fort 2009).

The corresponding spatial fluctuations of P are shown in figure A4.4. These color maps represent different configurations when the lake moves towards a catastrophic shift from clear to turbid water. Panels (A)–(C) correspond, respectively, to snapshots of the lake at $t = 250$ yrs (when still $F= H$), at $t = 437$ yrs (30 years before the peak in σ_s) and at $t = 467$ years (just at this peak of σ_s). For $t = 467$ typically there occur patches covering a wide scale of sizes (figure A4.4(C)). Notice that either far before or after the catastrophic shift, one of the two attractors dominates. We then conclude that the visualization of the onset of those patches, for example by aerial or satellite imaging of the lake surface, may be an effective way of anticipating a eutrophication transition.

Finally, let us have a look at the computed σ_t and compare it with σ_s. Figure A4.5 shows both variances. The temporal variance exhibits a temporal delay of around 40

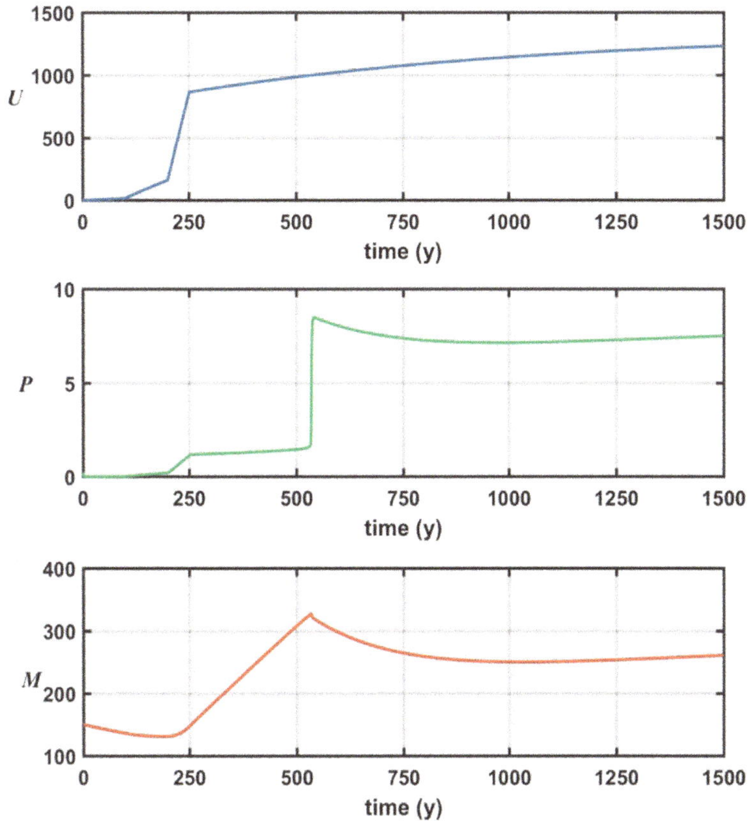

Figure A4.2. Phosphorus density (g m^{-2}) in soil (*U*), water (*P*) and sediment (*M*) versus time, in years, under the above four different historical stages I–IV.

years leaving much less margin for eventual remedial management actions. This delay can be easily understood since σ_t employs data for times where the fluctuations are still small. That is, when estimating the temporal variance, one must consider past values in the time series, which correspond to situations where the lake is far from undergoing a transition. The spatial variance considers only the present values; hence a signal announcing the shift is not blurred by averages including situations where these indications are absent.

In summary, using MLCA for describing the evolution of the P concentration, which dominates the eutrophication process in lakes, we have considered different possible early warnings for eutrophication shifts in lakes. In the case of σ_s, by measuring samples of P on a grid of points on the lake surface, we found that a significant growth in σ_s could serve as an early warning of an imminent transition. The spatial variance has an advantage over the temporal one, as σ_t is delayed with respect to σ_s.

As we have seen in chapter 4, the origin of the rise in σ_s is connected with the appearance of spatial patterns, in the form of clusters (in this case, of clear and

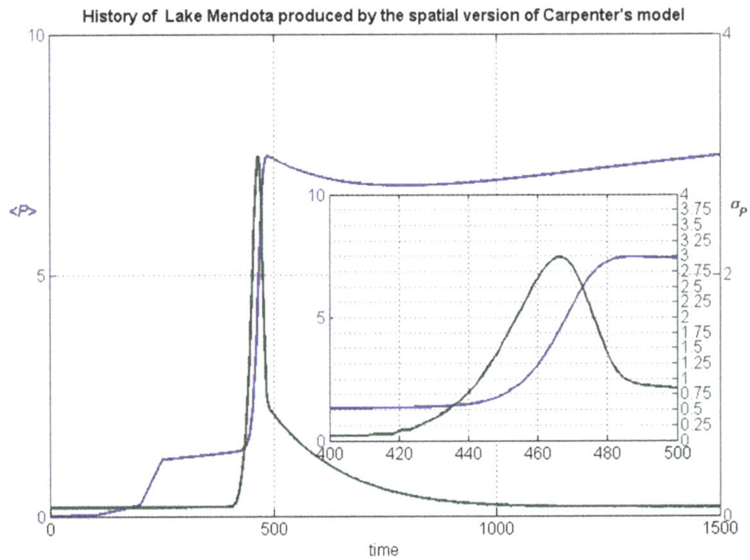

Figure A4.3. ⟨P⟩ (blue) and σ_s (green) versus time (in years). The inset is a zoom showing in detail the time region around the catastrophic transition, which shows clearly the delay between both quantities. Reprinted from Fort 2013 with permission from IntechOpen.

turbid water). Furthermore, as happened with the spatially explicit grazing model, the patchiness near the regime shift is very characteristic: patches of clear water tend to coexist with others of turbid water. In mathematical terms, close to the regime shift there are two competing attractors or alternative states (clear and turbid water) one occurs in a given set of cells and the other in the set of remaining cells. This explains the strong spatial fluctuations in the concentration of P.

Analyses presented here should be regarded as an illustration of what may happen, according to one parameter set and one model. Despite these uncertainties, it appears that slow dynamics of soil phosphorus can set the tempo of eutrophication for lakes in agricultural watersheds. Slow recovery of water phosphorus, in parallel with slow dynamics of soil phosphorus, occurs even in simple linear analyses of watershed-lake systems (Carpenter 2005). In this model the eutrophic state persisted for hundreds of years. The slow depuration of soil phosphorus plays a central role in retarding recovery from eutrophication.

The results do not depend strongly on the value of the two free parameters included in our CA model. However, the quantitative details of our conclusions depend on the choice of those values. For instance, the value of the diffusion parameter D affects the visual pattern of the patches of clear and turbid water at the transition between the *oligotrophic* (low nutrients density) and eutrophic conditions.

A4.5.2 Verdict: model validated, but…

Some notes of caution are required. First, this model used parameters from Lake Mendota, which is a thermally stratified lake. In such thermally stratified lakes two main layers can be distinguished. The *hypolimnion* is the dense, bottom layer of

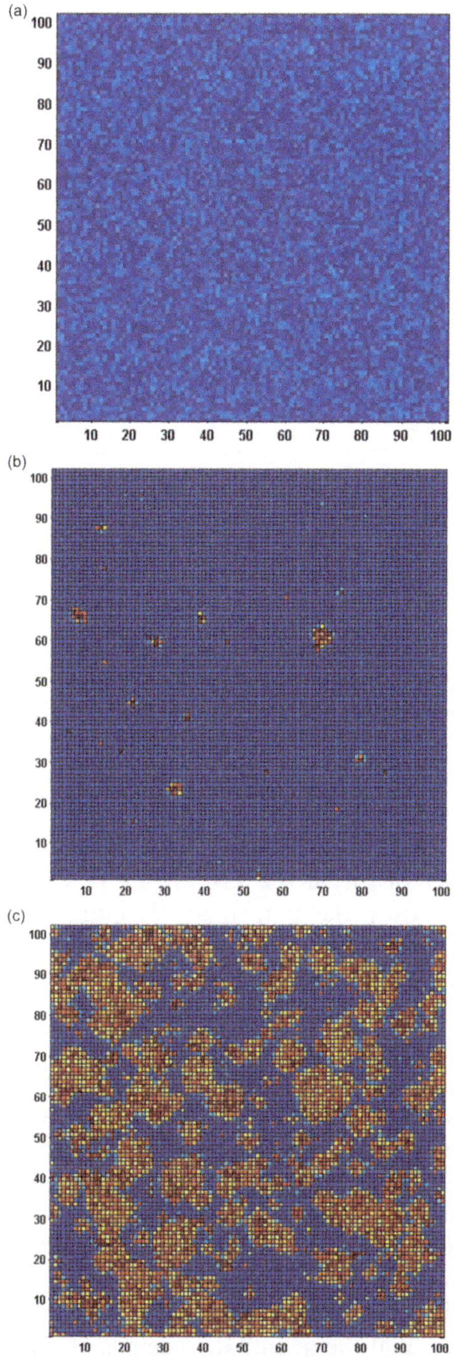

Figure A4.4. Maps of P density, i.e. values of $P(i,j;t)$ at each lattice cell (blue: low density, red: high density) for a 100 × 100 grid. (A) $t = 250$, well before the transition,. (B) $t = 437$. (C) $t = 467$, where σ_s has it maximum value. Reprinted from Fort 2013 with permission from IntechOpen.

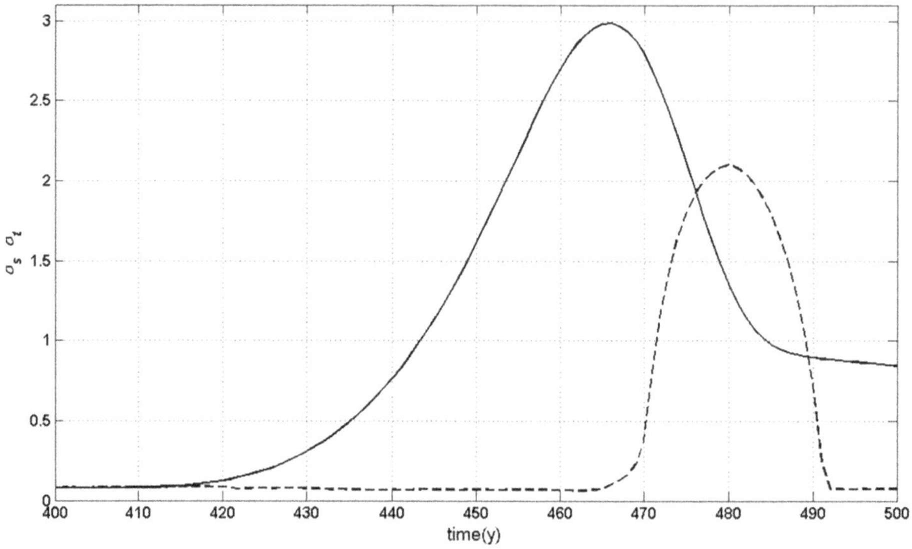

Figure A4.5. σ_s (solid line) versus σ_t (dashed line). Reprinted from Fort 2013 with permission from IntechOpen.

water, which is isolated from surface wind-mixing. The *epilimnion* is the top-most layer, occurring above the deeper hypolimnion and being exposed at the surface, it typically becomes turbulently mixed as a result of surface wind-mixing. The flow of phosphorus in recycling occurs first from sediment to and also from decomposition in the hypolimnion. Dissolved P in the hypolimnion tends to be rather well-mixed and homogeneous. Then vertical mixing of hypolimnetic and epilimnetic water occurs from time to time (roughly 10–15 day return time (Carpenter 2013)). Horizontal dispersion of dissolved P is fast after a vertical mixing event, within a day or so, smoothing over the patchiness structure in the epilimnion. As a consequence patches could never survive during a year, which is the unit of time in our model. Thus, instead of long lasting structures, these patches should be interpreted as frequent popping in and out structure along the course of one year. At any rate, since the spatial model we are analyzing is two-dimensional and there is no depth (the hypolimnion and the epilimnion are collapsed to a single layer), the above processes are not included. Hence, the model proposed here would be more appropriate to address the same effects of soil phosphorus in shallow lakes.

Second, some lakes show important spatial differences in terms of total phosphorus related to the spatial distribution of submerged plants or cyanobacteria or 'blue–green algae' blooms. The biomass tends to accumulate in several zones, by the hydrodynamic and wind actions, and determine strong differences of algal biomass as well as total phosphorous (Mazzeo 2013).

It is worth remarking that the quantitative details of our conclusions depend on the values of the model parameters. The reader can check that the qualitative behavior of results does not depend strongly on those values. Indeed, the model presented here is schematic in the sense that the quality of the lake's water is dependent on a single variable, the amount of phosphorus in solution. In real cases other environmental factors might be playing an important role in the dynamics of the water quality, and the model and its predictions could be much more complicated. Nevertheless, it appears that our main conclusions should hold in more realistic situations, namely the spatial variance of critical quantities provides an earlier warning signal than the temporal variance. Furthermore, the associated cluster structure of the patterns formed in the eutrophication process could be the fastest detectable warning of an ongoing catastrophic change for the water quality.

The CA presented here addresses an open problem in ecology, namely how to anticipate catastrophic shifts in ecosystems. This is a fundamental scientific question that goes beyond ecology and requires a major interdisciplinary effort and would have a significant impact on ecosystems management and conservation. This CA represents a minimal spatially explicit ecological model that should be regarded as an exploratory model, including the minimal number of adjustable parameters and some gross simplifications. More realistic assumptions and additional relevant parameters can be included in the model building (see next section).

A4.6 Usefulness of the early warnings

To determine the usefulness of the warning indicators presented in the previous section it is necessary to assess (1) their practicality and (2) whether they really allow the implementation of corrective actions to avoid the catastrophic shift.

Regarding the first problem of practicality, calculating variances over grids consisting of a large number of sites (e.g. 100×100 or 1000×1000) is easy on a computer. Nevertheless, performing such a large number of measurements, at different sites and on a regular basis, is an unfeasible task. Thus, to assess the practical feasibility of using σ^2_X as an early warning, we need to perform a sensibility analysis on the size of the sample grid L.

Figure A4.6 shows that the time at which the signal becomes appreciable depends on the number of points of the grid (or lattice size L). As expected, there is a trade-off between anticipation and cost. The larger the number of points involved in the statistics, the sooner will be this early warning signal, until, for grids above $100 \times 100 = 10\,000$ points, the curve doesn't change very much. For instance, for a 20×20 grid the early warning becomes noticeable 20 years later than when computed for the 100×100 grid.

In summary, monitoring the phosphorus concentration in water in a grid of some hundreds of points, typically separated by around 300 m, would be sufficient for the purpose of extracting an appropriate signal 40 years in advance. In fact, even for a small grid of only 25 points, σ^2_X still exhibits a noticeable peak and warns about the eutrophic shift 30 years before it happens.

Regarding possible remedial actions, in principle, and sometimes in practice, the eutrophic regime can be destabilized by management interventions, thereby changing the lake toward the clear-water regime (Carpenter 2005 and references therein). These interventions include, harvesting, herbicide/algicide application, oxygenation of water and chemical precipitation of phosphorous by the addition of iron or aluminum salts or calcium carbonate (Cooke *et al* 2005). Here we rather consider

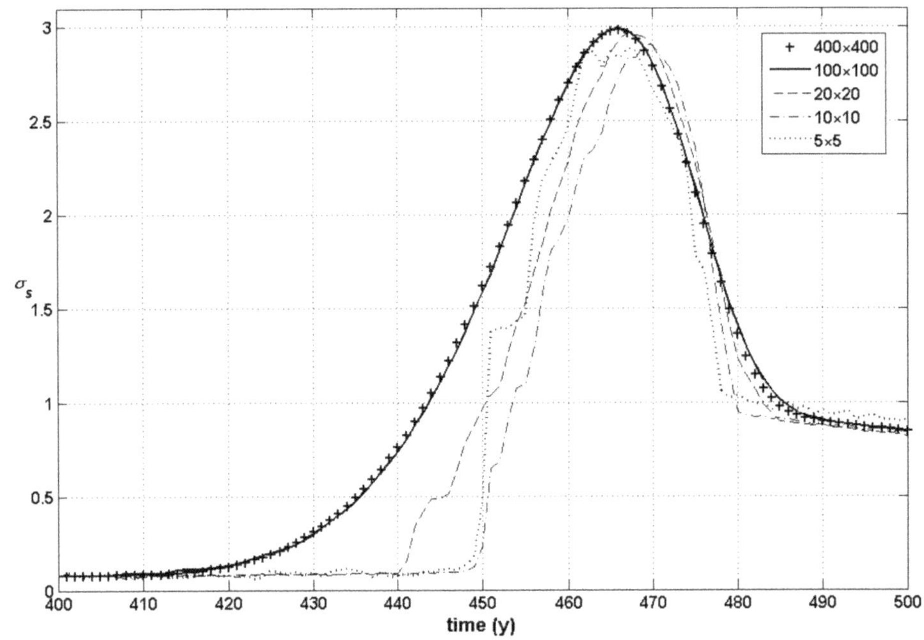

Figure A4.6. σ_s computed for different lattice sizes: $L = 5$, 10, 20, 100 and 400. The larger the lattice the earlier the signal. Reprinted from Fort 2013 with permission from IntechOpen.

contrasting long-term management policies for soil phosphorus which translates into a second simulation which coincides in stages (I) to (III) until year 250 with the one we previously presented. After year 250 this simulation includes, besides management to balance the phosphorus budget of agriculture, a slow decrease, from year 251–350, of W from W_D to W_0 (we have to be aware that natural runoff, which causes algal blooms in the wild, is common in ecosystems and should thus not reverse nutrient concentrations beyond normal levels) (figure A4.7). Notice this is enough to avoid the future large eutrophic jump appearing in the first simulation.

A4.7 Extensions, improvements and caveats

As we have seen, eutrophication results from a combination of natural processes and human impacts (Smol 2008). Natural processes may be isolated through *paleolimnological* analysis to estimate and connect historic input and runoff rates with temperature and humidity indices (García-Rodriguez *et al* 2004). Once we have calibrated these natural processes, like climate change (Sayer *et al* 2010), we can use cellular automata models, like the MLCA model, to simulate different managements to avoid severe eutrophication as well as possible remedial actions.

The model used here is a simple description of lake water quality. Indeed, in addition to the phosphorus concentration other variables may be also used to

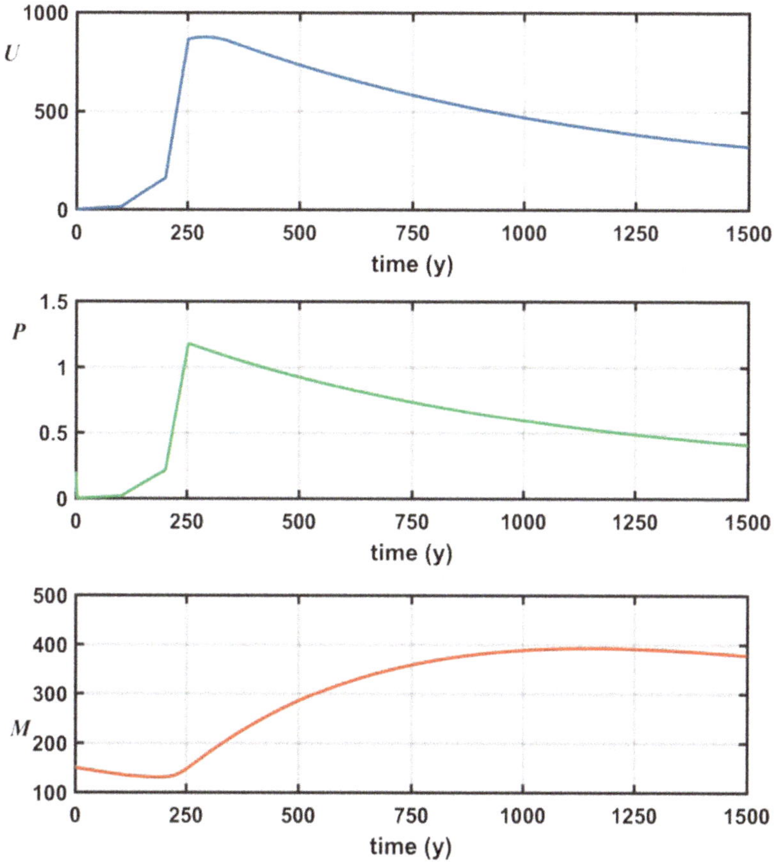

Figure A4.7. Phosphorus density (g m^{-2}) in soil (U), water (P) and sediment (M) versus time, in years, when performing a remedial action consisting in decreasing back in stage IV, from years 251 to 350, W from $W_D = 1.55$ to $W_0 = 0.147$.

describe the state of a water body, like turbidity, phytoplankton biomass, vegetation cover (Scheffer 1997).

A limitation of our CA model is that it is bi-dimensional and thus it ignores the lake depth. Indeed, P influx to the epilimnion via vertical fluxes (eddy diffusion, hypolimnetic entrainment, and sedimentation) can be substantially more important than external loading in stratified lakes (Carpenter 1992). Therefore, another interesting extension of this CA would be to take into account the lake depth and formulate it in three dimensions (rather than in just two). This would allow us to describe the flow of phosphorus in recycling from sediment and also from decomposition in the hypolimnion.

FORTRAN 77 code

```
      program mendota
c     mendota.f is based on carpenter05.f
c     Main difference: W+F-H varies with time as it varied for
c     the lake Mendota: 4 stages:
c          1) 0-100    y     -> small W, W=0.147, F=H
c          2) 101-200  y     -> large W, W=1.55,  F=H
c          3) 201-250  y     -> large W, W=1.55,  F-H=13
c          4) 251-1000 y     -> large W, W=1.55,  F=H
c     dU/dt=(W+F-H)-cU: as am=cU.
c     and
c     instead of r we have an effective reff(y,z) given by
c     reff=r*M(y,z), where M(y,z) is the P density in surface sediment
c     obtained from eq. 3:
c     dM/dt=sP-bM-reff*(x^8/(x^8+h^8))
      parameter dim=50
      implicit none
      real*8 x(-dim:dim,-dim:dim),xn(-dim:dim,-dim:dim), x00,xp0,xm0,x0p
     $      ,x0m,xt(5000),a(-dim:dim,-dim:dim),am,da,
     $       M(-dim:dim,-dim:dim),Mn(-dim:dim,-dim:dim), M00,Mp0,Mm0,M0p
     $      ,M0m,reff(-dim:dim,-dim:dim),c,U0,U(50000), tave,tave2,
     $       adot,s,o,r,h,b,dif,ran1,xave,xave2,delx,delt,Wf
      integer*4 l,lsample,y,z,t,tfin,iexp,istart,iseed, yp,ym,zp,zm,nbin
     $      ,tin,tout,imovie
      open(10,file='mendota.in',status='old')
      imovie=1 c This is to generate a movie
      if(imovie.eq.0) then
              open(20,file='mendota.m',status='unknown')
              open(30,file='mendotaMat.m',status='unknown')
              open(40,file='mendotaSigmat.m',status='unknown')
      else
              open(20,file='mendotaD06.m',status='unknown')
              open(30,file='mendotaMovieD06.m',status='unknown')
      endif

      read(10,*) c,U0,da,s,o,r,h,iexp,b,dif,l,tfin,iseed
      Wf=1.5
      write(20,*)'xs=['
c     Initialization
      lsample=1
      U(1)=U0
```

```
      do y=-1,l
          do z=-1,l
              a(y,z)=am+da*(1.-2.*ran1(iseed))
              x(y,z)=ran1(iseed)
            M(y,z)=300*ran1(iseed) ! i.e., initially an average of 150
c                         g/(m^2y)
          enddo
      enddo

c     Evolution
      do t=1,tfin
          am=c*U(t)
          do y=-1,l
             do z=-1,l
                 x00=x(y,z)
                 M00=M(y,z)
                 yp=y+1
                 if(y.eq.l) yp=-l
                    ym=y-1
                 if(y.eq.-l) ym=l
                    zp=z+1
                 if(z.eq.l) zp=-l
                    zm=z-1
                 if(z.eq.-l) zm=l
                 xp0=x(yp,z)
                 xm0=x(ym,z)
                 x0p=x(y,zp)
                 x0m=x(y,zm)
                 xn(y,z)=x00+a(y,z)-(s+o)*x00
     $                +r*M00*x00**iexp/(x00**iexp+h**iexp)+
     $                0.25*dif*(xp0+xm0+x0p+x0m-4*x00)
                 Mp0=M(yp,z)
                 Mm0=M(ym,z)
                 M0p=M(y,zp)
                 M0m=M(y,zm)
                 Mn(y,z)=M00+s*x00-b*M00
     $                     -r*M00*x00**iexp/(x00**iexp+h**iexp)
             enddo
          enddo


          if(imovie.eq.0) then
              if(t.eq.437) then
```

```
            write(30,*)'Mx437 = ['
                  do z=-dim,dim
              write(30,'(101f12.6)')(x(y,z), y=-dim,dim)
            enddo
            write(30,*)'];'
       elseif(t.eq.467) then
             write(30,*)'Mx467 = ['
                   do z=-dim,dim
              write(30,'(101f12.6)')(x(y,z), y=-dim,dim)
             enddo
             write(30,*)'];'
       elseif(t.eq.250) then
             write(30,*)'Mx250 = ['
                   do z=-dim,dim
              write(30,'(101f12.6)')(x(y,z), y=-dim,dim)
             enddo
             write(30,*)'];'
       endif

       elseif(imovie.eq.1.and.dif.le.0.5)then
        if(t.ge.420.and.t.le.480) then
             write(30,*)'P',t,' = ['
             do z=-dim,dim
              write(30,'(101f12.6)')(x(y,z), y=-dim,dim)
             enddo
             write(30,*)'];'
         endif
    elseif(imovie.eq.1.and.dif.gt.0.5)then
     if(t.ge.940.and.t.le.1000) then
          write(30,*)'P',t,' = ['
          do z=-dim,dim
              write(30,'(101f12.6)')(x(y,z), y=-dim,dim)
           enddo
          write(30,*)'];'
      endif
  endif

xave=0.
  xave2=0.
  do y=-1,1
     do z=-1,1
        x(y,z)=xn(y,z)
        M(y,z)=Mn(y,z)
```

```
        if(abs(y).le.lsample.and.abs(z).le.lsample) then
             xave=xave+x(y,z)
             xave2=xave2+x(y,z)**2
          endif
       enddo
    enddo
  xt(t)=x(0,0)
  xave=xave/(2*l+1)**2
  xave2=xave2/(2*l+1)**2
  delx=sqrt(abs(xave2-xave**2))
  write(20,*)c*U(t),xave,delx
  write(6,*)c*U(t),xave,delx
  if (t.le.100) then
      U(t+1)=U(t)+0.147-c*U(t)
  elseif (t.gt.100.and.t.le.200) then
      U(t+1)=U(t)+1.55-c*U(t)
  elseif (t.gt.200.and.t.le.250) then
      U(t+1)=U(t)+1.55+13-c*U(t)
  elseif (t.gt.250) then
      U(t+1)=U(t)+Wf-c*U(t)
  endif
    do y=-l,l
       do z=-l,l
          a(y,z)=am+da*(1.-2.*ran1(iseed))
       enddo
    enddo
 enddo
  do nbin=11,tfin
     tin=nbin-10
     tout=nbin
     tave=0.
     tave2=0.
     do t=tin,tout
        tave=tave+xt(t)
        tave2=tave2+xt(t)**2
     enddo
     tave=tave/11.
     tave2=tave2/11.
     delt=sqrt(abs(tave2-tave**2))
     write(40,*) tout,tave,delt
  enddo
  close(20)
  write(20,*)'];'
```

```
      write(20,*)'%plotyy(xs(:,1),xs(:,2),xs(:,1),xs(:,3)),grid'
      write(20,*)'plotyy([1:1500],xs(:,2),[1:1500],xs(:,3)),grid'
      close(20)
      stop
      end
      function ran1(idum)
      implicit none
           double precision ran1,am,eps,rnmx
      integer*4 idum,ia,im,iq,ir,ntab,ndiv
           parameter (ntab = 32)
           integer*4 j,k,iv(ntab),iy
      save iv,iy
      data iv /ntab*0/, iy /0/
      ia = 16807
           im = 2147483647
           am = 1.0D0/dfloat(im)
           iq = 127773
      ir = 2836
           ndiv = 1 + (im-1)/ntab
           eps = 1.2D-7
           rnmx = 1.0D0 - eps
           if ((idum.le.0).or.(iy.eq.0)) then
           idum = max(-idum,1)
      do 800 j = ntab+8,1,-1
      k = idum/iq
           idum = ia*(idum-k*iq) - ir*k
           if (idum.lt.0) idum = idum + im
           if (j.le.ntab) iv(j) = idum
800   continue
      if (j.eq.0) j = 1
              iy = iv(j)
           end if
      k = idum/iq
           idum = ia*(idum-k*iq) - ir*k
           if (idum.lt.0) idum = idum + im
           j = 1 + iy/ndiv
           iy = iv(j)
           iv(j) = idum
           ran1 = min(am*iy,rnmx)
           if(ran1.eq.1.) ran1=0.
     return
     end
```

References

Brock W A and Carpenter S R 2006 Variance as a leading indicator of regime shift in ecosystem services *Ecol. Soc.* **11** 9

Carpenter S R 1992 Destabilization of planktonic ecosystems and blooms of blue-green algae *Food Web Management: A Case Study of Lake Mendota* ed J F Kitchell (New York: Springer) 461–82

Carpenter S R 2003 *Regime Shifts in Lake Ecosystems: Pattern & Variation* (Oldendorf: Ecology Institute)

Carpenter S R 2005 Eutrophication of aquatic ecosystems: Bistability and soil phosphorus *Proc. Natl. Acad. Sci.* **102** 10002–5

Carpenter S R 2008 Phosphorus control is critical to mitigating eutrophication *Proc. Natl. Acad. Sci.* **105** 11039–40

Carpenter S R 2013 Personal communication

Carpenter S R and Brock W A 2006 Rising variance: A leading indicator of ecological transition *Ecol. Lett.* **9** 311–8

Chislock M F, Doster E, Zitomer R A and Wilson A E 2013 Eutrophication: causes, consequences, and controls in aquatic ecosystems *Nat. Educ.* **4** 10

Cooke G D, Welch E B, Peterson S A and Newroth P R 2005 *Restoration & Management of Lakes & Reservoirs* 3rd edn (Boca Raton, FL: CRC Press)

Dakos V, van Nes E H, Donangelo R, Fort H and Scheffer M 2010 Spatial correlation as leading indicator of catastrophic shifts *Theor. Ecol.* **3** 163–74

Dodds W K *et al* 2009 Eutrophication of U.S. freshwaters: analysis of potential economic damages *Environ. Sci. Technol.* **43** 12–9

Donangelo R, Fort H, Dakos V, Scheffer M and van Nes E H 2010 Early warnings of catastrophic shifts in ecosystems: Comparison between spatial and temporal indicators *Int. J. Bifurc. Chaos* **20** 315–21

Evans M R 2012 Modelling ecological systems in a changing world *Philos. Trans. R. Soc. B* **367** 181–90

Fernández A and Fort H 2009 Catastrophic phase transitions and early warnings in a spatial ecological model *J. Stat. Mech.* **2009** P09014

Fort H 2012 Two cellular automata designed for ecological problems: Mendota CA and Barro Colorado Island CA *Emerging Applications of Cellular Automata* ed A Salcido (Rijeka: Intech)

Fort H 2013 Statistical mechanics ideas and techniques applied to selected problems in ecology *Entropy* **15** 5237–76

García-Rodriguez F *et al* 2004 Holocene trophic state changes in relation to the sea level variation in Lake Blanca, SE Uruguay *J. Paleolimnol.* **31** 99–115

Gilmore R 1981 *Catastrophe Theory for Scientists and Engineers* (New York: Dover)

Lathrop R C 1992 Nutrient loadings, lake nutrients, and water clarity *Food Web Management: A Case Study of Lake Mendota* ed J F Kitchell (New York: Springer) 69–96

Mazzeo N 2013 Personal communication

National Research Council 1992 *Restoration of Aquatic Ecosystems: Science, Technology & Public Policy* (Washington, DC: National Academic Press)

Paerl H W 1988 Growth and reproductive strategies of freshwater blue-green algae (cyanobacteria) *Growth and Reproductive Strategies of Freshwater Phytoplankton* ed C D Sandgren (Cambridge: Cambridge University Press) 261–315

Sayer C D, Davidson T A, Jones J I and Langdon P G 2010 Combining contemporary ecology and palaeolimnology to understand shallow lake ecosystem change *Freshwater Biol.* **55** 487–99

Scheffer M 1997 *The Ecology of Shallow Lakes* (London: Chapman and Hall)

Scheffer M and Carpenter S R 2003 Catastrophic regime shifts in ecosystems: linking theory to observation *Trends Ecol. Evol.* **12** 648–56

Schindler D W 1977 Evolution of phosphorus limitation in lakes *Science* **195** 260–2

Schindler D W 1988 Experimental studies of chemical stressors on whole-lake ecosystems *Verh. Internat. Verein. Limnol.* **23** 11–41

Smol J 2008 *Pollution of Lakes and Rivers: A Paleoenvironmental Perspective* 2nd edn (Oxford: Blackwell)

Søndergaard M *et al* 2003 Role of sediment and internal loading of phosphorus in shallow lakes *Hydrobiologia* **506** 135–45

Thom R 1989 *Structural Stability and Morphogenesis: An Outline of a General Theory of Models* (Reading, MA: Addison-Wesley)

World Lake Database 2019 http://wldb.ilec.or.jp/Details/Lake/NAM-01

IOP Publishing

Ecological Modelling and Ecophysics
Agricultural and environmental applications
Hugo Fort

Appendix I

Equilibrium stability

A.1 Local and global stability

Let an ecological community of S species described by an autonomous continuous time model be of the form:

$$\frac{dn_i}{dt} = n_i f_i(n_1, n_2, \ldots, n_S) \equiv n_i f_i(\mathbf{n}) \qquad i = 1, \ldots, S, \tag{I.1}$$

where $f_1(\mathbf{n}), f_2(\mathbf{n}), \ldots, f_S(\mathbf{n})$ are continuous functions in the positive orthant. We can denote the product $n_i f_i(\mathbf{n})$ as $F_i(\mathbf{n})$.

Similarly, we use discrete time, and at time t let $N_i(t)$ denote the density of the ith species in an interaction among S species which we represent by a set of nonlinear difference equations:

$$n_i(t+1) = G_i(n_1, n_2, \ldots, n_S) \equiv G_i(\mathbf{n}) \qquad i = 1, \ldots, S, \tag{I.1$'$}$$

where, for convenience, we use n_i in place of $n_i(t)$; but in order to distinguish $n_i(t + 1)$ from $n_i(t)$ we shall retain the argument of $n_i(t + 1)$.

The simplest way to examine stability in a community model like equation (I.1) is by examining the eigenvalues of the so-called community matrix which is computed at an equilibrium of the model $\mathbf{n}^*$ that, by definition, verifies:

$$f_i(\mathbf{n}^*) = 0 \qquad i = 1, \ldots, S. \tag{I.2}$$

Similarly, for discrete time, we have that an equilibrium $\mathbf{n}^*$, by definition, must verify:

$$G_i(\mathbf{n}^*) = \mathbf{n}^* \qquad i = 1, \ldots, S. \tag{I.2$'$}$$

This community matrix is the Jacobian matrix, given by:

$$J_{ij} \equiv \frac{\partial\left(\frac{dn_i}{dt}\right)}{\partial n_j}\Bigg|_{\mathbf{n}^*} = \frac{\partial F_i(\mathbf{n})}{\partial n_j}\Bigg|_{\mathbf{n}^*} = \frac{\partial n_i}{\partial n_j} f_i(\mathbf{n})\Bigg|_{\mathbf{n}^*} + n_i{}^* \frac{\partial f_i(\mathbf{n})}{\partial n_j}\Bigg|_{\mathbf{n}^*} = f_j(\mathbf{n}^*) + \frac{\partial f_i(\mathbf{n})}{\partial n_j}\Bigg|_{\mathbf{n}^*} n_i{}^*, \quad (\text{I.3})$$

and then, by equation (I.2), we can simply write the community or Jacobian matrix J_{ij} as:

$$J_{ij} = \frac{\partial f_i(\mathbf{n})}{\partial n_j}\Bigg|_{\mathbf{n}^*} n_i^*. \quad (\text{I.4})$$

However, the problem with this method is that it can only establish **local** stability or instability. On the other hand, ecosystems in the real world are subject to large perturbations of the initial state, and continual disturbances on the system dynamics that may produce important departures from equilibrium. Since the equations of population biology are nonlinear, their solutions, which can be represented as an S-dimensional surface, can give rise to quite complicated 'landscapes'. And therefore neighborhood stability analysis may give a misleading representation of the full global stability of the system.

If the dynamical equations are linear, local and global stability are identical. Unfortunately, we have seen in chapter 1 that while the linear approximation is very useful for approaching many problems in physics, it is rarely a sensible approach in population biology. However, many biologically interesting models, although non-linear, produce relatively simple landscapes, with one valley or hilltop whose sides slope ever upward or downward, respectively. In this case the local stability analysis correctly describes the global stability. Such circumstances are characterized by the existence of a *Lyapunov function* and constitute the basis of a powerful analytical method for establishing that an equilibrium is **globally** stable, i.e. stable relative to finite perturbations of the initial state. This is the so-called *direct* or *second* method of Lyapunov (LaSalle and Lefschetz 1961, Gurel and Lapidus 1968, Willems 1970, Strogatz 1994). There are many methods for constructing Lyapunov functions (Schultz 1965, Gurel and Lapidus 1968). However, unfortunately, there is no general way of knowing whether a Lyapunov function exists, let alone a straightforward procedure to construct it if it does exist.

In any event, for a given model it is possible to use computer simulations to investigate the behavior of the model for finite perturbations of its initial state. But computer simulations cannot guarantee that an equilibrium does indeed have a finite region of attraction. Certainly, this procedure becomes increasingly worse as the number of species in a given community increases.

In the next section of this appendix we will consider the local stability for two-dimensional systems. In the third and final section we will return to local and global stability, review the two methods outlined above, the one based on the eigenvalues of the Jacobian matrix and Lyapunov's method.

A.2 Stability for two-dimensional systems

Stability theory addresses the stability of solutions of differential equations and of trajectories of dynamical systems under small perturbations of initial conditions. We discuss here the stability of a general autonomous ordinary differential bi-dimensional system of equations of the form

$$dx_1/dt = f_1(x_1, x_2),\ dx_2/dt = f_2(x_1, x_2), \tag{I.5}$$

where f_1 and f_2 are given functions or maps. This system can be written more compactly in vector notation as

$$d\mathbf{x}/dt = \mathbf{f}(\mathbf{x}), \tag{I.5$'$}$$

where bold denotes a column vector with two entries:

$$\mathbf{x} = \begin{bmatrix} x_1 \\ x_2 \end{bmatrix}, \tag{I.6a}$$

$$\mathbf{f}(\mathbf{x}) = \begin{bmatrix} f_1(\mathbf{x}) \\ f_2(\mathbf{x}) \end{bmatrix}, \tag{I.6b}$$

Thus $\mathbf{x}$ represents a point in the phase plane, and $d\mathbf{x}/dt$ is the velocity vector at that point, which is given by the vector field or bi-dimensional map $\mathbf{f}(\mathbf{x})$. By flowing along the vector field, a phase point traces out a solution $\mathbf{x}(t)$, corresponding to a trajectory or *phase curve* winding through the phase plane (figure I1).

However, what guarantee do we have that the general nonlinear system $d\mathbf{x}/dt = \mathbf{f}(\mathbf{x})$ actually *has* solutions? Fortunately, it turns out that there is an existence and uniqueness theorem for n-dimensional systems:

Existence and uniqueness theorem

Consider the initial value problem $d\mathbf{x}/dt = \mathbf{f}(\mathbf{x})$, $\mathbf{x}(0) = \mathbf{x}_0$. Suppose that $\mathbf{f}$ is ***continuously differentiable***, i.e. $\mathbf{f}$ is continuous and all its partial derivatives $\partial f_i/\partial x_j$, $i, j = 1,\ldots,n$, are continuous for $\mathbf{x}$ in some open connected set $\mathcal{D}$ contained in $\mathbf{R}^n$.

Then for $\mathbf{x}_0$ in $\mathcal{D}$, the initial value problem has a solution $\mathbf{x}(t)$ on some time interval $(-\tau, \tau)$ about $t = 0$, and the solution is unique.

Fixed or singular points

Phase curves or phase trajectories of equation (A1.1) are solutions of

$$\frac{dx_1}{dx_2} = \frac{f_1(x_1, x_2)}{f_2(x_1, x_2)}. \tag{I.7}$$

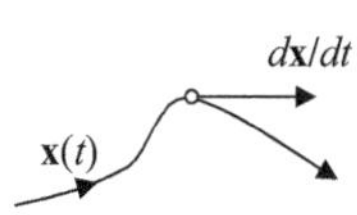

Figure I1. A trajectory or phase curve $\mathbf{x}(t)$ winding through the phase plane; the 'velocity' $d\mathbf{x}/dt$ is tangent to this curve.

We can imagine the entire phase plane as filled with such trajectories. In fact, through any point (x_1, x_2) there is a unique curve except at *fixed points* (x_1^*, x_2^*) where the vector field **f(x)** vanishes, i.e.

$$f_1(x_1^*, x_2^*) = f_2(x_1^*, x_2^*) = 0.$$

This is why fixed points are also called *singular points.*

It turns out that for systems of nonlinear equations in general it is impossible to find the trajectories analytically. Even when explicit formulas are available, they are often too complicated to provide much insight. However, something we can do is to determine the *qualitative* behavior of the solutions. That is, to find the system's phase portrait directly from the properties of the vector field **f(x)**. To do this we will use the ***linearization*** technique developed earlier for one-dimensional systems, namely a Taylor expansion around fixed points. The hope of this *linear stability analysis* is that we can approximate the phase portrait near a fixed point by that of a corresponding linear system, so that we can classify fixed points of *nonlinear* systems. More rigorously speaking, there is a theorem about the local behavior of dynamical systems in the neighborhood of a certain type of equilibrium point which asserts that linearization is effective in predicting qualitative patterns of behavior.

Linear stability analysis around fixed points, the linearization theorem of Hartman–Grobman

Suppose the map **f** is ***smooth***, i.e. it is at least differentiable everywhere (hence continuous) has an equilibrium state $\mathbf{x}^*$: that is, $\mathbf{f}(\mathbf{x}^*) = 0$. Then, the ***Hartman–Grobman*** *theorem* or ***linearization theorem*** states that the behavior of a dynamical system in a domain near a ***hyperbolic equilibrium point*** (we will define in a moment this kind of equilibrium) is qualitatively the same as the behavior of its linearization near this equilibrium point. Therefore, when dealing with such dynamical systems one can use the simpler linearization of the system to analyze its behavior around equilibria.

Just for simplicity of expression let us make the change of coordinates $x = x_1 - x_1^*$, $y = x_2 - x_2^*$, that moves the singular point to the origin $x = 0$ and $y = 0$. Then (0,0) is a singular point of the transformed equation (I.7′):

$$\frac{dx}{dy} = \frac{f_1(x, y)}{f_2(x, y)}. \tag{I.7′}$$

If f_1 and f_2 are analytic functions near (0,0), by definition of an analytic function, we can expand f_1 and f_2 in a Taylor series and, retaining only the linear terms, we get

$$\frac{dx}{dy} = \frac{f_{1x}x + f_{1y}y}{f_{2x}x + f_{2y}y}, \tag{I.8}$$

where the f_{ij} denote the partial derivative of the function f_i ($i = 1$ or 2) with respect to the direction $j = x$ or y evaluated at the origin, i.e. $f_{1x} = \frac{\partial f_1}{\partial x}|_{(0,0)}$, $f_{1y} = \frac{\partial f_1}{\partial y}|_{(0,0)}$, $f_{2x} = \frac{\partial f_2}{\partial x}|_{(0,0)}$, $f_{2y} = \frac{\partial f_2}{\partial y}|_{(0,0)}$. These four numbers define the ***Jacobian*** matrix **A**:

$$\mathbf{A} = \begin{pmatrix} f_{1x} & f_{1y} \\ f_{2x} & f_{2y} \end{pmatrix} \equiv \begin{pmatrix} \frac{\partial f_1}{\partial x} & \frac{\partial f_1}{\partial y} \\ \frac{\partial f_2}{\partial x} & \frac{\partial f_2}{\partial y} \end{pmatrix}_{(0,0)} . \tag{I.9}$$

Therefore, equation (I.5′) is equivalent, to first order (i.e. linear approximation) to:

$$\frac{d\mathbf{x}}{dt} = \mathbf{A}\mathbf{x}. \tag{I.10}$$

Let λ_1 and λ_2 be the eigenvalues of $\mathbf{A}$; given by equating the determinant of $\mathrm{A} - \lambda\,\mathbf{I}$:

$$\begin{vmatrix} f_{1x} - \lambda & f_{1y} \\ f_{2x} & f_{2y} - \lambda \end{vmatrix} = 0, \tag{I.11}$$

i.e. λ_1 and λ_2 are the roots of the second order ***characteristic equation***:

$$\lambda^2 - (f_{1x} + f_{2y})\lambda + f_{1x}f_{2y} - f_{1y}f_{2x} = 0, \tag{I.12}$$

which can be re-written as:

$$\lambda^2 - \mathrm{tr}\mathbf{A}\lambda + \det \mathbf{A} = 0, \tag{I.13}$$

where 'tr' denotes the trace of matrix $\mathbf{A}$ (the sum of diagonal elements) and 'det' its determinant $|\mathbf{A}|$. Therefore, we get

$$\begin{aligned} \lambda_1 &= 1/2\left(\mathrm{tr}\mathbf{A} + \sqrt{\mathrm{tr}\mathbf{A}^2 - 4\det \mathbf{A}}\right), \\ \lambda_2 &= 1/2\left(\mathrm{tr}\mathbf{A} - \sqrt{\mathrm{tr}\mathbf{A}^2 - 4\det \mathbf{A}}\right). \end{aligned} \tag{I.14}$$

In general, λ_1 and λ_2 are complex numbers. An equilibrium $\mathbf{x}^*$ is hyperbolic if no eigenvalue of the linearization has real part equal to zero. That is, hyperbolic equilibrium implies that $\mathrm{Re}(\lambda_1) \neq 0$ and $\mathrm{Re}(\lambda_2) \neq 0$.

The typical situation is for the eigenvalues to be distinct: $\lambda_1 \neq \lambda_2$. In this case, a theorem of linear algebra states that the corresponding eigenvectors $\mathbf{v}_1$ and $\mathbf{v}_2$ of $\mathbf{A}$ are linearly independent, and hence span the entire plane. In particular, any initial condition $\mathbf{x}_0$ can be written as a linear combination of eigenvectors, say

$$\mathbf{x}_0 = c_1\mathbf{v}_1 + c_2\mathbf{v}_2, \tag{I.15}$$

where c_1 and c_2 are arbitrary constants and the eigenvector $\mathbf{v}_i$ associated with the eigenvalue λ_i given by

$$\mathbf{v}_i = \left(1 + p_i^2\right)^{-1/2}\begin{bmatrix} 1 \\ p_i \end{bmatrix}, \quad p_i = \frac{\lambda_i - f_{1x}}{f_{1y}}, \quad f_{2x} \neq 0, \quad i = 1, 2. \tag{I.16}$$

This allows us to write down the general solutions of equation (I.10) simply as

$$\mathbf{x} = c_1\mathbf{v}_1 e^{\lambda_1 t} + c_2\mathbf{v}_2 e^{\lambda_2 t}. \tag{I.17}$$

This is a general solution because it is a linear combination of solutions to equation (A1.6), and hence is itself a solution. In addition, it satisfies the initial condition $\mathbf{x}(0) = \mathbf{x}_0$, and so by the existence and uniqueness theorem, it is the *only* solution.

If the eigenvalues are equal, i.e. equation (I.13) has a double root $\lambda_1 = \lambda_2 = \lambda$, the solutions are proportional to $(c_1 + c_2 t)\exp[\lambda t]$.

The mathematician Henri Poincaré distinguished four different singular points of differential equations. These are the ***node***, the ***saddle***, the ***focus*** and the ***center***. Figure I2 summarizes the possibilities in the so-called ***Poincaré diagram***, i.e. the (tr **A**, det **A**) parameter plane, which includes the parable $\Delta \equiv 1/4 \text{ tr } \mathbf{A}^2 - \det \mathbf{A} = 0$.

I. If det **A** < 0, then λ_1 and λ_2 are real and of opposite signs, regardless of the sign of tr **A**. Usually, solutions go to infinity as $t \to \infty$ so this case is considered to be unstable. Figure I2 shows the appearance of some trajectories near this kind of fixed point, denoted a **saddle point**. This type of behavior is found in the region below the horizontal axis of the (tr **A**, det **A**) parameter plane shown in the summary figure I2.

II. If det **A** > 0, then any of the following can happen:
(A) det **A** < 1/4 tr $\mathbf{A}^2$ (i.e. below the parable): In this case λ_1 and λ_2 are real. We then have two possibilities:

II(A).1. If tr **A** < 0: In this case $\lambda_1 < 0$ and $\lambda_2 < 0$. Solutions are both decreasing exponentials so that the *fixed point is stable*, denoted a **stable node** or **sink** (located in the Poincaré diagram between the horizontal axis and the parable, to the left-hand side of figure I2).

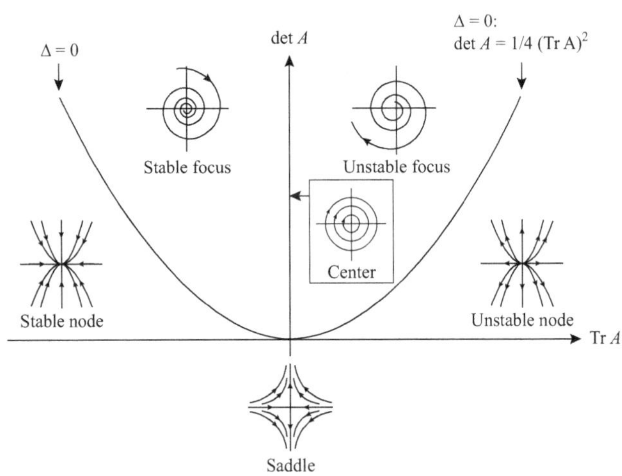

Figure I2. Poincaré diagram. Classification of phase portraits in the (tr **A**, det **A**)-plane. Author: Freesodas / Source: Gimp.

II(A).2. If tr **A** > 0: In this case $\lambda_1 > 0$ and $\lambda_2 > 0$. Solutions are both increasing exponentials so that the *fixed point is unstable*, denoted an **unstable node** or **source** (between the horizontal axis and the parable, to the rhs of figure I2).

(B) det **A** > 1/4 tr $\mathbf{A}^2$ (i.e. above the parable): In this case λ_1 and λ_2 are complex. We then have three possibilities:

II(B).1. If tr **A** < 0: In this case $\mathrm{Re}(\lambda_1) < 0$ and $\mathrm{Re}(\lambda_2) < 0$. Solutions are oscillations with decreasing amplitude so that the *fixed point* is a **stable focus** or **spiral sink** (located in the Poincaré diagram between the vertical axis and the parable, to the lhs of figure I2).

II(B).2. If tr **A** > 0: In this case $\mathrm{Re}(\lambda_1) > 0$ and $\mathrm{Re}(\lambda_2) > 0$. Solutions are oscillations with increasing amplitude so that the *fixed point* is an **unstable focus** or **spiral source** (located in the Poincaré diagram between the vertical axis and the parable, to the rhs of figure I2).

II(B).3. If tr **A** = 0: In this case $\mathrm{Re}(\lambda_1) = \mathrm{Re}(\lambda_2) = 0$. Solutions are periodic, with constant amplitude, and thus the phase curves are ellipses. This corresponds to a **center**, but is a marginal case. Centers are not stable in the usual sense, they are *neutrally stable*; a small perturbation from one phase curve does not die out in the sense of returning to the original unperturbed curve. The perturbation simply gives another solution. This implies that, in general, nonlinear terms will either stabilize or destabilize the system. In the case of center singularities, determined by the linear approximation to $f_1(x, y)$ and $f_2(x, y)$, we must look at the higher-order (than linear) terms to determine whether or not it is really a spiral and hence whether it is stable or unstable.

Summary

- Fixed points are **stable** when **the real part of λ_1 and λ_2 are negative.**
- There are four types of fixed points:
  1. A **node** if λ_1 and λ_2 are real, non null, with the same sign; if both λ_1 and λ_2 are negative (positive) the node is **stable (unstable)**.
  2. A **saddle point** if the signs of λ_1 and λ_2 are opposite.
  3. A **focus** when λ_1 and λ_2 are complex (with real part different from 0). Negative real parts for λ_1 and λ_2 imply a **stable focus**, whereas positive real parts for λ_1 and λ_2 mean an **unstable focus**.
  4. A **center** if λ_1 and λ_2 are purely imaginary the fixed point is. In this case we have to go beyond the linear stability analysis and look at the nonlinear terms to determine whether or not it is really a spiral and hence whether it is stable or unstable.

Limit cycles and Kolmogorov's theorem for predator–prey systems

An interesting question about trajectories that spiral outward from an unstable equilibrium is: do they spiral outward without bound until they intersect one of the axes and one of the species goes extinct? Or do they settle on a particular orbit which is itself stable? Such orbits are called stable **limit cycles**. A limit cycle is a closed trajectory such that neighboring trajectories are not closed; they spiral either toward (stable limit cycle) or away from the limit cycle (unstable limit cycle).

To elucidate the question of the fate of unstable spirals there are both negative theorems, which rule out closed orbit solutions in the phase plane, as well as the Poincaré–Bendixson theorem which establish that closed orbits exist under particular conditions. Before introducing these theorems, at the end of the next section, we will present a theorem by the Russian mathematician Andrei Kolmogorov for bi-dimensional predator–prey systems.

Kolmogorov's theorem

Given a bi-dimensional system like equation (I.5), if

$$\begin{aligned} f_1(x_1, x_2) &= x_1 f(x_1, x_2), \\ f_2(x_1, x_2) &= x_2 g(x_1, x_2), \end{aligned} \tag{I.18}$$

where $f(x_1, x_2)$ and $g(x_1, x_2)$ can be interpreted as the per capita growth rates for each species, provided that

(i) the functions f and g are continuous and differentiable in the domain $x_1 > 0$ and $x_2 > 0$

$$\begin{aligned}
&\textbf{(ii)}\quad \frac{\partial f}{\partial x_2} < 0 \\
&\textbf{(iii)}\quad \frac{\partial f}{\partial x_1} x_1 + \frac{\partial f}{d x_2} x_2 < 0 \\
&\textbf{(iv)}\quad \frac{\partial g}{\partial x_2} \leqslant 0 \\
&\textbf{(v)}\quad \frac{\partial g}{\partial x_1} x_1 + \frac{\partial g}{d x_2} x_2 > 0 \\
&\textbf{(vi)}\quad f(0, 0) > 0 \\
&\textbf{(vii)}\quad f(0, A) = 0 \\
&\textbf{(viii)}\quad f(B, 0) = 0 \\
&\textbf{(ix)}\quad f(C, 0) = 0 \\
&\textbf{(x)}\quad B > C
\end{aligned} \tag{I.19}$$

where A, B and C are three positive quantities, then this system has either a stable point of equilibrium or a stable limit cycle.

A.3 Some general theorems[1]

We will introduce some valuable theorems that we will accept without proving them (for proofs of these theorems see Goh 1980).

Local stability: the real parts of the eigenvalues of the Jacobian matrix must be negative

Suppose the autonomous system (I.1) has a positive equilibrium at $\mathbf{n}^*$ and let $x_i = n_i - n_i^*$ for $i = 1, 2, \ldots, S$ denote a small departure of each species density from its equilibrium value. Performing a first order Taylor expansion around the equilibrium we get:

$$\begin{aligned} n_i f_i(\mathbf{n}) &= n_i^* \left(f_i(\mathbf{n}^*) + \sum_{j=1}^{S} \frac{\partial f_i(\mathbf{n})}{\partial n_j} \bigg|_{\mathbf{n}^*} x_j + O(x^2) \right) \\ &= n_i^* \left(0 + \sum_{j=1}^{S} \frac{\partial f_i(\mathbf{n})}{\partial n_j} \bigg|_{\mathbf{n}^*} x_j + O(x^2) \right). \end{aligned} \tag{I.20}$$

Therefore, substituting equation (I.20) into (I.1) and using equation (I.4), the linearized dynamics is given by

$$\frac{dx_i}{dt} = \sum_{j=1}^{S} n_i^* J_{ij} x_j \qquad i = 1, \ldots, S. \tag{I.21}$$

It turns out that we have this valuable theorem for continuous time models:

Theorem 1. *The equilibrium* $\mathbf{n}^* = (n_1^*, n_2^*, \ldots, n_S^*)$ *of an autonomous continuum time system is locally stable if all the real parts of the eigenvalues of the Jacobian matrix* $J_{ij} \equiv \frac{\partial f_i(\mathbf{n})}{\partial n_j}\big|_{\mathbf{n}^*} n_i^*$ *are negative.*

Thus this theorem generalizes the two stability analysis for bi-dimensional systems of the previous section.

In the case of a Lotka–Volterra generalized linear model we have:

$$\frac{dn_i}{dt} = r_i n_i \left(1 + \sum_{j=1}^{S} a_{ij} n_j \right) \qquad i = 1, \ldots, S. \tag{I.22}$$

And thus we have,

$$J_{ij} = r_i a_{ij} n_i^* \tag{I.23}$$

[1] This section is mainly based on chapters 1, 3 and 5 of the thorough study on stability by Goh (1980) and chapter 2 of May (1974).

The equilibrium $\mathbf{n}^*$ is locally stable if all the real parts of the eigenvalues of the matrix $[r_i a_{ij} n_i^*]$ are negative. Note that in general the stability properties of the matrix $[r_i a_{ij} n_i^*]$ are different from those of the matrix $\mathbf{A} = [a_{ij}]$.

For the discrete time description (I.1′), to first order, we have:

$$x_i(t+1) = \sum_{j=1}^{S} \frac{\partial G_i(\mathbf{n})}{\partial n_j}\bigg|_{\mathbf{n}^*} x_j. \qquad \text{(I.21′)}$$

And, a theorem similar to theorem 1 but for discrete time models (Goh 1980) is:

Theorem 1′. *The equilibrium* $\mathbf{n}^* = (n_1^*, n_2^*, \ldots, n_S^*)$ *of a set of discrete difference equations is locally stable if the modulus of all the eigenvalues of the matrix* $\frac{\partial G_i(\mathbf{n})}{\partial n_j}\big|_{\mathbf{n}^*}$ *are less than one.*

Global stability: Lyapunov functions

There exists a method to determine whether a system is globally stable. It involves finding a function known as a *Lyapunov function*. The problem is that the existence of a Lyapunov function is often difficult to determine for multispecies models and, consequently, this approach has a limited utility. The discussion will be facilitated by considering physical systems analogous to the biological ones.

Consider an autonomous system of differential equations

$$dx_i/dt = f_i(\mathbf{x}) \qquad \text{(I.24)}$$

with a fixed point at $\mathbf{x}^* = (x_1^*, x_2^*, \ldots, x_S^*)$.

Definition: A *Lyapunov function*, for this system is a continuously differentiable, real valued function $V(\mathbf{x})$ with the following properties:

i. $V(\mathbf{x}) > 0$ for all $\mathbf{x} \neq \mathbf{x}^*$, and $V(\mathbf{x}^*) = 0$. (We say that V is *positive definite*.)

ii. $\frac{dV(x)}{dt} = \sum_{i=1}^{S} \frac{\partial V}{\partial x_i}\frac{dx_i}{dt} = \sum_{i=1}^{S} \frac{\partial V}{\partial x_i} f_i(\mathbf{x}) < 0$ for all $\mathbf{x} \neq \mathbf{x}^*$. (All trajectories flow 'downhill' toward $\mathbf{x}^*$.)

Theorem 2. *The equilibrium* $\mathbf{x}^*$ *of an autonomous continuum time system is locally asymptotically stable, i.e. for all initial conditions* $\mathbf{x}(t) \to \mathbf{x}^*$, *as* $t \to \infty$, *if there exists a Lyapunov function for* $\mathbf{x}^*$.

Intuitively, under conditions **i.** and **ii.**, all trajectories move monotonically down the graph of $V(\mathbf{x})$ toward $\mathbf{x}^*$ (figure I3).

For physical systems the direct method of Lyapunov generalizes the principle that a system, which continuously dissipates energy until it attains an equilibrium, is

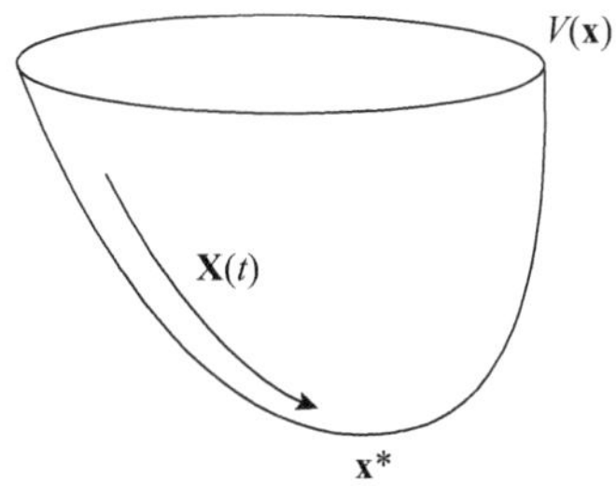

Figure I3. The solutions cannot get stuck anywhere else because if they did, V would stop changing, but by assumption, $dV/dt < 0$ everywhere except at $\mathbf{x}^*$.

stable. If a physical system, for example a vibrating spring and mass, dissipates energy over time and the energy is never restored then eventually the system must reach some final resting state. This final state is called the *attractor*. However, finding a function that gives the precise energy of a physical system can be difficult, and for biological systems, the concept of energy may not be applicable. Lyapunov's realization was that stability can be proven without requiring knowledge of the true physical energy, provided a Lyapunov function can be found to satisfy the above constraints.

Goh (1977) has discussed a Lyapunov function that fits all Lotka–Volterra models:

$$V(\mathbf{n}) = \sum_{i=1}^{S} \mathrm{k}_i(n_i - n_i^* - n_i \ln(n_i/n_i^*)), \qquad (\mathrm{I.25})$$

where the k_i are constants. If the k_i exist such that dV/dt is always negative except at $\mathbf{n}^*$ (where it is zero) then the system is globally stable.

The problem is how to obtain the k_i so that they satisfy condition $dV/dt < 0$. For simple examples this is easy, but for more complicated examples it is not.

Ruling out closed orbits

Closed orbits can be ruled out for the following systems[2]:

(A) ***Gradient systems***

That is, suppose the system can be written in the form $d\mathbf{x}/dt = -\nabla V$, for some continuously differentiable, single-valued scalar function $V(\mathbf{x})$ (this vector equality is written for each coordinate i as $dx_i/dt = -\partial V/\partial x_i$). Such a system is called a *gradient* with *potential function V*.

(B) **Systems with a Lyapunov function**

If a Lyapunov function exists, then closed orbits are forbidden

[2] For proofs we refer the reader to section 7.2 of Strogatz (1994).

Poincaré–Bendixson theorem

The Poincaré–Bendixson theorem is the main tool which historically has been used to show that a dynamical system has a stable cycle limit.

Poincaré–Bendixson Theorem

Suppose that:

(i) R is a closed, bounded subset of the plane;

(ii) $d\mathbf{x}/dt = \mathbf{f}(\mathbf{x})$ is a continuously differentiable vector field on an open set containing R;

(iii) R does not contain any fixed points;

(iv) There exists a trajectory $\mathcal{C}$ that is 'confined' in R, this means that it starts in R and stays in R for all future time.

Then either $\mathcal{C}$ is a closed orbit, or it spirals toward a closed orbit as $t \to \infty$. Thus, in either case, *R* ***contains a closed orbit***.

For a proof of this theorem we refer the interested reader to Coddington and Levinson (1955) *or* Wiggins (1990).

References

Coddington E A and Levinson N 1955 The Poincaré–Bendixson theory of two-dimensional autonomous systems *Theory of Ordinary Differential Equations* (New York: McGraw-Hill) pp 389–403

Goh B S 1977 Global stability in many species systems *Am. Nat.* **111** 135–43

Goh B-S 1980 *Management and Analysis of Biological Populations* (Amsterdam: Elsevier)

Gurel O and Lapidus L 1968 Stability via Liapunov's second method *Ind. Eng.Chem.* **60** 1–26

LaSalle J and Lefschetz S 1961 *Stability by Liapunov's Direct Method* (New York: AcademicPress)

May R M 1974 *Stability and Complexity in Model Ecosystems* (Princeton, NJ: Princeton University Press)

Schultz D G 1965 The generation of Liapunov functions *Advances in Control Systems* vol 2 ed. C T Leondes (New York: Academic)

Strogatz S H 1994 *Nonlinear dynamics and chaos. With Applications to Physics, Biology, Chemistry, and Engineering* (Reading, MA: Perseus)

Wiggins S 1990 *Introduction to Applied Nonlinear Dynamical Systems and Chaos* (New York: Springer)

Willems J L 1970 *Stability Theory of Dynamical Systems* (London: Nelson)

IOP Publishing

Ecological Modelling and Ecophysics
Agricultural and environmental applications
Hugo Fort

Appendix II

Fermi problems or back-of-the-envelope calculations

A **Fermi problem, Fermi question, or Fermi estimate** is an estimation problem based on heuristic methods. Named for the 20th century physicist Enrico Fermi, such problems typically involve making justified guesses about quantities that seem impossible to compute given limited available information.

Fermi was known for his ability to make good approximate calculations with little or no data. One well-documented example is his estimate of the yield of the atomic bomb detonated during the Trinity test, based on the distance traveled by pieces of paper dropped from his hand during the blast. Fermi's estimate of 10 kilotons of TNT was remarkably close to the now-accepted value of around 20 kilotons.

Let us consider a first Fermi problem:

'How much oil (in barrels) is consumed in the United States per year?' To answer this question we will split the problem into two simpler quantities to estimate:

First, let us estimate how much oil is used by cars every year.

Secondly, we will increase the estimate to account for non-automotive uses.

A typical solution might include the following assumptions and estimations:

1. There are approximately 330 000 000 people in the United States.
2. On average, each person owns a car, so let us say the number of cars $N_c \sim 3 \times 10^8$.
3. What about the number of gallons consumed per capita per day or year? Let us estimate this quantity in two different ways. (A) Thinking how frequently you refill the tank of your car at a gas station; e.g. every ten days. This implies an average consumption of 13 gallons every ten days or more roughly one gallon per day. (B) Using the annual average mileage of a car; maybe around 10 000 miles. A typical value for the miles per gallon (mpg) for a car is around 25. This gives 10 000/25 = 400 gallons per year or, once again, roughly one gallon per day.

4. A rough and simple estimation of the fraction of oil used by cars is 1/2; the other half is used for other means of transportation (trucks, buses, trains, boats, planes, etc), for heating and cooling and for manufacturing plastics and chemicals, as well as many lubricants, waxes, tars, asphalts, pesticides and fertilizers.

From these assumptions:

$$\text{Oil used by cars per day} = (3 \times 10^8 \text{ cars}) \times (1 \text{ gallon per car per day}) = 3 \times 10^8 \text{ gallons d}^{-1}.$$

How many gallons does a barrel contain? The answer is 42. But suppose we do not know. We can estimate it using that a barrel costs around $50 and the average US price of regular-grade gasoline is $2.50. If we assume that half of the price of a gallon of gas ($1.25) is the cost required to produce the oil, we get that a barrel contains roughly 50/1.25 = 40. Not so bad! Thus we have:

$$\text{Oil used by cars per day} = 3 \times 10^8 \text{ gallons d}^{-1}/40 \text{ gallon barrel}^{-1} = 7.5 \times 10^6 \text{ barrels d}^{-1}$$

Therefore, we have a total consumption of oil of:

$$\text{Oil used in US per day} = 2 \times 7.5 \times 10^6 \text{ barrels d}^{-1} = 15 \text{ million barrels d}^{-1}$$

Or equivalently,

$$\text{Oil used in US per year} = 365 \times 15 \times 10^6 \text{ barrels d}^{-1} \approx 5 \times 10^9 \text{ barrels y}^{-1}$$

Let us perform a second estimation:

The world consumes around 100 millions of oil barrels per day. The US GDP represents between 15% (measured in purchasing power parity PPP) and 25% (measured in US $) of the world GDP. Let us assume that the same proportionality holds true for the oil consumption. Therefore, we get 15–25 million barrels d^{-1}. If we take the midpoint we get 20 million barrels d^{-1}.

Notice the lower estimate is exactly the number we estimated before using a different procedure! But this is just a coincidence. Actually, overall, there were an estimated 272.48 million vehicles registered in 2017. The figures include passenger cars, motorcycles, trucks, buses, and other vehicles (Statista 2019a). So if we remove the number of trucks and buses it turns out we overestimated the number of cars by more than 10% (and probably more than 15%). However, our estimation of the fraction of the gasoline consumed by cars of 1/2 was quite accurate: in 2018, consumption of finished motor gasoline averaged about 9.33 million b/d (392 million gallons per day), which was equal to about 45% of total US petroleum consumption according to the US Energy Information Administration (EIA 2019a). Additionally, we also had luck when estimating the number of gallons in a barrel with an error of only 5%. Nevertheless, notice that had we performed this same estimation five years ago, when the price of the oil barrel was around $100, we would have gotten a much worse estimate of 80 gallons per barrel!

The true value for 2018 was 20.5 million barrels of petroleum per day, or a total of about 7.5 billion barrels of petroleum per year (EIA 2019b). Thus, we underestimated the number of barrels consumed in the US by 25% according to our first estimation and by less than 3% according to our second estimation. Not so bad, remember that the goal of Fermi problems is to estimate quantities with scarce information within one order of magnitude from its actual value. Therefore, overall we can assume our guesses were not quite off the mark.

Another Fermi problem could be: 'How many shopping malls S there are in the USA?'

The total number of customers C is smaller but comparable with the population of the US, N, say $C = 2/3 \times N$, then we will estimate S by dividing the total amount of money all the shopping malls receive from customers by the average net profits all the owners of stores of an average shopping mall make.

Let us call the average percentage of net profits of a retail store, p.

Now, owning a business means risking your money, dealing with employees, etc. Therefore, we assume that, on average, the owner makes more money than employees. This implies that the mean profits of a store > mean income (i); say $3 \times i$.

An average consumer spends, say a percentage q of her/his income in shopping malls.

If we denote by n the average number of stores in a shopping mall, thus we have

$$3 \times i \times S \times n = C \times i \times (q/100) \times (p/100).$$

So we see that i cancels out because it multiplies in both sides of the equality, so we get:

$$S = 2/3 \times N \times q \times p/(3 \times n \times 10^4)$$

Now we have to put numerical values for the percentages p and q, say: $p = 10\%$, $q = 20\%$ and for n, say, on average, $n = 20$ stores per shopping mall. Thus we finally get:

$$S = 2 \times 3.3 \times 10^8 \times 20 \times 10/(9 \times 20 \times 10^4) = 73\,333.$$

According to Statista (2019b) there are 116 000 shopping malls in the US.

According to the US Bureau of Labor Statistics (2020), an average consumer spends \$11 185 in 'other goods and services' from the \$67 241 average income after taxes (this number does not include other expenses like food away from home, etc). Therefore, we can estimate q as 16.6%.

According to this NYU Stern database for more than 7000 US companies (Stern 2020) in many different industries, **the average profit margin is $p = 7.9\%$ for all companies and 6.9% for more than 6000 companies excluding financials**.

Additional Fermi problems

Try your hand at the following problems. Remember, you are not expected to get the 'right' answer, and you will not be rewarded for extra accuracy, but you do need to clearly show your reasoning at each step. After solving these problems by estimation, search the Web for the 'true' answer.

I. Estimate the total number of cattle in the world.
II. How much milk is produced in the US each year?
III. Estimate the total number of hairs on your head.
IV. How many commercial planes are flying simultaneously over the US?

References

EIA 2019a https://www.eia.gov/energyexplained/oil-and-petroleum-products/use-of-oil.php

EIA 2019b https://www.eia.gov/tools/faqs/faq.php?id=33&t=6

Statista 2019a https://www.statista.com/statistics/183505/number-of-vehicles-in-the-united-states-since-1990/

Statista 2019b https://www.statista.com/statistics/208059/total-shopping-centers-in-the-us/

Stern 2020 http://pages.stern.nyu.edu/~adamodar/New_Home_Page/datafile/margin.html

US Bureau of Labor Statistics 2020 https://www.bls.gov/news.release/wkyeng.nr0.htm

IOP Publishing

Glossary

Adiabaticity or quasistatic evolution	A slow process in which all time derivatives are very small.
Attractor	In dynamical systems, an attractor is a set of numerical values toward which a system tends to evolve, for a wide variety of starting conditions of the system. System values that get close enough to the attractor values remain close even if slightly disturbed, in such a way that all trajectories not contained in that region will eventually wind up in the region. An attractor may be a point or a cycle that is an equilibrium and generates transients that return to the equilibrium state after perturbation. It may also be an attractive region that has no individual equilibrium points or cycles (a chaotic or strange attractor).
Autonomous dynamical system	A system of ordinary differential equations which does not explicitly depend on the independent variable. When the independent variable is time, they are also called time-invariant systems.
Basin of attraction	For each attractor, its basin of attraction is the set of initial conditions leading to long-time behavior that approaches that attractor. That is, the collection of points that converge on a particular attractor.
Bifurcation	A bifurcation occurs when a small smooth change made to the parameter values (the 'bifurcation' parameters) of a system causes a sudden 'qualitative' or topological change in its behavior.
Bifurcation diagram	A graph of the attractors of a system as a function of some parameter (the 'bifurcation' parameter). It shows the values visited or approached asymptotically (fixed points, periodic orbits, or chaotic attractors) of a system as a function of this bifurcation parameter in the system.
Bifurcation, local	A local bifurcation occurs when a parameter change causes the stability of an equilibrium (or fixed point) to change. In continuous systems, this corresponds to the real part of an

doi:10.1088/978-0-7503-2432-8ch7

eigenvalue of an equilibrium passing through zero. In discrete systems (those described by maps rather than ODEs), this corresponds to a fixed point having an eigenvalue with modulus equal to one. In both cases, the equilibrium is **non-hyperbolic** (at least the real part of one eigenvalue becomes zero) at the bifurcation point. The topological changes in the phase portrait of the system can be confined to arbitrarily small neighborhoods of the bifurcating fixed points by moving the bifurcation parameter close to the bifurcation point (hence 'local'). By contrast, global bifurcations cannot be revealed by eigenvalue degeneracies.

Bifurcation, normal form of In mathematics, the normal form of a dynamical system is a simplified form that can be useful in determining the system's behavior. Normal forms are often used for determining local bifurcations in a system. All systems exhibiting a certain type of bifurcation are said to be locally (around the equilibrium) topologically equivalent to the normal form of the bifurcation.

Bifurcation point A point of structural instability in which a single equilibrium condition is split into two.

Carrying capacity The maximum attainable size of a population, usually symbolized as K.

Catastrophe or Imperfect bifurcation A catastrophe occurs when the stability of an equilibrium breaks down, causing the system to jump into another state. This jump could be truly catastrophic for the equilibrium of a bridge or a building or a species that extinguishes. Catastrophes can be also regarded as *imperfect bifurcations*, often described by the addition of an imperfection parameter to the **normal form of a bifurcation**.

Competitive exclusion principle Sometimes referred to as Gause's law, is the proposition that two species competing for the same limiting resource cannot coexist at constant population values. When one species has even the slightest advantage over another, the one with the advantage will dominate in the long term. This result can be derived from the Lotka–Volterra competition equations: if interspecific competition between two species is sufficiently large, the equilibrium of both species coexisting is unstable.

Density dependence The condition in which the rate at which a population increases or decreases is a function of its density (in contrast with density independence)

Dynamical system A means of describing how one state develops into another state over the course of time in terms of a system of equations. These equations describe the time dependence of a point's position in its ambient (geometrical) space. *Dynamical systems theory* brings a qualitative and geometrical approach to the analysis of ordinary differential equations (ODEs), addressing the existence, stability, and global behavior of sets of solutions, rather than seeking exact or approximate expressions for individual solutions.

Equilibrium point The value of a variable that does not change under the rules of a dynamical system. An equilibrium point may be stable (in which case it is commonly referred to as an **attractor**) or unstable (in which case it is commonly referred to as a **repeller**).

Euler's constant Approximately 2.7183, the base of natural logarithms, normally symbolized by a lowercase *e*.

Facultative mutualism Mutualism in which one species can survive without its mutualist but performs better with it.

Ferromagnetism The basic mechanism by which certain materials, such as iron and nickel, form permanent magnets. Microscopically the ferromagnetism is explained in terms of the electrons contained in the material. Specifically, one of the fundamental properties of an electron is that it has a magnetic dipole moment, i.e. it behaves itself as a tiny magnet. When these tiny magnetic dipoles are aligned in the same direction, their individual magnetic fields add together to create a measurable macroscopic magnetic field.

Functional response In consumer–resource (predator–prey) equations, the function that stipulates how the per capita consumption rate (or predation rate) changes with changes in resource density.

Gause principle See **Competitive exclusion principle.**

Hamiltonian The mathematical descriptor for the energy of a given interaction. The total Hamiltonian describes all energies of all the interactions that affect the system.

Intraspecific competition The competitive interaction among individuals in the same population.

Intrinsic rate of natural increase The growth of a population under the theoretical state of extremely low population density, usually symbolized as r.

Isocline or Nullcline In population dynamics, the term isocline refers to the set of population sizes at which the rate of change for one population in a pair of interacting populations is zero. More generally, for a dynamical system, the set of all points for which one of the variables does not change, so that the time derivative is equal to zero.

Limit cycle An oscillatory system that can be either stable (an oscillatory attractor) or unstable (an oscillatory repeller).

Logistic equation Sometimes called the Verhulst model or logistic growth curve, is a model of population growth first published by Pierre Verhulst (1845). The model is continuous in time, but a modification of the continuous equation to a discrete quadratic recurrence equation known as the logistic map is also widely used.

Logistic population growth Population growth that appears qualitatively exponential at low population density but approaches an asymptote as the population becomes larger; population growth that follows the logistic equation.

Malthus equation The simplest population equation describing an exponentially growing population, introduced by Thomas R Malthus in 1798.

Mean-field approximation (MFA) — In physics and probability theory, the mean-field approximation consists in approximating a random (stochastic) model by a simpler model that results from averaging over degrees of freedom. Such models consider many individual components that interact with each other. The effect of all the other individuals on any given individual is approximated by a single averaged effect, thus reducing a **many-body problem** to a **one-body problem**.

Metastability — In physics, metastability is a stable state of a dynamical system other than the system's state of least energy. In isolation the state of least energy is the only one the system will inhabit for an indefinite length of time, until more external energy is added to the system. That is, the system will spontaneously leave any other state (of higher energy) to eventually return (after a sequence of transitions) to the least energetic state. A ball resting in a hollow on a slope is a simple example of metastability. If the ball is only slightly pushed, it will settle back into its hollow, but a stronger push may start the ball rolling down the slope.

Nullcline = Isocline = Zero growth line

Obligate mutualism — Mutualism in which one species is unable to survive without its mutualist.

Ordinary differential equations (ODEs) — A differential equation containing one or more functions of one independent variable and the derivatives of those functions. The term *ordinary* is used in contrast with the term partial differential equation (PDE) which may be with respect to *more than* one independent variable.

One-dimensional map — A function f that projects a single variable x_t through discrete time t, $x_{t+1} = f(x_t)$. For example, in the logistic map $f(x_t) = rx_t(1 - x_t)$.

Partial differential equation (PDE) — A differential equation that contains several unknown variables and their partial derivatives (i.e. the derivative with respect to one of those variables, with the others held constant). PDEs are used to formulate problems involving functions of several variables, typically space coordinates and time. A special case is ordinary differential equations (ODEs), which deal with functions of a single variable and their derivatives.

Population — A group of individual items. In the context of population ecology, a population is a group of individual living organisms.

Repeller — A point or cycle that is theoretically an equilibrium but generates transients that deviate from the equilibrium position when perturbed.

Separatrix — The boundary between two basins of attraction.

Simulation — A numerical simulation is a calculation that is run on a computer following a program that implements a mathematical model for a physical system. Numerical simulations are required to study the behavior of systems whose mathematical models are too complex to provide analytical solutions, as in most nonlinear systems.

Strange attractor A chaotic attractor. A region of space that attracts all trajectories but contains no attractive points or cycles.

Structural stability A higher-level stability concept in which the qualitative nature of a system is unchanged when the parameters of the system are varied.

Structured models Models that do not assume that all individuals in the population are identical. E.g. models can be spatially-structured (spatial heterogeneous environment), age-structured or sex-structured.

Vector field The set of vectors that determine the behavior of a dynamic system.

Zero growth line = Nullcline = Isocline

Lightning Source UK Ltd.
Milton Keynes UK
UKHW051348211121
394305UK00002B/51